STUDENT WORKBOOK
Dee U. Silverthorn

Richard D. Hill

Lawrence D. Brewer

HUMAN An Integrated Approach
PHYSIOLOGY

S I L V E R T H O R N

PRENTICE HALL, Upper Saddle River, NJ 07458

Executive Editor: David Brake
Special Projects Manager: Barbara A. Murray
Production Editor: Mindy DePalma
Supplement Cover Manager: Paul Gourhan
Supplement Cover Designer: Liz Nemeth
Manufacturing Buyer: Ben Smith
Editorial Project Manager: Byron D. Smith

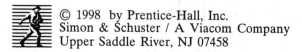 © 1998 by Prentice-Hall, Inc.
Simon & Schuster / A Viacom Company
Upper Saddle River, NJ 07458

Printed in the United States of America

10 9 8 7 6 5 4 3 2 1

ISBN 0-13-267543-9

Prentice-Hall International (UK) Limited, *London*
Prentice-Hall of Australia Pty. Limited, *Sydney*
Prentice-Hall Canada, Inc., *London*
Prentice-Hall Hispanoamericana, S.A., *Mexico*
Prentice-Hall of India Private Limited, *New Delhi*
Prentice-Hall of Japan, Inc., *Tokyo*
Simon & Schuster Asia Pte. Ltd., *Singapore*
Editora Prentice-Hall do Brazil, Ltda., *Rio de Janeiro*

CONTENTS

CHAPTER

OWNER'S MANUAL

This workbook is divided into sections to help you study the chapters in the text.

Summary
For each chapter in the textbook, you will find a brief list that points out the key learning tasks for the chapter, along with a short narrative summary.

Teach Yourself the Basics
This section is organized using the section headers from the chapter. Teach Yourself the Basics has a series of questions about each section, with page and figure references so that you can refer back to the book if you can't remember the answer. There are two ways to use Teach Yourself the Basics. You can fill it in as you read, using the workbook to direct your reading and notetaking. This is an excellent method for making sure that you are getting the important information out of each section of the chapter. Or you can wait until you have studied the chapter, then see if you can answer the questions without referring to the textbook.

$\int$ This symbol marks information from other chapters than can be integrated into the current chapter.

☛   These POINTERS give you interesting facts or helpful ways to remember information.

Talk the Talk is a vocabulary list of the important terms from the chapter. Use this list to quiz yourself.

Errata is a section that points out potentially confusing typographical errors in the book.

Quantitative Thinking includes quantitative problems and shows you how to think about them and work them.

Practice Makes Perfect is a set of questions that deal with material in the chapter. They range from simple "memorization" questions to difficult "application" questions. The answers to these questions are contained in an appendix at the back of the workbook.

Beyond the Pages contains additional material that is related to the chapter.
　　　Clinical Correlates point out the clinical applications of the basic physiology.
　　　Try It sections include activities that you may want to try, such as mini-experiments and demonstrations or interesting web sites.
　　　Reading includes books or articles that relate to material covered in the chapter.
　　　Viewing suggests movies and videos of interest.

A Note on Conventions Used in the Workbook
Ions in the workbook are written in the following format : Na^+, Ca^{2+}, Cl^-, K^+ . Pi or PO_4^{3-} are used to represent phosphate. If you are not familiar with the concept of ions, read p. 22 in Chapter 2. Body fluid compartments are abbreviated as follows:
　　　ICF = intracellular fluid
　　　ECF = extracellular fluid
　　　IF = interstitial fluid

Study Hints for Physiology Students

There are several differences you will notice when studying and learning science:

• There is a large volume of unfamiliar and frequently intimidating vocabulary.

• Science textbooks have a different writing style and need to be read differently than textbooks in the humanities.

• The thought process for science requires linear thinking and the ability to trace a process in some detail from beginning to end. Humanities students who are used to analyzing interrelationships may tend to think too broadly.

Here are some suggestions to use in class and when studying physiology.

◆ Notetaking

Do not try to write down every word the instructor says. Listen particularly for vocabulary, concepts, points the instructor emphasizes. Develop your own shorthand so that you can get down more information. Example: Use up and down arrows for "increase" and "decrease."

Stop the instructor when a major point is unclear, or if the instructor has gone too fast.

If you can't ask a question about something or if you get behind in your notetaking, put a question mark (?) in the margin so that you know that your notes in that section are lacking. Check with a friend or the instructor to clear up what you missed.

Develop one or two "study buddies" with whom you can compare your notes. If you have to miss class, try to get notes from more than one person. Notes are a memory aid rather than a verbatim copy of the lecture, so two different sets of notes are more likely to give you a complete overview of the lecture.

◆ Vocabulary

Most scientific words sound terribly complicated and difficult to remember. However, you can learn some common prefixes, suffixes, and roots that will help you to remember the words. In the textbook after some vocabulary words, there will be a note in brackets [ ] that shows the roots and their meanings. In this workbook, there is a list of some of the most common roots you will encounter in physiology. You can also use a dictionary to find out the origin of a word. Start a list in your own notebook of other suffixes, prefixes, and roots.

◆ Reading the textbook

Find out from your instructor if you are responsible for material that is in the text but has not been covered in lecture. If you are, you will need to add the extra information from the text to the information in your class notes. If you are not, then you need to be familiar with your notes so that you can pay less attention to material in the text that is not relevant.

Science texts are written with important facts in every sentence. You cannot speed-read a science text. Go slowly and analyze each sentence as you read it. Ask yourself if you understand the concepts in that sentence, and how that sentence relates to the facts presented in previous sentences.

Use the charts and diagrams in the book. Read the captions to the figures as they will frequently explain what is in the diagram. Sometimes the figures in a physiology text provide a summary of material in a section.

◆ Organizing your studying

As soon as possible after class, you should glance over your notes and mark the points that are unclear or where your notes may be lacking.

Some students sit down and spend a lot of time rewriting their notes. Usually there are other, more profitable ways of spending your study time. You will still be writing down the information in your notes, but you will also be reorganizing it into study notes in a form that you can remember. The following are some possible ways to do this. You probably won't have time to do all of them, so experiment until you find the method that works best for the way you learn.

1. **Mark up your original notes with colored pens.**

Use colored pens to see if you can divide the notes into levels and sublevels for an outline. Assign a different color to each level of organization. Underline or draw a box around the word(s) that fits into that heading. For example:
SUBJECT HEADING: Purple
I. Major topic = red
      A. Secondary topic = green
            1. Facts under that topic = turquoise

Give vocabulary words and concepts their own color, such as yellow.

Once you have marked your notes this way, go back and make a skeleton outline using the words you marked. It probably won't be a perfect outline (Topic I may have A but not B), but don't let that worry you. Use this outline to give yourself an overview of the material covered and the progression of the ideas or concepts covered.

2. **Make a working vocabulary list.**

• Take a sheet of lined notebook paper and fold it in half lengthwise.
• Down the left-hand side, list all the words and concepts you have marked in yellow in your notes.
• Down the right-hand side (on the other side of the fold), list an abbreviated definition.
• Study with the paper open so that you can see both sides. When you think you have learned the material, fold the paper in half so that you see only the list of words. Test yourself by going down the list and saying the definitions to yourself. If you don't know a definition, keep going. When you reach the end of the list, go back and look at the definitions of the words you missed.
• Now turn the paper over so that you are looking at the definitions. Read the definitions and see if you can say and spell the word that fits each definition.

3. Get the big picture as well as the details.

In physiology, for each system studied, you should make an outline, study sheet, or chart that answers the following questions:

• What is the anatomical structure of the system? Can you trace a molecule involved in the system through all the parts? Example: trace a drop of blood from the aorta to various parts and back through the heart. What kinds of tissues or cells make up this system? What kind of muscle? Is there some structural entity that we can call the "functional unit"?

• What is the function(s) of the system? Which parts carry out which function? How are the functions carried out?

• How is the system regulated? Consider control by the nervous system and the endocrine system. Are there any reflexes? Know where any pertinent hormones are secreted and what controls their release.

• How is the circulatory system involved with this system?

• Certain themes will keep popping up throughout the chapters. Make note of them. They include:
      Movement of molecules across membranes
      Pressure and flow
      Biomolecules: carbohydrates, fats, proteins. Their roles, transport, and metabolism.
      Ions: Na^+, K^+, H^+, HCO_3^-
      Gases: oxygen and carbon dioxide
      Energy use and storage

4. Make charts, diagrams, flow charts, and concept maps.

One advantage of a chart is that it also allows you to compare and contrast different concepts that at first glance may not seem to have much of a relationship.

To make a chart:

Divide your paper into columns and rows. Across the top, write the topics you want to compare (Example: male and female reproduction). Down the left side, label the rows or blocks with the points you want to compare (Example: name of gamete, name of gonad, hormones). Go back and fill in the chart.

One technique some students have used is to try to condense everything they have learned about a system onto a piece of poster paper. One effective way to do this is to make a giant drawing of the structure (anatomy) of the system and then add in all the physiological processes at or near the appropriate structure.

Example: Make a poster of the respiratory system.
      On the board you might draw a large upper body with the upper and lower respiratory systems diagram labeled. In the head you would also include the neurological control of ventilation. Add an enlarged cluster of alveoli just below the lung, and draw in the circulatory system going to a single cell.

5. Practice higher level thinking.

One objective of many physiology courses is to teach students how to use what are called higher level cognitive (factual knowledge) processes. The table below* shows one scheme for classifying the types of learning that people do. How many levels do you usually use when you study?

Lowest

1. Knowledge	Requires that you recognize or recall information
2. Comprehension	Requires that you think on a low level such that the knowledge can be reproduced or communicated without a verbatim repetition
3. Application	Requires that you solve or explain a problem by applying what you have learned to other situations and learning tasks
4. Analysis	Requires that you solve a problem through the systematic examination of facts or information
5. Synthesis	Requires that you find a solution to a problem through the use of original, creative thinking
6. Evaluation	Requires that you make an assessment of good or not so good, according to some standards

Highest

To use the higher thinking skills, you must master the first two levels; in other words, you must have a memorized database of information upon which to act. Once that background information is in place, you can begin to analyze and apply it.

You can recognize questions that require use of higher level thinking by the following phrases:

How would you (solve some problem)?	[synthesis]
Predict what would happen if...?	[application]
What inference would you make?	[synthesis]
What is more important...?	[analysis]
Compare and contrast...	[analysis]

* *A Taxonomy of Educational Objectives: The Classification of Educational Goals, Handbook I: Cognitive Domain.* Benjamin S. Bloom, editor. McKay Publishers, New York, 1956.

Ten Tasks for Students in Classes That Use Active Learning

Written by Marilla Svinicki, Ph.D.
Director, University of Texas Center for Teaching Effectiveness

1. Make the switch from an "authority-based" conception of learning to a "self-regulated" conception of learning. Recognize and accept your own responsibility for learning.

2. Be willing to take risks and go beyond what is presented in class or the text.

3. Be able to tolerate ambiguity and frustration in the interest of understanding.

4. See errors as opportunities to learn rather than failures. Be willing to make mistakes in class or in study groups so that you can learn from them.

5. Engage in **active** listening to what's happening in class.

6. Trust the instructor's experience in designing class activities and participate willingly, if not enthusiastically.

7. Be willing to express an opinion or hazard a guess.

8. Accept feedback in the spirit of learning rather than as a reflection of you as a person.

9. Prepare for class physically, mentally, and materially (do the reading, work the problems, etc.).

10. Provide support for your classmate's attempts to learn. The best way to learn something well is to teach it to someone who doesn't understand.

<u>Dr. Dee's Eleventh Rule</u>

DON'T PANIC! Pushing yourself beyond the comfort zone is scary but you have to do it in order to improve.

Word Roots for Physiology

a- or an-	without; absence
anti-	against
-ase	signifies an enzyme
auto-	self
bi-	two
brady-	slow
cardio-	heart
cephalo-	head
cerebro-	brain
contra-	against
-crine	a secretion
crypt-	hidden
cutan-	skin
-cyte or cyto-	cell
de-	without, lacking
di-	two
dys-	difficult, faulty
-elle	small
endo-	inside or within
exo-	outside
extra-	outside
-emia	blood
epi-	over
erythro-	red
gastro-	stomach
-gen, -genic	produce
gluco-, glyco-	sugar or sweet
hemo-	blood
hemi-	half
hepato-	liver
homo-	same
hydro-	water

hyper-	above or excess
hypo-	beneath or deficient
inter-	between
intra-	within
-itis	inflammation of
kali-	potassium
leuko-	white
lipo-	fat
lumen	inside of a hollow tube
-lysis	split apart or rupture
macro-	large
micro-	small
mono-	one
multi-	many
myo-	muscle
oligo-	little, few
patho-, -pathy	related to disease
para-	near, close
peri-	around
poly-	many
post-	after
pre-	before
pro-	before
pseudo-	false
re-	again
retro-	backward or behind
semi-	half
sub-	below
super-	above, beyond
supra-	above, on top of
tachy-	rapid
trans-	across, through

MAPPING STRATEGIES FOR PHYSIOLOGY

Introduction

Mapping is a technique to improve a student's understanding and retention of subject material. It is based on the theory that each person has a memory bank of knowledge organized in a unique way based on prior experience. Learning occurs when you attach new ideas to your preexisting framework. By actively interacting with the information and by organizing it in your own way before you load it into memory, you will find that you remember the information longer and can recall it more easily.

Mapping is a non-linear way of organizing material, closely related to the flow charts used to explain many physiological processes. A map can take a variety of forms but usually consists of terms or concepts linked by explanatory arrows. The map may include diagrams or figures. The connecting arrows can be labeled to explain the type of linkage between the terms (structure/function, cause/effect) or may be labeled with explanatory phrases ("is composed of").

You will find a number of maps in the text that you can simply memorize, but the real benefit from using maps occurs when you create the maps yourself. By organizing the material yourself, you question the relationships between terms, organize concepts into a hierarchial structure, and look for similarities and differences between items. Such interaction with the material ensures that you process it into long-term memory instead of simply memorizing it for a test. Teaching you how to map is an important part of the process, as you may not know where to begin.

Key Elements of Maps

A map has only two parts: the concepts and the linkages between them. A concept is an idea, event, or object. Concepts do not exist in isolation; they have associations to other relevant concepts. An example is the sentence "The heart pumps blood." Heart and blood are two concepts related by the verb pumps. A map consists of a group of related terms that are hierarchically ranked and linked by explanatory arrows. In this Student Workbook, we have provided some groups of words to be mapped. You will probably want to develop your own groups of words to fit the material you are studying.

How to Make a Map

1. Choose the concepts to map. Begin at either the top or the center with the most general, important, or overriding concept from which all the others naturally stem. If this is a reflex pathway, you would start with the stimulus. Next, use the other concepts to break down this one idea into progressively more specific parts or to follow the reflex pathway. Use horizontal cross-links to tie branches together. The downward development of the map may reflect the passage of time if the map represents a process or increasing levels of complexity if the map represents something like a cell.

If you are trying to map a large number of terms, you might try writing each term on a small piece of paper. You can then lay out the papers on a table and rearrange them until you are satisfied with your map. Even an experienced physiologist may draw a map several times before being satisfied that it is the best representation of the information.

2. Think about the type of association between two concepts. Arrows will point the direction of the linkage, but you should also label the kind of linkage. You may label the line with linking words or

by the type of link, such as CE (cause or effect). Color is very effective on maps. You can use colors for different types of links or for different sections.

3. Once you have your map, sit back and think about it. Are all the items in the right place? You may want to move them around once you see the big picture. Revise your map to expand the picture with new concepts or to correct wrong linkages. Review the information in the map by beginning with recall of the main concept and then moving to the more specific details. Ask yourself questions like, "What is the cause? effect? parts involved? main characteristics?" to jog your memory.

4. The best way to study with a map is to trade maps with your study partner and see if you can understand each other's maps. You may want to find an empty classroom, put your maps on the blackboards, then step back and compare them. Did one of you put in something the other forgot? Did one of you have an incorrect relationship between two items?

Practice making maps. The study questions in each chapter of your textbook will give you some ideas of what you should be mapping. Your instructor can help you if you do not know how to get started.

LIBRARY RESEARCH

Resources in the library come in two forms: books and journals. Books are the best resource for general background information. If you are trying to learn about a new subject, they can be an excellent place to start, especially older books that have simpler background information. However, unless the book has a recent publication date, it may not be the most up-to-date source of information. One exception is the book that is a compilation of published papers from a **symposium** [a meeting or conference held to discuss a certain topic]. An effort is made to see that these books are published with a minimum of delay.

JOURNALS: Scientific journals are usually sponsored by a scientific organization and consist of contributed papers that describe the **original scientific research** of an individual or group. When a scientist speaks of writing "a paper," he/she is usually referring to the scientific paper published in a journal. Many journals will publish **review articles**. A review article is a synopsis of recent research on a particular topic and is an excellent place to begin a search for information, since it usually contains more up-to-date information than a book on the same topic.

CITATION FORMAT: Citation formats for papers will vary but will usually include the following elements somewhere:

Title [Brackets around the title indicate an English translation of a foreign language paper.]
Year the paper was published.
Name of author(s) Within a body of work, a multiauthor paper is usually cited as **first author, et al.** Et al. is the abbreviation for the Latin *et alii* meaning "and others," and indicates that there are additional authors.
Journal abbreviation, **volume (issue)**: inclusive **pages**. A **volume** number is usually given to all issues published in one calendar year (six months for weekly journals). **Issue** 1 would be the first issue published in a volume, Issue 2 would be the second, etc.

Example: Horiuchi, M., Nishiyama, H., and Katori, R. (1993) Aldosterone-specific membrane receptors and related rapid non-genomic effects. [Review] *Trends Pharmacol Sci* 14(1):1-4.

In many citations, the name of a journal is abbreviated. Here is a list of commonly used abbreviations.

Adv	advances	**Am**	American
Ann	annals	**Annu**	annual
Appl	applied	**Arch**	archives
Assoc	association	**Behav**	behavior
Biochem	biochemistry	**Biol***	biology or biological
Biophys	biophysics	**Br**	British
Can	Canadian	**Chem**	chemistry or chemical
Clin	clinical	**Commun**	communications
Curr	current	**Dev**	developmental
Dis	disease	**Eur**	European
Exp	experimental	**Gen**	general
Hum	human	**Int**	internal
Intl	international	**J**	journal
Med	medicine or medical	**Monogr**	monograph
Nat	natural	**Natl**	national

Pharm	pharmacy	**Physiol***	physiology or physiological
Proc	proceedings	**Q**	quarterly
Res	research	**Rev**	review
Sci	science	**Soc**	society or social
Surg	surgery or surgical	**Symp**	symposium
Ther	therapy		

* Most words ending with -ology or -ological will be abbreviated by stopping after the "l".
Titles of one word such as "Nature" are never abbreviated.

Citing Sources Published on the World Wide Web

Searching the World Wide Web has become a standard method for gathering information. However, a word of caution is in order. Most articles published in scientific journals have gone through a screening process known as peer review, in which the article is read and critiqued by other specialists in a particular field. In some cases, articles that are submitted are rejected by the journal editor, and in many cases the authors must make revisions to the article before it can be published. This process acts as a safeguard against the publication of poorly done research. However, anyone can create a web page and publish information on the Web. There is no screening process, so the reader must decide how valid the information is. Web sites that are published by recognized universities and not-for-profit organizations are likely to have good information. But an article on vitamins on the web page of a health food store should be viewed with a skeptical eye unless the article cites published research.

Citing sources from the Web requires a different format. Here is one suggested format:

Author (year; month and day if appropriate) Title. Source name [online]. Available: (electronic address sufficient for retrieval)

Examples:

Long, Chris. (1995, August 14) Teachers hope to use Net to make science fun. *Austin American Statesman* [online], City/State section, p. B2. Available: NEXIS: News NEXIS File:AAS.

English, Peter. (1997, Nov. 10) Birds of the Ecuadorian Rainforest. Internet.
http://www.utexas.edu/depts/grg/gstudent/grg394k/spring97/english/english.html

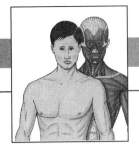

CHAPTER 1

INTRODUCTION TO PHYSIOLOGY

SUMMARY

What should you take away from this chapter?

• What are the different levels of organization for living organisms?
• Be able to name the physiological systems of the human body.
• What is the difference between a teleological and mechanistic approach to science?
• List six key themes in physiology.
• Be able to describe how scientists design and execute experiments.

Physiology is the study of how organisms function and adapt to a constantly changing environment.

The human body is comprised of ten organ systems. However, these organ systems do not act as isolated units. Instead they communicate and cooperate to maintain homeostasis, a relatively stable internal environment composed of the extracellular fluid that baths the cells. Physiologists use a wide variety of techniques to study how the human body operates. Some of these techniques examine the activities of molecules and cells, while others focus on the response of entire organ systems. Physiology is usually approached from a functional, or mechanistic, viewpoint. Physiological events can also be explained in terms of their significance, which is considered a teleological approach to physiology. Scientific experimentation includes formulation of a hypothesis, observation and experimentation, and data collection and analysis. Experiments using human subjects are difficult to perform and analyze because of tremendous variability within human populations and because of ethical problems.

TEACH YOURSELF THE BASICS

LEVELS OF ORGANIZATION

• List the ten levels of organization, starting with atoms and ending with the biosphere. (p. 2)

_____

_____

• List the ten human organ systems. (p. 2; Table 1.1, p. 3)

_____

_____

PHYSIOLOGY IS AN INTEGRATIVE SCIENCE

• What do we mean when we say that physiology is an integrative science? (p. 3-4)

_____

_____

PROCESS AND FUNCTION

• What is the difference between a teleological approach to physiology versus a mechanistic approach? Use the pumping of blood by the heart as an example. (p. 5)

_____

_____

THE EVOLUTION OF PHYSIOLOGICAL SYSTEMS

• Humans are animals adapted to a terrestrial environment. What is the primary challenge of life on land? (p. 5)

_____

_____

• What is the "external environment" for the individual cells of the body? (p. 5)

_____

_____

• Define homeostasis. (p. 6)

_____

_____

THEMES IN PHYSIOLOGY

• List five key themes of physiology in addition to homeostasis. (p. 6)

_____

_____

THE SCIENCE OF PHYSIOLOGY

• List the key steps a scientist goes through in a scientific inquiry. (p. 7)

_____

_____

• In an experiment, which are independent variables and which are dependent variables? (p. 7)

_____

_____

• Why should every experiment have a control? (p. 7)

_____

_____

• How does a scientific theory differ from a hypothesis? (p. 7)

_____

_____

• Why is a crossover study better than a study in which the experimental and control groups are composed of different organisms? (p. 9)

_____

_____

• What advantage is gained by having a blind study? A double-blind study? (p. 10)

_____

_____

TALK THE TALK

Aristotle
blind study
cell
cell to cell communication
circulatory system
concept map
control
crossover study
data
dependent variable
digestive system
double-blind crossover study
double-blind study
endocrine system
energy
external environment
extracellular fluid
fertilization
fetal development
Hippocrates
homeostasis
hypothesis
immune system
independent variable

integration of body systems
integumentary system
internal environment
law of mass balance
lumen
mass flow
mechanistic approach
musculoskeletal system
nervous system
organ system
pH
physiology
placebo
placebo effect
reproductive system
respiratory system
salinity
scientific method
skin
teleological approach
temperature regulation
tissue
urinary system

ERRATA

p. 7, left column, third text block: should read ... "infusion of glucose containing **50** grams of glucose per liter."
p. 12, question 7: should be Figure 1-1, not 1-10.

PRACTICE MAKES PERFECT

1. How does physiology differ from anatomy?

_____

2. A scientist wants to study the effects of the cholesterol-lowering drug pravastatin in rats.

a) What might be the hypothesis in this experiment?_____

_____

b) The scientist decides to use four different doses of pravastatin. What would be an appropriate control for this experiment?

_____

3. Write one or two sentences that summarize the graph below.

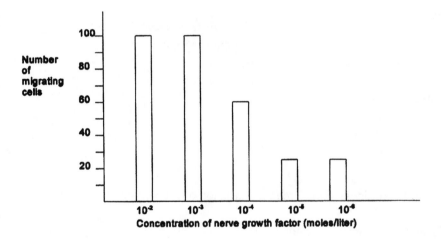

4. Glucose is transported into cells from the extracellular fluid. Construct a graph using the following data and label the axes. Summarize these results in the legend. Identify and distinguish between the independent and dependent variables.

Extracellular concentration of glucose (mM)	Intracellular concentration of glucose (mM)
150	100
140	101
130	100
120	91
110	83
100	75
90	61
80	52

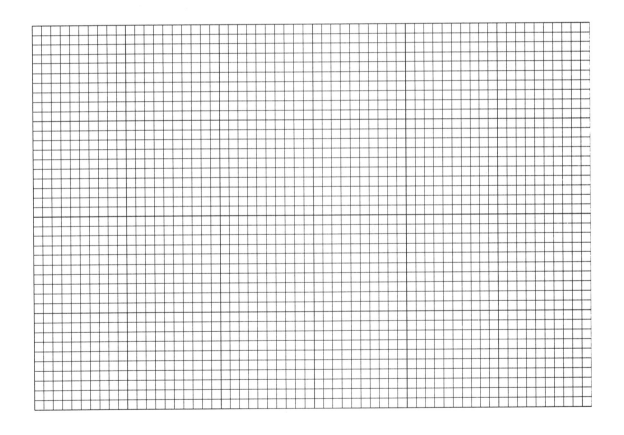

5. Construct a map showing the relationships between the ten organ systems. Include key words that illustrate functional interactions between systems. For example, the circulatory system transports wastes to the urinary system.

As you encounter additional information about these physiological systems, consider adding more information to this map. Maps are an excellent way to illustrate the functional integration of tissues, organs, and organ systems.

BEYOND THE PAGES

GRAPHS

Find examples of good and bad graphs in your textbooks or in newspapers and magazines. *USA Today* is an excellent source of misleading graphs.

READING

Clarke, Geoffrey M. (1994), *Statistics and experimental design: An introduction for biologists and biochemists* (3rd ed.). Halsted Press, New York.

Franklin, Kenneth J. (1949) *A short history of physiology* (2d ed.). Staples Press, New York.

Fye, W. Bruce. (1987) *The development of American physiology: Scientific medicine in the nineteenth century.* Johns Hopkins University Press, Baltimore.

Goldstein, M. and Goldstein, I. F. (1978). *How we know: An exploration of the scientific process.* Plenum Press, New York.

Goodfield, June. (1960). *The growth of scientific physiology; physiological method and the mechanist-vitalist controversy, illustrated by the problems of respiration and animal heat.* Hutchinson, London.

Hall, Thomas S. (1969) *Ideas of life and matter; studies in the history of general physiology, 600 B.C.-1900 A.D.* University of Chicago Press, Chicago. (Published in 1975 under title: History of general physiology, 600 B.C. to A.D. 1900.)

Hodgkin, A. L. (ed.) (1977) *The pursuit of nature: informal essays on the history of physiology.* Cambridge University Press, New York.
    These essays were written as part of the celebrations of the centenary of the Physiological Society in 1976.

Pease, Craig M. and Bull, J.J. (1992, April) *Is science logical?* Bioscience 42 (4): 293-298.

Rothschuh, Karl E. (1973) *History of physiology.* Translated and edited by Guenter B. Risse. R. E. Krieger Pub. Co., Huntington, N.Y.

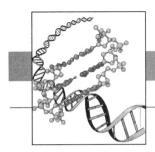

CHAPTER 2

ATOMS, IONS, AND MOLECULES

SUMMARY

What should you take away from this chapter?

• Be able to describe the general structure of atoms.
• Be able to discuss how the different subatomic particles affect the nature of atoms.
• You should have a general knowledge of the four types of bonds.
• What is the difference between polar and nonpolar molecules?
• Spend some serious time learning all about solutions and solutes.
• Become familiar with molarity and osmolarity, equivalents, and percent solutions.
• Understand the basic differences between the biomolecules.

Atoms are composed of protons, neutrons, and electrons, but it is the number of protons that distinguishes one element from another element. Atoms of a particular element that have differing numbers of neutrons are called isotopes. Electrons are located in shells that circle the nucleus. Covalent, ionic, and hydrogen bonds are formed by specific types of interactions between electrons of different atoms. Van der Waals forces are very weak bonds formed by non-specific attractions and repulsions between atoms.

Water is known as the universal solvent. Molecules that easily dissolve in water are called hydrophilic and molecules that do not dissolve are called hydrophobic. The concentration of a solution can be expressed in molar, equivalent, and percent solutions. Concentration is important to many body functions. Understanding concentration allows you to predict water movement and cell volume changes. The body has many homeostatic mechanisms, to be discussed later in the book, that are specifically designed to maintain concentration and water balance.

The concentration of hydrogen ions (H^+) determines body pH. Acids are molecules that contribute H^+; bases bind H^+. The body contains specific buffers that act as bases to maintain a normal body pH of 7.4.

There are four basic groups of biomolecules: carbohydrates, lipids, proteins, and nucleotides. Carbohydrates are structurally divided into monosaccharides, disaccharides, and polysaccharides. The most common lipid, the triglyceride, is composed of three fatty acids linked to a glycerol. Other lipid-related molecules include phospholipids, steroids, and eicosanoids. Proteins are composed of hundreds of amino acids that are arranged into chains, spirals, sheets, and complex globular configurations. These molecules have a wide variety of functions ranging from structural components to carriers and enzymes. Lipoproteins, glycoproteins, and glycolipids are frequently associated with cell membranes. Nucleotides consist of one or more phosphate groups, a five-carbon sugar, and a nitrogenous base. Examples of nucleotides include ATP, ADP, cAMP, NAD, FAD, DNA, and RNA.

TEACH YOURSELF THE BASICS

ELEMENTS AND ATOMS

Atoms Are Composed of Protons, Neutrons, and Electrons

• List the subatomic particles and describe their characteristics. (p. 14)

_____

_____

_____

_____

• What is found in the nucleus of an atom? (Fig. 2-1, p. 14)

_____

An Element Is Distinguished by the Unique Number of Protons in Its Nucleus

• Know how to read a periodic table. (Fig. 2-2, p. 15)

• What is the significance of essential elements? (p. 14)

_____

• What determines the atomic number of an element? (p. 16)

_____

• What determines the atomic mass of an element? (p. 16)

_____

Isotopes of An Element Are Atoms with Different Numbers of Neutrons

• Describe isotope characteristics. Use hydrogen as an example. (Fig. 2-4, p. 17)

_____

_____

• What are some of the uses of isotopes? (p. 17)

_____

Electrons Form Bonds between Atoms and Capture Energy

• Describe how electrons are arranged in an atom. (p. 17)

_____

_____

• What is a high-energy electron? (p. 18)

_____

• What role do high-energy bonds play in the cell? (p. 18)

_____

MOLECULES AND BONDS

• What is a molecule? (p. 19)

_____

What Does the Chemical Formula of a Molecule Tell Us?

• Know how to calculate the molecular weight of a molecule. (p. 19)

• What does a chemical formula tell you about the atoms in a compound? What does it not tell you? (p. 19)

_____

_____

• List the four common types of bonds. (p. 19)

_____

Covalent Bonds Are Formed When Adjacent Atoms Share Electrons

• Explain how covalent bonds are formed. (p. 19)

_____

• How can you determine the number of covalent bonds that an atom can form? (pp. 19-20)

_____

Functional groups
• Know the most common functional groups in biological molecules. (Table 2-1, p. 21)

• What are the characteristics of a functional group? (p. 21)

_____

Molecular shape
• Describe some of the shapes found in biological molecules (Fig. 2-7, p. 21; Fig. 2-16, p. 31; Fig. 2-18, p. 33)

_____

Polar and nonpolar molecules
• What are the properties of a polar molecule? Give at least one example. (p. 21)

_____

• What are the properties of a nonpolar molecule? Give at least one example. (p. 22)

_____

Ionic Bonds Form When Atoms Gain or Lose Electrons

• What is an ion? (p. 22)

_____

• What is a cation? What is an anion? Give an example of each. (p. 22)

_____

• How is an ionic bond different from a covalent bond? (p. 22;   p. 19)

_____

Hydrogen Bonds Are Weak Bonds between Molecules or Regions of a Molecule

• How are hydrogen bonds formed and what elements form hydrogen bonds? (p. 22)

_____

• What property of water is associated with hydrogen bonds between water molecules? (p. 22)

_____

Van der Waals Forces Are Weak Attractions between All Molecules

• Define van der Waals forces. (p. 23)

_____

• Name a function of van der Waals forces. (p. 23)

_____

SOLUTIONS AND SOLUTES

• What is a solute? A solvent? A solution? (p. 24)

_____

Solubility in Aqueous Solutions

• What is the difference between a hydrophilic and a hydrophobic molecule? (p. 24)

_____

• What property of a molecule determines its solubility? (p. 24)

_____

The Concentration of Solutions

• Concentration is defined as: (p. 25)

_____

• When describing the concentration of a solution, (p. 25)

      Weight is usually expressed in what units? _____

      Number of solute molecules is usually expressed in what units? _____

      Solute ions can also be expressed in what units? _____

      Volume is usually expressed in what units? _____

Moles and molarity
• 1 mole   =   _____ atoms (p. 25)

• Define a molar solution. (p. 25)

_____

• Sodium chloride (NaCl) has a molecular weight of 58.5. How would you make a 1 molar solution of NaCl? (p. 25; answer in workbook appendix)

_____

Equivalents
• Define an equivalent. (p. 25)

_____

• One mole of magnesium ions (Mg^{2+}) contains _____ equivalents. (p. 25; answer in workbook appendix)

Weight/volume and percent solutions
• How would you make 250 mL of a 10% solution of NaCl? (p. 25; answer in workbook appendix)

_____

• If a solution contains 200 mg NaCl/dL, what is its concentration in g/L? (p. 25; answer in workbook appendix)

_____

The Concentration of Hydrogen Ions in the Body Is Expressed in pH Units

• Molecules that ionize and release an H^+ are called _____. (p. 25)

• Molecules that combine with H^+ or produce OH^- are called _____. (p. 25)

• Define pH in words. (p. 25)

_____

• Define pH mathematically: (Appendix A, p. 739)

_____

• How do buffers in the body help maintain normal pH? (p. 26)

_____

BIOMOLECULES

• What is an organic molecule and what elements are most commonly found in these molecules? (p. 27)

_____

• Name the four basic classes of biomolecules. (p. 27)

_____

Carbohydrates Are the Most Abundant Type of Biomolecule

• What is the basic formula of a carbohydrate? (p. 27)

_____

• What are monosaccharides? Give some examples. (p. 27)

_____

• What are disaccharides? Give some examples. (p. 27)

_____

• What are polysaccharides? What are some of their functions? Give examples. (p. 27)

_____

_____

• In what form do animals store complex sugars? (p. 27)

_____

Lipids Are Made Mainly of Hydrogen and Carbon

• What elements are lipids made of? How do lipids differ from carbohydrates? (p. 27)

_____

• Name three types of lipid-related molecules. Give at least one function for each type. (pp. 27-29)

_____

• What is the difference between glycerol and a fatty acid? (p. 27; Fig. 2-14, p. 29)

_____

• What is the difference between a saturated, a mono-unsaturated, and a polyunsaturated fatty acid? (p. 27)

_____

Proteins Are the Most Versatile of the Biomolecules

• What biomolecules are the building blocks of proteins? (p. 29)

_____

• What is an essential amino acid? (p. 29)

_____

• Describe the organization of proteins, starting with amino acids and ending with the quaternary structure. (Fig. 2-16, p. 31)

_____

_____

_____

• What functions do proteins serve in the body? (pp. 30-31)

_____

Some Molecules Combine Carbohydrates, Proteins, and Lipids

• What is a conjugated protein? (p. 31)

_____

• Name three different types of conjugated proteins and tell where in the body they are found. (p. 31)

_____

Nucleotides Are Responsible for the Transmission and Storage of Energy and Information

• List the components of a nucleotide. (p. 32)

_____

• List as many examples of nucleotides as you can. (p. 33)

_____

• Name the bases that make up the DNA molecule. How does DNA differ from RNA? (p. 33)

_____

_____

TALK THE TALK

acid
adenine
ADP
alkaline
alpha-helix
amino acid
amino group
anion
atom
atomic mass
atomic mass unit
atomic number
ATP
base
bicarbonate ion
biomolecule
buffer
carbohydrate
carbonic acid
carboxyl group
cation
cellulose
chemical formula
compound
concentration
conjugated protein
covalent bond
Crick, F.H.C.
cyclic AMP
cytosine
dalton
deciliter (dL)
deoxyribose
dextrose
disaccharide
DNA
eicosanoid
electron

element
equivalent
essential amino acid
essential element
fat
fatty acid
fibrous protein
fructose
galactose
globular protein
glucose
glycerol
glycogen
glycolipid
glycoprotein
guanine
hydrogen bond
hydrophilic
hydrophobic
hydroxyl group
ion
ionic bond
isotope
lactose
lipid
lipoprotein
maltose
molarity
mole
molecular weight
molecule
monosaccharide
neutron
nonpolar molecule
nucleotide
orbital
organic compound

peptide
percent solution
pH
phosphate group
phospholipid
pleated sheet
polar molecule
polymer
polypeptide
polysaccharide
primary structure of a protein
protein
proton
purine
pyrimidine
quaternary structure of a protein
radiation
radioisotope
ribose
RNA
saturated fatty acid
secondary structure of a protein
solute
solution
solvent
starch
steroid
sucrose
surface tension
tertiary structure of a protein
thymine
trace element
triglyceride
unsaturated fatty acid
uracil
van der Waals force
Watson, J.D.

ERRATA

p. 14, Running problem: Yes, we really do know that football teams don't have forwards!

p. 22-23, Running problem: Chromium ions are cations, not anions as shown. The valence is correct in the conclusion on p. 38.

p. 29, Fig. 2-14: Palmitate has one CH_2 too many. It and oleate should be named palmitic and oleic acids, as they are not in the anion form as indicated by the suffix -ate.

QUANTITATIVE THINKING

In the laboratory, you may be asked to make up or mix solutions. The first example below shows you how to make up a molar solution. The second example shows how to calculate the concentration when you mix two different solutions.

<u>Task 1</u>: Make 500 mL of a 150 millimolar (mM) NaCl solution.

<u>Step 1</u>: Calculate the molecular weight of NaCl.

Atomic weight of Na is 23 × 1 atom of Na =	23
Atomic weight of Cl is 35.5 × 1 atom of Cl =	<u>35.5</u>
sum	58.5

One mole of a substance contains its molecular weight in grams; this is known as its gram molecular weight. One mole of NaCl therefore weighs 58.5 grams.

<u>Step 2</u>: Molar solutions are expressed in moles per liter of solution. You want to make a solution with 0.150 moles per liter (= 150 mM). However, you cannot weigh out moles of a compound. So you must calculate how many grams of NaCl are equal to 0.150 moles. You know that one mole weighs 58.5 grams.

Set up a ratio as shown:

(a) $\dfrac{58.5 \text{ g NaCl}}{1 \text{ mole}}$ $=$ $\dfrac{? \text{ g NaCl}}{0.15 \text{ moles}}$

(b) $\dfrac{58.5 \text{ g NaCl} \times 0.15 \text{ moles}}{1 \text{ mole}}$ $=$ $? \text{ g NaCl}$

(c)   ? = 8.8 g NaCl

Therefore 0.15 moles of NaCl weighs 8.8 grams, and a 0.15 M solution has 8.8 g/liter of solution.

<u>Step 3</u>: You know from step 2 how much NaCl to use to make one liter of 0.15 M NaCl. But you have been asked to make up 500 mL, not one liter. Again, set up a ratio:

$\dfrac{8.8 \text{ g NaCl}}{1000 \text{ mL}}$ $=$ $\dfrac{? \text{ g}}{500 \text{ mL}}$

The answer is 4.4 grams of NaCl is needed to make up 500 mL of a 150 mM solution.

The next example shows you how to calculate the concentration when you mix two different solutions.

<u>Task 2</u>: Mix 2 liters of 3 M NaCl with 1 liter of 6 M glucose. What is the concentration of glucose in the mixed solution? What is the concentration of NaCl?

First figure out the <u>amount</u> of solute that you have in the two starting solutions:

    3 moles NaCl/L × 2 L   =   6 moles NaCl

    6 moles glucose/L × 1 L   =   6 moles glucose

Now put those amounts in the total volume formed when the solutions are added to each other:

    <u>6 moles NaCl + 6 moles glucose</u>

           3 L total volume

    6 moles/3 liters = 2 moles/liter for NaCl and 2 moles/liter for glucose

Therefore the NaCl concentration is 2 M and the glucose concentration is 2 M.

What is the <u>total</u> concentration of the mixed NaCl/glucose solution? Show your work.

<u>Task 3</u>: Make 600 mL of a 300 mM glucose solution. (Mol. wt. of glucose = 180)

PRACTICE MAKES PERFECT

1. Draw a carbon atom and an oxygen atom with the appropriate number of protons, neutrons, and electrons.

2. A carbon dioxide molecule consists of covalent bonds between one carbon and two oxygen atoms. Draw how their valence (outer shell) electrons are arranged using an electron-dot model.

3. Matching:

_____ atomic mass   A. atoms with different numbers of neutrons

_____ atomic number   B. protons plus neutrons

_____ isotopes   C. capture and transfer energy

_____ electrons   D. number of protons

_____ radioisotopes   E. atoms with different numbers of protons

_____ radiation   F. unstable isotopes

_____ atomic mass unit   G. energy emitted by radioisotope

                                            H. dalton

4. Using what you know about atomic number, atomic weight, protons, neutrons, and electrons in atoms, fill in the table below. There is no way to predict the number of neutrons in the different isotopes of an element, but you can guess the number of neutrons in the most common isotope by using the atomic weight. Check your answers against the periodic table on p. 15 of the text.

element	symbol	atomic number	protons	electrons	neutrons*	atomic weight
Calcium		20			20	40.1
Carbon			6	6	12.0	
Chlorine			17		18	35.5
	Co			27	27	58.9
Hydrogen			1	0		
Iodine	I	53		74		
Magnesium			12	12	12	
	N	7		7		14
Oxygen	O		8			16.0
Sodium			11			23
Zinc			30		35	
Copper				29	35	
	Fe	26				55.8
Potassium		19	19		20	

* Number of neutrons in the most common isotope.

5. What is the molecular weight of water? Indicate the proper units.

6. Matching:

____ anion   A. two or more atoms that share electrons
____ cation   B. a pair of electrons shared by two atoms
____ covalent bond   C. molecule that contains more than one element
____ ion   D. bond that shares protons or neutrons
____ molecule   E. positively charged ion
____ compound   F. negatively charged ion
                         G. an atom that gains or loses one or more electrons from/to another atom

7. Which of the following structural formulas for the amino acid leucine ($C_6H_{13}O_2N$) is (are) correct? Be able to explain your reasoning.

8. Answer the following questions about the properties of water.

a. A single water molecule is held together by what type of bond? _____

b. Many individual water molecules are held together by what type of bond? _____

c. Why is water called the universal solvent?

d. When a salt crystal of sodium chloride is dropped into water, how do the molecular properties of water and the ionic properties of the salt interact?

9. What happens to an oxygen atom if it gains a proton?

10. When potassium and chloride form an ionic bond, which ion gains an electron and which ion loses an electron? Which ion(s) becomes stable? Explain your reasoning.

11. Why are polar molecules hydrophilic and nonpolar molecules hydrophobic?

12. What is the weight of a half mole of sodium chloride?

13. How would you make a 500 mL NaCl solution that has a molarity of 0.5 M?

14. A 0.1 M solution is equal to how many millimoles per liter? _____

15. How many grams of glucose are in 100 mL of a 50 mM glucose solution? _____

16. How would you make 100 mL of a 3% glucose solution? What would be the molarity of this solution? (The molecular weight of glucose is 180.)

17. If you mix 1 liter of 0.4 M glucose with 1 liter of 0.8 M NaCl:
    a) What is the total concentration of the mixed solution? _____

    b) What is the NaCl concentration in the mixed solution? _____

    c) What is the glucose concentration in the mixed solution? _____

18. The plasma concentration of Na^+ is 142 mEq/L. What is the concentration of Na^+ in millimoles per liter?

19. The plasma concentration of Ca^{2+} is 5 mEq/L. What is the concentration of Ca^{2+} in millimoles per liter?

20. Matching. (Blanks may have more than one correct answer.)

____ acid   A. concentration of hydrogen ions

____ base   B. prevents pH changes

____ pH   C. molecule that ionizes and donates H^+

____ alkaline   D. molecule that combine with free H^+

____ buffer   E. molecule that produce hydroxide ions (OH^-)

                             F. solution with a low concentration of H^+

21. Amines are organic compounds that act as acids and bases. Identify the acids and bases in the reaction below.

$$CH_3-\overset{\overset{\displaystyle CH_3}{|}}{N}-H \ + H-OH \ \rightarrow \ CH_3-\overset{\overset{\displaystyle CH_3}{|}}{\underset{\underset{\displaystyle H}{|}}{N^{\pm}}}-H \ + \ OH^-$$

22. The reaction below occurs in red blood cells. Identify the acids and bases in this reaction.

$$CO_2 \ + \ H_2O \ \rightleftarrows \ H_2CO_3 \ \rightleftarrows \ H^+ \ + \ HCO_3^-$$

23. Which HCl solution is more acidic, a 50 mM or a 0.5 M? _____

24. Matching:

____ triglycerides   A. large molecules with repeating units

____ fatty acids   B. long carbon chains with terminal carboxyl groups

____ carbohydrates   C. fatty acid chains linked to a glycerol

____ polymers   D. polar, hydrophilic molecules

____ steroids   E. $(CH_2O)n$

                             F. lipid-related molecules with four carbon rings

25. Matching. (Blanks may have more than one answer, and answers may be used more than once.)

____ essential amino acid   A. made from amino acids

____ protein   B. a molecule that has proteins plus lipids

____ conjugated protein   C. a molecule that has proteins plus carbohydrates

____ glycoprotein   D. amino acid not made by the body

____ phospholipid   E. may act as a membrane receptor

                             F. major component of membranes

26. Use the following terms to create a map of biomolecules. Be sure to put labels on your linking arrows. The term "biomolecule" is the most general term and should appear at the top of your map.

amino acids
carbohydrates
cellulose
disaccharides
fats
fatty acids
glucose
glycogen
glycolipids

lipids
lipoproteins
maltose
monosaccharides
peptide
phospholipids
polymer
polypeptide
polysaccharides

proteins
saturated fatty acids
starch
steroids
sucrose
triglycerides
unsaturated fatty acids

BEYOND THE PAGES
READING

Breast cancer imaging. (1996, January/February) *Science & Medicine.*

The discovery of X-rays. (1993, November) *Scientific American.*

Electromagnetic fields and power lines. (1995, July/August) *Science & Medicine.*

Health effects of ozone. (1997, May/June) *Science & Medicine.*

Magnesium: Growing in clinical importance. (1994, January 15) *Patient Care.*

The photon radiosurgery system. (1995, November/December) *Science & Medicine.*

Xenon-enhanced CT of cerebral blood flow. (1995, September/October) *Science & Medicine.*

SUMMARY

What should you take away from this chapter?

- Familiarize yourself with the general organization of a cell. Learn the organelles and their functions.
- Be able to describe differences between the tissue types and subtypes: epithelia, connective tissue, nerve, and muscle.
- What is the importance of the extracellular matrix?

The study of cytology has greatly benefited from the use of sophisticated light and electron microscopes. These instruments combined with other techniques have illuminated the structure and function of cellular organelles. This chapter discusses the basic components of a cell, how cells are held together, and the different types of tissues formed by highly differentiated cells.

An extracelluar matrix composed mostly of glycoproteins and protein fibers helps hold cells together and provides a structural base for cell growth and migration during development. A variety of cell junctions formed by proteins and glycoproteins also hold cells together and participate in cell to cell communication. Adhesive junctions, such as desmosomes, resist physical stresses while tight junctions prevent the leakage of material between cells. Gap junctions are pores or channels between cells that allow substances such as ions to pass directly between cells.

The four basic tissue types are epithelial, connective, muscle, and neural. Epithelia can be functionally divided into five types: exchange, transporting, ciliated, protective, and secretory. The seven types of connective tissue are loose connective tissue, dense regular and dense irregular connective tissue, adipose, blood, cartilage, and bone. These connective tissues consist of an extracellular matrix that contains one or more of the following protein fibers: collagen, elastin, fibrillin, and fibronectin. There are three types of muscle: skeletal, cardiac, and smooth. Neural tissue and muscle are considered to be excitable tissues. Organs are composed of collections of tissues that perform certain functions.

TEACH YOURSELF THE BASICS

STUDYING CELLS AND TISSUES

- What is the difference between cytology and histology? (p. 40, 53)

_____

- What can light microscopes do that electron microscopes cannot? What can a scanning electron microscope do that a transmission electron microscope cannot? (p. 41)

_____

_____

CELLULAR ANATOMY

• What happens to a cell when it differentiates? (p. 41)

_____

• What is extracelluar fluid? (p. 43)_____

The Cell Membrane

• Describe cell membrane structure. (p. 43; Fig. 3-5, p. 46)

_____

• What are the two primary functions of the cell membrane? (p. 43)

_____

The Cytoplasm

• What is the difference between cytosol and cytoplasm? (p. 43)

_____

• Based on their structure, organelles are divided into what two groups? (p. 43)

_____

Nonmembranous Organelles

• Name the two groups of nonmembranous organelles. (p. 43)

_____

• Ribosomes are made of _____. (p. 43; Fig. 3-6, p. 46)

• Their function is to _____. (p. 43)

• Distinguish between free ribosomes, fixed ribosomes, and polyribosomes. (p. 43)

_____

• List the four sizes of cytoplasmic protein fibers. (p. 43)

_____

• List four types of proteins that make up these fibers. (p. 43)

_____

• Describe the structure of the cytoskeleton. (p. 47; Fig. 3-7)

_____

• List five functions of the cytoskeleton. (p. 47)

_____

• What are microvilli? (p. 47) _____

• Describe the structure and function of centrioles. (pp. 47; Fig. 3-8b, p. 48)

_____

• Compare the structure of cilia and flagella. (pp. 47-48)

_____

• Compare the functions of cilia and flagella and the cell types on which they are found. (pp. 47-48)

_____

Membranous Organelles

• What functional advantage do membranous organelles have? (p. 48)

_____

• List five membranous organelles. (p. 48)

_____

• Describe the structure of a mitochondrion. Be sure to include the following words: matrix, cristae, intermembrane space. (pp. 48-49)

_____

• What is the primary function of the mitochondria? (p. 48)

_____

• What two unusual characteristics do mitochondria possess? (p. 49)

_____

• What are the anatomical and functional differences between rough and smooth endoplasmic reticulum or ER? (p. 49; Fig. 3-10, p. 50; p. 27)

_____

_____

• Describe the relationship between the Golgi apparatus and the rough endoplasmic reticulum (RER). (pp. 50-51; Fig. 3-11, p. 50)

_____

• Distinguish between secretory vesicles, transport vesicles, storage vesicles, and lysosomes. (pp. 49-51)

_____

_____

• How is lysosomal enzyme activity regulated? (p. 51)

_____

• What happens to cells if lysosomes fail to function? If they leak enzymes into the cytosol? (p. 51)

_____

_____

• How do peroxisomes differ from lysosomes? (p. 52)

_____

The Nucleus

• Describe the organization and components of the nucleus. (p. 52; Fig. 3-13, p. 32)

_____

_____

• What is the difference between the nucleus and a nucleolus? (p. 53)

_____

TISSUES OF THE BODY

• What is a tissue? (p. 53)

_____

• List five physical features that histologists use to describe tissues. (p. 53)

_____

_____

• Name the four primary tissue types and give some distinguishing characteristics of each type. (Table 3-1, p. 54)

_____

_____

_____

Extracellular Matrix Helps Support Tissues

• Describe the structure and function of the extracellular matrix.

_____

Cell Junctions Hold Cells Together to Form Tissues

• How do adhesive junctions, tight junctions, and gap junctions differ in their structure and function? (p. 54; Fig. 3-14, p. 55)

_____

_____

_____

Adhesive junctions:
The second type of adhesive junction is the adherens junction. **Adherens junctions** are similar to desmosomes but are not quite as strong.

Epithelia

• List three general functions of epithelial tissues. (p. 56)

_____

• What is the basal lamina and what is its function? (p. 56)

_____

• Name the five functional types of epithelia and give examples of where each type can be found in the body. (p. 56; Table 3-2, p. 58; Fig. 3-16, p. 57)

_____

_____

• Explain how the function of each type of epithelium is reflected by its structure.

_____

_____

_____

_____

• Explain the significance of these terms as they relate to a transporting epithelium: lumen [p. 2], apical membrane, microvilli, basolateral membrane, tight junction, extracellular fluid. (p. 58; Fig. 3-17, p. 59)

_____

_____

• Define secretion. (p. 60)

_____

• Distinguish between an endocrine gland and an exocrine gland. (p. 60; Fig. 3-20, p. 61)

_____

• What is the difference between a serous secretion and a mucous secretion? (p. 60)

_____

Connective Tissues

• What is the distinguishing characteristic of connective tissues? (p. 61)

_____

• Name the three basic components of matrix. (pp. 61-62; Table 3-3, p. 62)

_____

• Name the four main protein fibers of matrix and describe their key properties. (p. 62)

_____

• Name the seven types of connective tissue and tell where they are found. (pp. 62-63; Table 3-3, p. 62)

_____

_____

Muscle and Nerve

• What is meant by the term "excitable tissue"? (p. 63)

_____

• What is the primary characteristic of muscle? (p. 64)

_____

• List the three types of muscle tissues. (p. 64)

_____

• What is the primary function of neural tissue? (p. 66)

_____

ORGANS

• What is an organ? (p. 67)

_____

TALK THE TALK

actin
adherens junction
adhesive junction
adipose tissue
apical membrane
basal body
basal lamina
basement membrane
basolateral membrane
blood
bone
cartilage
cell or plasma membrane
centriole
ciliated epithelium
cilium ‡
collagen
connective tissue
crista ‡
cytology
cytoplasm
cytoskeleton
cytosol
dense connective tissue
desmosome
differentiation
elastin
endocrine gland

epithelium ‡
exchange epithelium
exocrine gland
extracellular fluid
extracellular matrix
fibrillin
fibronectin
fixed ribosome
flagellum ‡
gap junction
goblet cell
Golgi apparatus
histology
Hooke, Robert
hydrogen peroxide
inclusion
intermediate filament
intermembrane space
keratin
loose connective tissue
lysosome
matrix ‡
membranous organelle
microfilament
microtubule
microvilli *
mitochondria *
mucous secretion

myosin
neurofilament
nonmembranous organelle
nuclear envelope
nucleolus
nucleus ‡
organ
organelle
peroxisome
polyribosome
pore
protective epithelium
ribosome
rough endoplasmic reticulum
secretion
secretory epithelium
secretory vesicle
serous secretion
smooth endoplasmic reticulum
spot desmosome
storage vesicle
thick filament
tight junction
transport epithelium
transport vesicle
van Leeuwenhoek, Antonie

* These words are in the plural form. What are their singular forms?
‡ These words are in the singular form. What are their plural forms?

ERRATA

p. 47, right column: Callout for Fig. 3-8a should read 3-8b; Callout for 3-8b should read 3-8a.

p. 55: In the description of adhesive junctions, a sentence was dropped:
   Adherens junctions are similar to desmosomes but are not quite as strong.

RUNNING PROBLEM: Pap Smear

Neuromedical Systems, Inc., the developers of PAPNET, ran an informational ad in the 9/2/96 edition of *People* magazine, suggesting that women request PAPNET for their pap smears "for added peace of mind" and about $40 over the usual cost of reading a pap smear. For information, call 1-88-PAPNET-4.

About benefits and costs: Cervical screening as defensive medicine. Science & Medicine, November/December 1995.

PRACTICE MAKES PERFECT

1. Which organelle if separated from its cell would have the highest probability of existing and evolving into a functional life form? Explain your reasoning.

2. What is the advantage of having the nucleus partially separated from the cytoplasm?

3. If intestinal cells lacked a cytoskeleton, how would their structure and function be affected?

4. Why are epithelia more susceptible to developing cancer than most other tissues?

5. A histological examination of a tissue shows that it has tight junction. Would you expect the solutions normally found on either side of the epithelium to be the same or different? Explain.

6. What anatomical and functional qualities of the epidermis (skin) help regulate body temperature and prevent dehydration?

7. Matching:

_____ ground substance
_____ fibroblast
_____ collagen
_____ elastin
_____ fibrillin
_____ fibronectin

A. exceptionally strong, inelastic protein fiber
B. cell that breaks down extracellular matrix
C. matrix of glycoproteins, water, protein fibers
D. cell that secretes extracellular matrix
E. a coiled protein fiber with elasticity
F. protein fiber that connects cells to extracellular matrix
G. a very thin fiber that combines with elastin

8. Choose the type of protein fiber(s), if any, that is associated with a particular type of connective tissue.
a) collagen   b) elastin   c) fibrillin   d) fibronectin   e) none

_____ under the skin
_____ sheaths that surround nerves and muscles
_____ cartilage
_____ adipose

_____ tendons and ligaments
_____ blood
_____ lungs and blood vessels
_____ bones

9. Put the letter of the correct answer in front of the phrase below:
a) rough endoplasmic reticulum   b) Golgi apparatus   c) both   d) neither

_____ packages proteins into vesicles
_____ modifies proteins by adding or subtracting fragments
_____ a series of interconnected hollow tubes or sacs
_____ proteins are synthesized in this location

10. Be sure you can distinguish between the following pairs or groups. Give examples when appropriate.

a) desmosome, intermediate junction, gap junction, tight junction, junctional complex

b) cilia and flagella

c) microfilaments, microtubules, microvilli, and intermediate filaments

d) exchange and transporting epithelia

e) intracellular, extracellular, and intercellular compartments

f) cristae and matrix of mitochondria

g) lysosomes and peroxisomes

h) rough and smooth endoplasmic reticulum

i) epithelium and endothelium

j) apical and basolateral sides of an epithelium

k) movement of substances across a tight epithelium and a leaky epithelium

l) endocrine and exocrine glands

m) serous and mucous secretions

n) fibroblasts, melanocytes, adipocytes, macrophages, and mast cells

o) collagen and elastin

p) tendons and ligaments

11. Create a map of "the cell," including as many cellular components as you can. When you have completed your map, check the table on p. 45 of the text to be sure you have not omitted any organelles.

BEYOND THE PAGES
READING

A is for ...: Cadherins. (1996, March/April) *Science & Medicine*.

A is for...: Integrin.(1995, January/February) *Science & Medicine*.

Annexins. (1997, March) *Trends in Cell Biology* (7):87ff.

Artificial organs. (1995, September) *Scientific American*. Engineering artificial tissue.

Bioengineered skin substitutes. (1997, July/August) *Science & Medicine*.

The birth of complex cells. (1996, April) *Scientific American*. A model of the evolution of eukaryotic cells.

Cell adhesions - spreading frontiers, intricate insights. (1997, March) *Trends in Cell Biology* (7):107ff.

The centromere: Hub of chromosomal activities. (1995, December 8) *Science* 270: 1591-1594.

The extracellular matrix. (1995, May/June) *Science & Medicine*.

Extracellular matrix: The cellular environment. (1994, June) *News in Physiological Sciences* 9:110-114.

Extracellular matrix: The central regulator of cell and tissue homeostasis. (1997, January) *Trends in Cell Biology* (7):40ff.

Fibronectins. (1986, June) *Scientific American*.

Liposomes. (1996, May/June) *Science & Medicine*.

Mechanism and function of vacuolar acidification. (1993, February) *News in Physiological Sciences* 8:24-29.

Membrane polarity in epithelial cells: protein sorting and establishment of polarized domains. (1997) *American Journal of Physiology* 272 (*Renal Physiol.* 41): F425-F429.

The microtubule as an intracellular engine. (1987, February) *Scientific American*.

Putting the actin cytoskeleton into perspective. (1997) *American Journal of Physiology* 272 (*Renal Physiol.* 41): F430-F433.

Small wonder: Delicate and protean, the cell's "skeleton" shapes life far beyond the cell wall. (1993, May/June) *The Sciences*, pp.21-25.

Sunlight and skin cancer. (1996, July) *Scientific American*.

Tissue culture in microgravity. (1997, May/June) *Science & Medicine*.

Tubulin and microtubules. (1995, January/February) *Science & Medicine*.

VIEWING

Rent the movie *Lorenzo's Oil* about a family whose child has an inherited peroxisomal disorder.

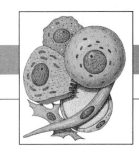

SUMMARY

The key themes to take from this chapter are:

- Energy is required to do work. There are two types of energy: kinetic and potential.
- There are many different types of work. [See Appendix A for more discussion.]
- Chemical reactions transfer energy. You should take away a general understanding of activation energy, free energy, endergonic/exergonic reactions, reversible reactions, and coupled reactions.
- Enzymes exhibit specificity, competition, and saturation. These three characteristics are observed in all protein-substrate interactions. [∫ Ch. 5 and membrane transporters]
- Cells use a series of chemical reactions to release the energy stored in the bonds of biomolecules. Energy from these bonds is transferred to the bonds of ATP and other high energy compounds. The high-energy compounds are used by the body to do work. Ask your instructor which details you should learn about cellular metabolism at this point.
- Take some time to become very familiar with protein synthesis. This concept is key to understanding many aspects of physiology, and the time you spend learning it will be well spent.

This chapter discusses how cells obtain and store energy in the form of chemical bonds. The potential energy stored in chemical bonds is later used as kinetic energy to perform work. In biological systems, this work relates to chemical, transport, and mechanical work. The types of work performed by chemical bond energy include protein synthesis, transport of substances across membranes, and muscle contraction.

Chemical reactions are of two general types, exergonic or endergonic. During exergonic reactions, the free energy of the products is less than that of the reactants. In endergonic reactions, the opposite occurs. Many exergonic and endergonic biological reactions are coupled and use nucleotides (e.g., ATP) to capture, store, and transfer energy. Some of these reactions are irreversible, but many are reversible. Most reversible reactions require the aid of an enzyme to overcome their activation energy.

Enzymes are involved in many biological reactions. Enzymatic activity depends in part upon the ability of a substrate to bind to the active site of the enzyme. Numerous other factors affect enzyme activity including proteolytic activation, co-factors or co-enzymes, modulation, and the concentration of the enzyme, substrate, or product. The reaction rate of enzymes is measured by how fast products are synthesized. This rate is determined by modulators, enzyme and substrate concentrations, and the ratio of substrates to products (Law of Mass Action). Types of biological reactions catalyzed by enzymes include oxidation-reduction, hydrolysis and dehydration, and addition-subtraction-exchange reactions.

Metabolic reactions are either catabolic (net breakdown) or anabolic (net synthesis). The activity of metabolic pathways can be altered in a variety of ways, including mechanisms that regulate enzyme activity and ATP concentrations.

Glucose, fatty acids, glycerol, and amino acids are used as energy sources to produce ATP. During glycolysis (an anaerobic pathway), one glucose molecule yields 2 pyruvate molecules, 2 ATP, 2 NADH, 2 H^+, and 2 H_2O molecules. In the absence of sufficient oxygen, pyruvate is converted to lactate. However, during aerobic metabolism pyruvate is converted to acetyl CoA, which enters the citric acid cycle (CAC). The yield from CAC metabolism of two pyruvate molecules is 8 NADH, 2 $FADH_2$, 2 ATP, and 6 CO_2 molecules. NADH and $FADH_2$ enter the electron transport system, and donate their electrons and H^+ to produce 32-34 ATP, H_2O, and heat. Therefore, the net reaction for aerobic metabolism is: glucose + 6 O_2 + 36 P_i + 6 H_2O $\rightarrow$ 6 CO_2 + 48 H_2O + 36 (or 38) ATP.

Glucose is not the only fuel source for making ATP during anaerobic and aerobic respiration. Glycogen can be converted to glucose-6-phosphate. Amino acids can be converted through deamination to pyruvate, acetyl CoA, or other intermediates of the CAC. Glycerol can enter glycolysis, and fatty acids can be converted to acetyl CoA.

Many fuel sources can be synthesized by the body. For example, glucose can be synthesized from glycogen, glycerol or amino acids whenever there is not enough glucose for ATP production. Most glucose is synthesized in the liver and released into the bloodstream for transport to other cells. Nerve cells normally only use glucose for fuel, so the body always tries to maintain adequate glucose concentrations in the blood. Lipids can be synthesized from glucose, acetyl CoA, glycerol, and fatty acids.

Proteins are synthesized by an elaborate processes that includes transcription and translation. Genes are selected for transcription by initiation factors that bind to specific promoter regions on DNA. DNA is transcribed into mRNA, and mRNA is translated into a protein. These proteins are sorted, modified, packaged, and directed to a specific destination.

TEACH YOURSELF THE BASICS

ENERGY IN BIOLOGICAL SYSTEMS

• Where do plants get energy and in what form(s) do they store excess energy? (p. 73)

_____

• Where do animals get energy and in what form(s) do they store excess energy? (pp. 73-74)

_____

Energy Is Used to Perform Work

• Define energy. (p. 74)

_____

• List three kinds of work in biological systems and give an example of each. (p. 74)

_____

_____

The Two Major Categories of Energy Are Kinetic Energy and Potential Energy

• What is the difference between kinetic energy and potential energy? (p. 74)

_____

• How is potential energy stored in biological systems? (p. 74)

Energy Can Be Transformed from One Type to Another

• When potential energy is converted to kinetic energy, what happens to the energy released? (p. 74)

Thermodynamics Is the Study of How Energy Is Converted to Work

• State the first law of thermodynamics. (p. 75)

• State the second law of thermodynamics. (p. 75)

• What is entropy? (p. 75)

CHEMICAL REACTIONS

• What is bioenergetics? (p. 75)

Energy Is Transferred between Molecules during Reactions

• Combine these words into a sentence that explains their relationship: reaction, product, molecule(s), substrate, reactant. (p. 75)

• How do we measure the rate of a chemical reaction? (pp. 75-76)

• Define and give an example of free energy. (p. 76)

• List three fates for chemical bond energy that is released during a reaction. (p. 76)

• Define and give an example of activation energy. (p. 76)

_____

• If a reaction proceeds spontaneously, what does that tell you about its activation energy? (p. 76)

_____

• What is the difference between exergonic reactions and endergonic reactions? (p. 76)

_____

• Compare the free energy of the products in an exergonic reaction to the free energy of the products in an endergonic reaction. (p. 76; Fig. 4-4, p. 77)

_____

• What is the difference between reversible and irreversible reactions? (p. 77)

_____

• Define metabolism. (p. 78)

_____

• What is the advantage of coupling exergonic and endergonic reactions? (p. 78)

_____

ENZYMES

Enzymes Lower the Activation Energy of Reactions

• What is the function of an enzyme? (p. 78)

_____

• How does an enzyme increase the rate of reaction? (p. 78)

_____

Enzymes Bind to Their Substrates

• How do enzymes and substrates interact? (p. 79)

_____

• What is an active site on an enzyme? (p. 79)

_____

• What is the difference between the lock-and-key model of enzyme activity and the induced fit model? (p. 79)

_____

_____

• What do we mean by the specificity of an enzyme? (p. 79)

_____

• Give an example of an enzyme, state the reaction it catalyzes, and explain its specificity. (p. 79; Table 4-4, p. 87)

_____

• What suffix is often found on enzyme names? (p. 79) _____

• The first part of an enzyme name usually refers to _____

_____. (p. 79)

• What are isozymes? (p. 80)

_____

Many Factors Affect Enzyme Activity

• Why are some enzymes manufactured as inactive precursors? (p. 81)

_____

• How are inactive precursor enzymes activated? (p. 81)

_____

• What suffix is used to indicate an inactive enzyme? (p. 81) _____

Some Enzymes Require Cofactors or Coenzymes

• What is the difference between a cofactor and a coenzyme? (p. 81)

_____

_____

• List three different types of modulators that can influence enzyme activity. (p. 82)

_____

• What is the difference between a competitive inhibitor and an allosteric inhibitor of an enzyme? (p. 83)

_____

• How do covalent modulators work? Give an example of one. (p. 84)

_____

Enzyme and Substrate Concentration Affect Reaction Rate

• List four factors that can affect the rate of an enzymatic reaction. (p. 85)

_____

• What happens to the rate of a reaction if the amount of enzyme present is increased? (p. 85)

_____

• What happens to the rate of a reaction as the amount of substrate decreases? (p. 85)

_____

• Define enzyme saturation. (p. 85)

_____

• Explain the concept of equilibrium as it applies to a reversible reaction ($A + B \rightleftarrows C + D$). (p. 86)

_____

• If the reaction above is at equilibrium, what happens as you add more substrate? (Fig. 4-16b, p. 86)

_____

• Define the Law of Mass Action. (p. 86)

_____

Types of Enzymatic Reactions

☛ MNEMONIC: OIL RIG = Oxidation Is (electron) Loss, Reduction Is Gain

• Fill in the blanks in the following table (p. 86-87; Table 4-4, p. 87)

REACTION TYPE	WHAT HAPPENS	TYPE OF ENZYME
hydrolysis of lipids		
	add a phosphate	
	gain electrons	
		dehydrogenase
hydrolysis		
exchange	phosphate	
		deaminase
oxidation		
	subtract a phosphate	
exchange	amino group	
add	amino group	

METABOLISM

• Define metabolism. (p. 87)

_____

• Distinguish between catabolic and anabolic reactions. (p. 87)

_____

• Define a kilocalorie (kcal). (p. 87)

_____

• What is a metabolic pathway? (pp. 87-88)

_____

Regulation of Metabolic Pathways

• List five basic ways that cells regulate metabolism. (p. 88)

_____

_____

• Explain feedback inhibition, using the pathway L → M → N → P as an example. (p. 88)

_____

• What is the advantage of having a reaction that is regulated by two enzymes (one for the forward direction, the other for the reverse direction) ? (p. 89)

_____

• What advantage does a cell gain by isolating some enzymes within specific intracellular compartments (organelles)? (p. 89)

_____

• If the ATP:ADP ratio decreases, what do you expect to see happen to the metabolic pathways that synthesize ATP? (p. 89)

_____

ATP Transfers Energy between Reactions

• What is the biological significance of ATP? (p. 89)

_____

• What is the difference between aerobic and anaerobic pathways? (p. 89)

_____

• Compare ATP production by aerobic and anaerobic pathways. (p. 89)

_____

ATP PRODUCTION

• Name two major pathways for high-energy electron production. (p. 90)

_____

• What biomolecules can be used by these pathways? (p. 90)

_____

• What pathway transfers energy from high-energy electrons to ATP? (p. 90)

_____

Glycolysis Converts Glucose and Glycogen into Pyruvate

• Where in the cell does glycolysis take place? (p. 90) _____

• What are the possible end products of glycolysis and what is their fate?

_____

Pyruvate Is Converted into Lactate in Anaerobic Metabolism

• Under what conditions is lactate converted into pyruvate? (p. 91)

_____

• What is the net energy yield from the conversion of one glucose to two lactate? (p. 91)

_____ ATP   _____ NADH   _____ FADH$_2$

Pyruvate Enters the Citric Acid Cycle in Aerobic Metabolism

• In aerobic metabolism, pyruvate is converted into _____. (p. 91)

• Where in the cell does this reaction take place? (p. 91) _____

• What are the end products of the citric acid cycle and what is their fate? (pp. 92-93)

_____

• The net energy for one pyruvate completing the citric acid cycle is ___ ATP, ___ NADH, ___ FADH$_2$

The Electron Transport System Transfers Energy from NADH and FADH$_2$ to ATP

• What is the relationship between the electron transport system (ETS) and oxidative phosphory-lation? (p. 93)

_____

• In the ETS, NADH and FADH$_2$ donate _____ and _____. (p. 93)

ATP Production Is Coupled to Movement of Hydrogen Ions across the Inner Mitochondrial Membrane

• In the ETS, how do electrons move through the inner mitochondrial membrane? (p. 93)

_____

• Explain how the ETS makes ATP and water. (pp. 93-94)

_____

_____

The Net Energy Yield of One Glucose Molecule Is 36 to 38 ATP

• Total energy production from 1 glucose in aerobic metabolism = _____ ATP. (Table 4-5, p. 92)

Conversion of Large Biomolecules to ATP

• When glycogen is broken down for energy, it is converted to _____ or _____.
   (p. 95)

• The first step in protein catabolism is the removal of a/an _____ group through the process

   known as _____. (p. 95)

• When amino acids are converted into organic acids, the latter can enter which metabolic pathways? (p. 95)

   _____

• In lipolysis, lipids are converted into _____ and _____. (p. 95)

• What happens to fatty acids that undergo beta oxidation? (p. 96)

   _____

• Liver cells convert some fatty acids into _____ rather than into acetyl CoA. (p. 96)

SYNTHETIC PATHWAYS

Glycogen Can Be Made from Glucose

• What two organs have the largest glycogen stores? (p. 96) _____

• Through what pathway(s) can glycogen be made? (p. 97) _____

Glucose Can Be Made from Glycerol or Amino Acids

• What is gluconeogenesis? (p. 97) _____

• What organ is the primary site of gluconeogenesis? (p. 97) _____

Acetyl CoA Is an Important Precursor for Lipid Synthesis

• What two-carbon precursor can be used to synthesize fatty acids? (p. 97) _____

• Can the body make cholesterol or must it come from the diet? (p. 97) _____

Proteins Are Necessary to Many Cell Functions

• What property of proteins allows them to be more variable and specific than any other biomolecule? (p. 98)

_____

• What is a codon? (p. 98) _____

• What relationship do codons of DNA have to the codons of mRNA? (p. 98) _____

_____

• What is the difference between transcription and translation? (pp. 98-100)

_____

• Describe the process by which transcription is initiated. (p. 99)

_____

• What is the role of RNA polymerase? (p. 99) _____

• Explain the relationship between introns, exons, mRNA processing, and alternative splicing. (pp. 99-100)

_____

_____

• What is the difference between mRNA, rRNA, and tRNA? (pp. 99-101)

_____

_____

• What do ribonucleases do? (p. 102) _____

• What is a signal sequence? (p. 102) _____

• What happens to proteins as they come off the ribosome? (p. 102)

_____

• Trace the path of a newly synthesized protein from the ribosome to a secretory vesicle. (p. 102)

_____

TALK THE TALK

acetyl coenzyme A
activation energy
active site
aerobic respiration
aerobic metabolism
allosteric modulator
alternative splicing
amination
anabolism
anaerobic metabolism
anticodon
antisense strand
ATP synthase
beta-oxidation
catabolism
catalyst
chemical work
chemical reaction
chemiosmotic theory
citric acid cycle
codon
coenzyme
cofactor
compartmentation
competitive inhibitor
concentration gradient
coupled reaction
covalent modulator
deamination
dehydration
denatured
DNA
electron transport system
endergonic reaction
energy
entropy

enzyme
exergonic reaction
exon
$FADH_2$
feedback inhibition
free energy
gene
gluconeogenesis
glucose-6-phosphate
glycogen metabolism
glycogenolysis
glycolysis
high-energy phosphate bond
hydrolysis
induced fit model
initiation factor
intermediate
intron
irreversible reaction
isozyme
ketone bodies
key intermediate
kilocalorie
kinase
kinetic energy
Krebs cycle
lactate
lactate dehydrogenase
Law of Mass Action
laws of thermodynamics
lipolysis
lock-and-key model
matrix (of mitochondria)
mechanical work
metabolism
modulator

mRNA
mRNA processing
NADH
oxidation
oxidative pathway
oxidative phosphorylation
pathway
peptidase
phosphorylation
potential energy
product
promoter
protease
protein sorting
proteolytic activation
pyruvate
reactant
reaction rate
reduction
reversible reaction
ribonuclease
ribosomal RNA (rRNA)
RNA polymerase
saturation
sense strand
signal sequence
specificity
substrate
transamination
transcription
transition vesicle
translation
transport work
work

RUNNING PROBLEM: Tay Sachs Disease

Tay-Sachs Disease — Carrier screening, prenatal diagnosis, and the molecular era. An international perspective, 1970 to 1993. (1993, Nov. 17) *Journal of the American Medical Association* 270 (19):2307-2315.

FOCUS ON PHYSIOLOGY: The Law of Mass Action

The **Law of Mass Action** is a simple relationship that holds for chemical reactions ranging from those in a test tube to those in the blood. This law says that the ratio of the concentrations of the molecules is a constant number, known as the *equilibrium constant*, K.*

$$K = \frac{[substrates]}{[products]}$$

*Mathematically:

$$K = \frac{[A]^m\,[B]^n}{[C]^p\,[D]^q}$$

where m, n, p, and r represent the number of molecules of A, B, C, and D respectively in the balanced equation.

In very general terms, the Law of Mass Action says that when a reaction is at equilibrium, the ratio of the substrates to the products is always the same. If the concentration of a substrate (A or B) increases, the equilibrium will be disturbed. To restore the proper substrate/product ratio, some of the added substrates will convert into product. Conversely, if the amount of product (C or D) decreases, such as would occur when they are used up in a different reaction, some substrate will convert into product to replace that which was lost. This has the effect of decreasing the amount of A and B but keeping the ratio at the value needed for K.

The equilibrium constant is important in physiology because as the concentrations of substrates change, the change will be reflected in the concentrations of the products. This is most easily illustrated by equating the chemical reaction to a scale, with the number of blocks on each side representing the concentrations of the substrates and products. The reaction is reversible. The equilibrium constant, K, is indicated by the position of the triangular base of the scale. For simplicity, our example has an equilibrium constant of 1, so that the reaction is in equilibrium when there are equal numbers of blocks on each side: 1 = [S]/[P]. The figure below shows the reaction at equilibrium, with substrate and product concentrations balanced at the equilibrium constant (K) for the reaction.

(a)   (b)   (c)

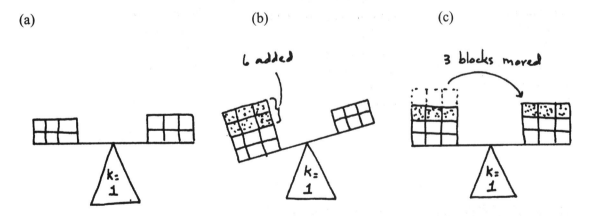

In part (b) of the figure, some outside force has added substrate to the left side of the reaction, sending the reaction out of equilibrium. Now, without removing any blocks from the scale, how can you get the reaction back to equilibrium? Easy...move some of the added blocks over to the product side. Then each side still has an equal number of blocks (the 1:1 ratio that we set up as our equilibrium constant) and the reaction is at equilibrium. Notice that BOTH product and substrate have increased, but by an equal amount. You cannot change the K since this is a constant for any given reaction.

The next figure shows a different reaction that is at equilibrium when the product is twice the substrate (d). That means the K for this reaction is fi. Notice that the balance point has moved to compensate for the K value: 1/2 = [S]/[P]. In part (e) of this figure, 9 units have been added to the substrate side. How can you rearrange the blocks so that you balance the scale at the 1/2 ratio? Try this on your own, using part (f) .

(d)   (e)   (f)

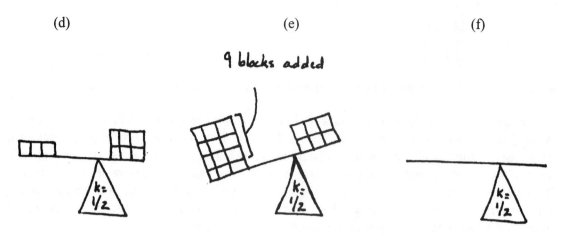

In the figures below, you again have the reaction with a K value of 1/2 (g). In part (h), 3 units of product are removed by another metabolic reaction. Without adding or subtracting any new units to the scales, move the units between the scales until the reaction is at equilibrium again (i).

(g)   (h)   (i)

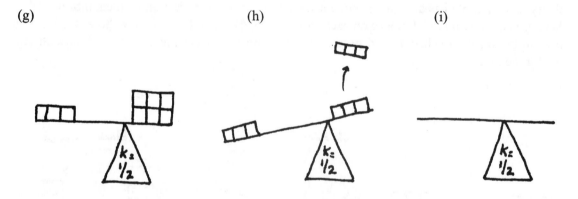

Now, instead of blocks on a scale, assume that the reaction is: A ↔ C + D →E

If more substrate A is added to the reaction, what happens to the amounts of the following when the reaction reaches equilibrium?

A _____   C _____   D _____

If C and D are converted into E, when the system again reaches equilibrium, what has happened

to the amount of A in the system? _____

PRACTICE MAKES PERFECT

1. Matching. (More than one letter may apply to each number.)

1. _____ entropy
2. _____ potential energy
3. _____ kinetic energy
4. _____ exergonic
5. _____ endergonic

A. ATP
B. a muscle contracting
C. reaction that converts glucose to ATP
D. second law of thermodynamics
E. free energy
F. reaction that releases a lot of energy
G. reaction that requires a lot of energy

2. Most breakfast cereals would not be very nutritious unless they were supplemented with vitamins and ions such as Mg^{2+}. How do some of these ions and vitamins affect enzyme activity?

3. In the human body, carbonic anhydrase (CA) has maximum activity at pH 7.4. Some other animals have an isozyme that has maximum activity at about 6.4. Draw a curve of the activity of human carbonic anhydrase and its isozyme as a function of pH, using the graph below.

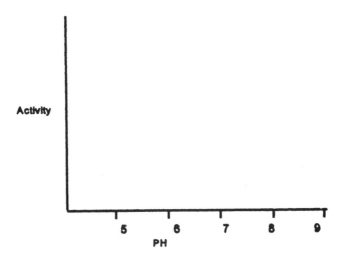

4. Answer the questions using the two graphs below.

a. Which graph represents an exergonic reaction and which graph represents an endergonic reaction?

b. Which reaction is more likely to proceed in the forward direction? Why?

c. Which reaction products yield more free net energy?

d. Which reactions below belong with graph A and which belong with graph B?
(Reactions 2, 3, and 4 are from the glycolysis pathway.)

1. ___ $6 CO_2 + 6 H_2O + \text{sunlight} \rightarrow C_6H_{12}O_6 + 6 O_2$

2. ___ glucose + ATP $\rightarrow$ glucose-6-phosphate + ADP

3. ___ 3-phosphoglyceraldehyde + P_i + NAD^+ $\rightarrow$ 1,3 diphosphoglycerate + NADH

4. ___ 1,3 diphosphoglycerate + ADP $\rightarrow$ 3-phosphoglycerate + ATP

A.   **B.**

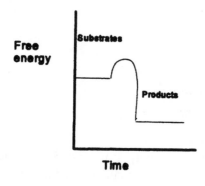

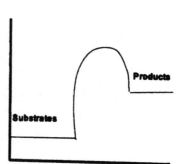

5. Matching:

_____ hydrolysis

_____ dehydration

_____ oxidation

_____ reduction

_____ kinase

_____ deamination

_____ transamination

A. lose electron or gain H^+

B. removal of water

C. -OH is added to one molecule and -H to another

D. gain electron or lose H^+

E. loss of activity

F. transfer of an amino group

G. enzyme that transfers a phosphate group

H. removal of an amino group

6. Ribonuclease is a digestive enzyme secreted by the pancreas. This enzyme catalyzes the hydrolysis of RNA.

a. A 100 mM solution of ribonuclease is added to a series of test tubes with RNA concentrations ranging from 10% to 70%. Plot the reaction rate as a function of substrate concentration on the graph below. Assume that the reaction rate is maximum at a 50% substrate concentration.

b. What is the relationship between the substrate and the enzyme when the reaction rate is maximal?

c. If the enzyme concentration was only 50 mM, indicate on the graph below how the reaction rate would change as a function of substrate concentration.

Reaction Rate

10   20   30   40   50   60   70

RNA Concentration (%)

7. During anaerobic metabolism, the reaction shown below occurs. Which molecules have been reduced and which have been oxidized?

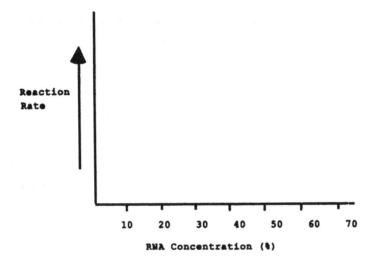

$NADH + H^+$ NAD^+

$CH_3 - C - COO^-$ ———————→ $CH_3 - CH - COO^-$
$\quad\quad\quad ||$ *lactate dehydrogenase* $\quad\quad\quad |$
$\quad\quad\quad O$ $\quad\quad\quad OH$
pyruvate   **lactate**

8. During glycolysis, the reaction shown below occurs. What type of reaction is this?

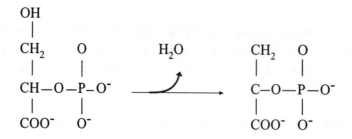

2-phosphoglycerate   phosphoenolpyruvate

9. Matching:

_____ chemiosmotic theory

_____ Sir Hans A. Krebs

_____ NADH and FADH$_2$

_____ glycolysis

_____ beta-oxidation

_____ glycogen

_____ gluconeogenesis

_____ glycogenolysis

_____ acetyl coenzyme A

_____ deamination

A. described citric acid cycle

B. conversion of proteins or fats to glucose

C. converted to glucose-6-phosphate

D. using H$^+$ gradient to make ATP

E. derived from pyruvate and the vitamin pantothenic acid

F. pathway that converts glucose to pyruvate

G. donates electrons and H$^+$

H. glycogen breakdown

I. remove NH$_3$ from amino acid to make organic acid that enters aerobic metabolic pathway

J. converts fatty acid or glycerol into pyruvate

K. converts fatty acids into acetyl CoA

10. Explain the function of the following in the electron transport system:

a. NADH -

b. FADH$_2$ -

c. oxygen -

d. ATP synthase -

e. H$^+$ -

f. complex inner membrane proteins -

11. The three major food groups, proteins, carbohydrates, and fats, are used as fuel sources to produce ATP. Create a map using the list of terms (and any terms you wish to add) to show:

    a) the important or key intermediates
    b) where the breakdown products of these food groups enter the pathway
    c) the major products for each part of the pathway. What are the fates of these products?
    d) both aerobic and anaerobic pathways
    e) where these reactions occur in the cell

acetyl CoA	cytoplasm	glucose	lactic acid
$ADP + P_i$	electron transport	glycerol	liver
aerobic	system	glycogen	mitochondria
anaerobic	electrons	glycogenesis	NADH
ATP	fats	glycolysis	oxygen
carbohydrates	fatty acids	H^+	proteins
citric acid cycle	gluconeogenesis	H_2O	pyruvate

12. Design a map that outlines the events of DNA transcription and mRNA processing using the following terms (plus any terms you wish to add):

alternative splicing	initiation factor	RNA polymerase
codon	introns	sense strand
complementary base pairing	mRNA	splicing enzyme
DNA double helix	mRNA processing	stop codon
exons	promoter	unwind
gene		

13. Design a map that outlines the events of translation and protein sorting. Include in your answer the following terms (plus any terms you wish to add):

amino acid	mRNA	rough endoplasmic reticulum
anticodon	peptide bond	signal sequence
codon	polypeptide	transition vesicles
Golgi complex	ribosomes	tRNA

BEYOND THE BASICS

Ethics in Medicine

Suppose that both Sarah and David were carriers of the Tay Sachs gene. What parenting options are available for such a couple?

A couple who are both carriers has a one-in-four chance of conceiving a child who will be born with the disease. These are the basic options that the genetic counselor presents to them:
1. The couple can decide not to have children.
2. They can decide to adopt children.
3. They can elect to use an egg or sperm donor who does not carry the gene.
4. If they decide to have their own children, they can take the one-in-four chance that their child will be born with the disease. Screening techniques are available to test the genetic makeup of a fetus as early as six weeks after conception to determine whether it will be born with Tay Sachs disease. If the couple's personal convictions allow, they can screen for the disease in the fetus and elect to abort a fetus who would be born with the disease.

READING

A is for ...: Gene. (1996, July/August) *Science & Medicine*.

A is for...:Polymerase chain reaction. (1994, September/October) *Science & Medicine*.

A is for...: Transgenic animal. (1994, November/December) *Science & Medicine*.

About benefits and costs: Genetic testing for Huntington's disease. (1995, March/April) *Science & Medicine*

Adaptation of mitochondrial gene expression to changing cellular energy demands. (1997, August) *News in Physiological Sciences* 12: 178-183.

Doubly labeled water measures energy use. (1994, March/April) *Science & Medicine*.

Molecular modeling of protein structure. (1997, January/February) *Science & Medicine*.

Protein translocation at the ER membrane. (1997, March) *Trends in Cell Biology* 7:90-94.

Quality control of protein folding: participation in human disease. (1997, August) *News in Physiological Sciences* 12: 162-165.

RNA editing. (1995, March/April) *Science & Medicine*.

Signal transduction by intramitochondrial Ca^{2+} in mammalian energy metabolism. (1994, April) *News in Physiological Sciences* 9: 71-76.

Why introns? (1995, April) *News in Physiological Sciences* 10:98-99.

CHAPTER 5
MEMBRANE DYNAMICS

SUMMARY

This chapter discusses some key themes that recur throughout the remainder of the book. Learn them well now to make your life easier as the class progresses!

- Cell membranes are fluid-mosaics consisting of a phospholipid bilayer with proteins and carbohydrates.
- What are the differences between membrane-spanning proteins and associated proteins?
- Membrane proteins have various functions. Learn the different groups of membrane proteins and understand how they participate in cellular activities.
- The body fluids are compartmentalized. Learn the compartments and what keeps them separate.
- Several factors affect diffusion. You should be familiar enough with these factors to explain them verbally and express them mathematically.
- There are two types of active transport: primary and secondary. You should be able to compare and contrast these processes.
- Large particles use different forms of vesicular transport to enter or leave the cell. Learn the distinctions between phagocytosis, endocytosis, and exocytosis.
- Cells can be polarized. This is an important concept for transepithelial transport. Understand how polarization allows unidirectional transport.
- Cell membranes create a selectively permeable barrier that allow cells to maintain electrochemical disequilibrium across the membrane. Learn how this disequilibrium is maintained so that you can understand its roles in physiological processes to be discussed in later chapters.
- Osmolarity and tonicity are very important concepts! It's important that you understand the difference between them and learn the terminology for comparing solutions.

SUMMARY

Cell membranes act as a selectively permeable barrier between the cells and their external environment, mediating transport and communication between the two compartments. In addition, membranes provide structural support. A cell membrane consists of a phospholipid bilayer with proteins that are inserted partially or entirely through the membrane. Many of these proteins are mobile while others have restricted movement. Membrane proteins form channels, carriers, receptors, or enzymes and also have a structural role.

The ability of a substance to cross a cell membrane depends upon the properties of the membrane and the size and lipid solubility of the substance. Small lipid-soluble substances can cross a membrane by passive diffusion. Channels formed by transmembrane proteins allow the passage of water and smaller ions. Larger molecules and ions are transported by protein carriers. One type of carrier-mediated transport, primary active transport, uses ATP while other forms of carrier-mediated transport use the energy stored in concentration gradients. Large molecules and particles move into and out of cells by vesicles that fuse with the membrane. Epithelial cells are polarized, with different types of protein-mediated transporters on the apical and basolateral membranes. This allows the one-way transport of certain molecules.

Water is able to move freely across most membranes, moving primarily through open water-filled channels. The movement of water from a region of lower solute concentration to one of higher solute concentration is called osmosis. Osmolarity describes the number of particles in a solution; tonicity is a comparative term that describes how the size of a cell changes by osmosis when exposed to a solution.

Ions and most solutes do not diffuse freely across cell membranes and therefore a state of chemical disequilibrium exists between the cell and its environment. All cells have a resting membrane potential resulting from chemical and electrical gradients of ions. Depending upon the type of cell, resting membrane potentials are usually between -50 to -90 mV relative to the extra-cellular fluid. This resting potential is established by a membrane that is more permeable to K^+ than to Na^+ or other ions. Active transport maintains this potential by transporting Na^+ and K^+ ions across the membrane.

TEACH YOURSELF THE BASICS

CELL MEMBRANES (Fig. 5-10)

• List the four general functions of membranes. (p. 108)

_____

Membranes Are Mostly Lipid and Protein

• What biomolecules form the three main components of biological membranes? (p. 109)

_____

• Diagram a typical membrane based on the fluid mosaic membrane model. (p. 109; Fig. 5-2, p. 110)

• Why is the membrane described by the word "fluid"? (p. 111)

_____

Membrane Lipids Form a Barrier between the Cytoplasm and Extracellular Fluid

• Describe the structure and polarity of a phospholipid molecule. (p. 109)

_____

• Distinguish between a micelle, liposome, and phospholipid bilayer. (p. 109; Fig. 5-3, p. 110)

_____

Membrane Proteins May Be Tightly or Loosely Bound to the Membrane

• Name the two anatomical classifications of membrane proteins (p. 109; Fig. 5-2, p. 110)

_____

• Describe the characteristics and regions of membrane-spanning proteins. (pp. 110-111)

_____

_____

• Describe the characteristics of associated proteins. (p. 111)

_____

Membrane Proteins Function as Structural Proteins, Enzymes, Receptors, and Transporters

• List four functions of membrane proteins. (p. 111)

_____

• What are the two major roles of structural membrane proteins? (p. 111)

_____

• What are microvilli and what function do they serve on a cell? (∫ p. 47)

_____

• Give one example each of an intracellular and an extracellular membrane enzyme. (p. 112)

_____

• How does the role of a membrane receptor differ from the role of a membrane transporter? (p. 112)

_____

_____

• What is a ligand?

_____

• Name t3

_____

Membrane Carbohydrates Attach to Both Lipids and Proteins

• What is the glycocalyx and where in the cell is it found? (p. 114)

_____

• Name one important function of membrane glycoproteins. (p. 114)

_____

BODY FLUID COMPARTMENTS

• What is the external environment of a cell? (p. 114) _____

• What separates extracellular fluid from intracellular fluid? (p. 114) _____

• What separates interstitial fluid from the plasma? (p. 115) _____

• What is endothelium and what kind of epithelium is it? ($\int$ p. 57)

_____

MOVEMENT ACROSS MEMBRANES

• What two properties of a molecule determine whether it can diffuse across a membrane? (p. 116)

_____

• If a molecule cannot move into a cell, we say that the cell is _____ to the molecule. (p. 116)

• What is the difference between active transport and passive transport? (p. 116)

_____

Diffusion Uses Only the Energy of Molecular Movement

• Define diffusion. (p. 117)

_____

• List six properties of diffusion. (pp. 117-118)

_____

_____

_____

Lipophilic Molecules Can Diffuse through the Phospholipid Bilayer

• List four factors that influence the rate of simple diffusion of molecules across the phospholipid bilayer. As the factor increases in magnitude, does the rate of diffusion increase or decrease? (p. 119)

_____

_____

• Fick's Law describes simple diffusion. Write the mathematical expression for Fick's Law in words:

diffusion rate $\propto$

• Do ions obey the rules for simple diffusion? Explain. (p. 119)

_____

Mediated Transport Uses Membrane Proteins

• List and describe the three properties of mediated transport. (pp. 120-121)

_____

_____

_____

• What is the difference between a competitor for a transport protein and a competitive inhibitor? (p. 120)

_____

_____

Facilitated Diffusion Is Diffusion That Uses Membrane Proteins

• How is facilitated diffusion similar to simple diffusion? (p. 121)

_____

• How is facilitated diffusion different from simple diffusion? (p. 121)

_____

Active Transport Requires the Input of Energy from ATP

• How does active transport differ from facilitated diffusion? (p. 121)

_____

• How is active transport similar to facilitated diffusion? (p. 121)

_____

• Distinguish between a cotransporter, a symport protein, and an antiport protein. (p. 122)

_____

• Define and give an example of primary active transport. (p. 122; Fig. 5-19, 5-20, p. 123)

_____

• How does secondary active transport differ from primary active transport? (p. 124)

_____

• What happens to secondary active transport if a cell's production of ATP is stopped? (p. 125)

_____

Vesicular Transport Across Membranes

• What distinguishes molecules that move by vesicular transport from molecules that move by mediated transport? (p. 126)

_____

• How is phagocytosis similar to endocytosis and how is it different? (p. 126)

_____

_____

• What is the difference between pinocytosis and receptor-mediated endocytosis? (p. 127)

_____

• What role do coated pits play in receptor-mediated endocytosis? (p. 127)

_____

• Explain the process of membrane recycling. (p. 127)

_____

• Describe the steps of exocytosis and give some examples of molecules released by this method. (p. 127)

_____

_____

Molecules Move Across Epithelia Using Passive and Active Transport

• What distinguishes the apical and basolateral membranes of a transporting epithelium from each other? (p. 128)

_____

• Name the three transporters involved in the transepithelial transport of glucose. (pp. 128-129)

_____

• Where does the Na^+-glucose symporter get the energy to move glucose against its concentration gradient? (p. 128)

_____

• Name the three steps of transcytosis. (p. 130)

_____

• What kinds of molecules are likely to move by transcytosis? (p. 130) _____

DISTRIBUTION OF WATER AND SOLUTES IN THE BODY

Living Cells Use Energy to Maintain a State of Chemical and Electrical Disequilibrium

• Explain what is meant by "a state of chemical disequilibrium in the body." (p. 131)

_____

• Is water in a state of disequilibrium in the body? (p. 132) _____

Water Distributes throughout the Body

• Name the three body compartments. (p. 132)

_____

• Which compartment contains the most water? (p. 132) _____

• What is the average volume of each compartment in a male weighing 70kg? (p. 132)

_____

Water Crosses Membranes by Osmosis

• Distinguish between osmosis, osmotic pressure, osmolarity, and osmolality. (pp. 132-134)

_____

_____

_____

_____

• Sulfuric acid, H_2SO_4, dissociates into three ions when placed in water. What is the osmolarity of a 1 mM solution of sulfuric acid? (p. 134) _____

• If solution A has more particles per liter than solution B, then A is said to be _____osmotic to solution B, and solution B is said to be _____osmotic to solution A.

The Tonicity of a Solution Describes How the Size of a Cell Would Change If It Were Placed in the Solution

• Solution C has 100 mosmoles/L and solution D has 200 mosmoles/L. Can we describe the tonicities of these two solutions relative to each other? Explain your reasoning. (p. 135)

_____

• What is the one factor that determines the tonicity of a solution relative to a cell? (p. 135)

_____

☛ *HELPFUL HINT: To remember what happens to a cell in hypotonic solutions versus hypertonic solutions, look at the letter directly preceding the "t" in each word. For hypotonic, the "o" can be a visual reminder that cells swell in hypotonic solutions. The "r" in hypertonic can be a visual reminder that cells shrink in hypertonic solutions.*

Resting Membrane Potential

• List the major intracellular ions and the major extracellular ions. (Fig. 5-28, p. 131; Table 5-9, p. 137)

_____

• Describe the Law of Conservation of Electric Charges. (p. 137)

_____

• What distinguishes a conductor from an insulator? Which of these terms describes the cell membrane? (p. 137)

_____

• What forces create the electrical disequilibrium that exists between the intracellular and extracellular compartments? (p. 138)

_____

• Define the resting membrane potential of a cell. (p. 139)

_____

• If the resting membrane potential of a cell is -70 mV and the extracellular fluid is assumed to have a charge of zero millivolts, what happened to the extra positive charges that must exist according to the Law of Conservation of Electric Charges? (pp. 138-139)

_____

• If a cell was freely permeable to all charged molecules, would it have a resting membrane potential? Explain. (p. 140)

_____

_____

• What is the equilibrium potential for an ion? (p. 140)

_____

• Which ion contributes most to the resting potential of most cells? (p. 141) _____

• Why is the Na^+/K^+-ATPase considered to be an electrogenic pump? (p. 141)

_____

TALK THE TALK

active transport	fluid mosaic model	osmotic pressure
antiport	freezing point depression	passive transport
apical	gated channels	penetrating solute
associated proteins	glycocalyx	permeable
atherosclerosis	hyperosmotic	phagocytosis
basolateral	hypertonic	phospholipid bilayer
cell (plasma) membrane	hyposmotic	pinocytosis
chemical disequilibrium	hypotonic	plasma
chemical equilibrium	impermeable	polarized
chemical gradient	interstitial fluid	primary (direct) active transport
chemically gated channel	isosmotic	receptor-mediated endocytosis
cholesterol	isotonic	resting membrane potential
coated pit	ligand	restricted diffusion
competitive inhibitor	liposome	saturation
co-transporter	low density lipoprotein	secondary (indirect) active
diffusion	mechanically gated channel	transport
electrical gradient	mediated transport	selectively permeable
electrochemical gradient	membrane recycling	sodium-potassium ATPase
electrogenic pump	membrane spanning protein	specificity
endocytosis	micelle	symport
equilibrium	nonpenetrating solute	tonicity
equilibrium potential	osmolality	transcytosis
exocytosis	osmolarity	transepithelial transport
extracellular fluid	osmosis	vesicular transport
facilitated diffusion	osmotic equilibrium	voltage gated channel

ERRATA

Answer to concept check on p. 136 should read: "Both 5% dextrose and 0.9% NaCl are almost isosmotic, so neither would affect body osmolarity significantly."

Table 5-8, p. 137: 5% dextrose is only slightly hyposmotic and it is often considered to be isosmotic.

QUANTITATIVE PHYSIOLOGY

Volumes of Distribution

The techniques by which the volumes of the body compartments are determined provides us with interesting insight into experimental design. The determination of a volume of liquid can be estimated using a **dilution technique**. There are several requirements for this experiment. (1) The volume to be measured must be well-mixed, so that any marker that is given will distribute evenly throughout the compartment. (2) The marker must be able to be measured. If the system is a living organism, the marker must also be non-toxic, not metabolized or excreted, and must distribute only into the compartment being measured.

To carry out a dilution experiment, a known amount of marker in a known volume is administered. It is allowed to distribute, then a sample of the compartment liquid is removed and analyzed for its marker content. Because the concentration of marker in the sample is the same as the concentration throughout the compartment, the following ratio is used to calculate the compartment volume:

$$\frac{\text{amount marker in sample (known)}}{\text{volume of sample (known)}} = \frac{\text{total amount of marker (known)}}{\text{total volume of compartment (unknown)}}$$

For example, 1 gram of dye is put into a giant vat of water. The dye is stirred until it is evenly distributed throughout the vat. A 1 mL sample of water is taken out, analyzed, and found to contain 0.02 mg of dye. What is the volume of the vat?

0.02 mg dye/1 mL = 1000 mg dye/ χ mL

χ = 1000 mg dye • 1 mL/0.02 mg dye

= 50,000 mL or 50 L

The dye distributed into a volume of 50 L. For this reason, the volumes calculated by this method are known as *volumes of distribution*.

The determination of body compartment volumes in humans has been done using a variety of markers. The marker must be restricted to the compartment being measured in order to give accurate results. **Total body water** has been estimated using water with radioactive isotopes of hydrogen. Both deuterium oxide (D_2O or heavy water) and tritium oxide can be used. Calculating the volume of distribution for the **extracellular fluid (ECF) volume** requires a molecule that can move freely between the plasma and the interstitial fluid but that cannot enter the cells. Sucrose, the disaccharide commonly known as table sugar, fits this requirement as does a molecule called *inulin*. Inulin is a plant polysaccharide extracted from the roots of dahlias. It is not metabolized by humans but is excreted in the urine, so this must be taken into account when estimating ECF volume. The final compartment that can be directly measured is the **plasma volume**. This measurement requires a large molecule that distributes in the plasma but cannot cross the leaky epithelium to the interstitial fluid. Since endogenous plasma proteins meet this requirement, researchers found a dye, Evans blue, that binds to plasma proteins and therefore distributes only in the plasma.

There are no markers for the interstitial fluid and the intracellular compartment, but we have been able to accurately estimate those volumes as well. Using the information in the paragraph above, can you explain how to calculate interstitial fluid and the intracellular volumes? (Answer in appendix)

PRACTICE MAKES PERFECT

1. According to Einstein, the time required for a molecule to diffuse from point A to point B is proportional to the square of the distance. Fill in the table below and graph the relationship:

Distance	Time
1 mm	
2 mm	
3 mm	
4 mm	

2. A woman with pneumonia has an accumulation of fluid in her lungs. This has an effect similar to that of increasing membrane thickness by a factor of 2. How does this affect the rate of diffusion? Give a quantitative and qualitative answer. Assume that all other parameters do not change.

3. Complete the table below.

METHOD	MOVEMENT RELATIVE TO CONCENTRATION GRADIENT	ENERGY SOURCE	WHAT AFFECTS RATE?	THROUGH MEMBRANE BILAYER OR THROUGH PROTEIN TRANSPORTER?	EXHIBITS SPECIFICITY ?	EXHIBITS COMPETITION ?	EXHIBITS SATURATION ?	EXAMPLES
SIMPLE DIFFUSION								
FACILITATED DIFFUSION								
DIRECT ACTIVE TRANSPORT								
INDIRECT ACTIVE TRANSPORT								

4. Match the transporter with all the descriptions that apply.

_____ Na^+/K^+-ATPase   A. symport

_____ Na^+-glucose transporter   B. antiport

_____ Ca^{2+}-ATPase   C. active transport

_____ $Na^+/K^+/2$ Cl^-- transporter   D. cotransport

_____ Na^+/H^+ -transporter   E. secondary active transport

5. After completing physiology, you become so fascinated by biology that you go on to become a world-famous zoologist. In the year 2045 you are part of a scientific expedition to investigate newly discovered life forms on the planet Zwxik in another solar system. You are assigned to study a single-cell organism found in the aqueous swamps of Zwxik. In one of your first studies using a radioisotope of calcium, you observe that calcium moves freely in and out of the cell during daylight hours but is unable to get into or out of the cell in the dark. What is your hypothesis about how calcium is crossing the cell membrane in this organism? Be as specific and as scientific as possible.

6. In the compartments below, diagram the osmotic, chemical, and electrical equilibrium/disequilibrium that exists in the living body. For ions, use large symbols where concentrations are high and small symbols where concentrations are low.

Key: Na^+, K^+, Ca^{2+}, Cl^-, proteins, osmolarity (mOsm), plasma, interstitial fluid, cells

Impermeable   **Na^+/K^+-ATPase**

7. Construct a map of the mechanisms used to transport substances across membranes using the following terms, plus any terms you chose to add.

active transport
antiport
ATP
cell membrane
coated pit
concentration gradient
diffusion

mediated transport
endocytosis
exocytosis
facilitated diffusion
passive transport
phagocytosis
pinocytosis

primary active transport
receptor-mediated endocytosis
secondary active transport
sodium-potassium ATPase
symport
vesicular transport

8. The addition of solute to water disrupts the hydrogen bonds of water and interferes with the crystalline lattice formation of ice. As a result, the water will not freeze until cooled below the freezing point for pure water, a phenomenon known as freezing point depression. For an ideal solution, its freezing point will drop 1.86° Celsius for each osmole of solute/liter of water. A plasma sample from a patient shows a freezing point depression of 0.55° C. What is the osmolarity of this patient's plasma? What is the intracellular osmolarity in this patient?

9. You put 5 grams of glucose into a giant beaker of water and stir. You then take a 1 mL sample and analyze it for glucose. If the 1 mL sample contains 1 mg of glucose, how much liquid is in the beaker?

10. You are monitoring the absorption of a new drug, Curesall, across the intestine of a rat *in vitro*. After numerous experiments, you get data that give you the graph below. The only difference between experiments A and B is that in experiment A, the apical solution contains 150 mM Na$^+$ and in experiment B the apical solution contains 50 mM Na$^+$. What conclusion(s) can you draw about Curesall's absorption based on the graph? Based on what you have learned about different kinds of transport, hypothesize about the type of transporter that carries Curesall.

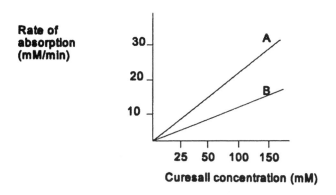

11. What is the osmolarity of a 0.9% NaCl solution? Assume complete dissociation of NaCl. (To review molar solutions, and percent solutions, see p. 25 of the text. For atomic weights, see p. 15 of text.)

12. What is the osmolarity of a 5% dextrose (= glucose, $C_6H_{12}O_6$) solution?

13. An artificial cell with zero glucose inside is placed in a solution of glucose. The amount of glucose inside the cell is measured at different times. The two graphs below show the results of the experiments.

Graph #1

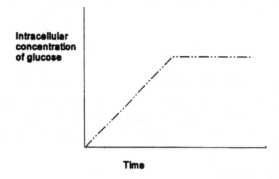

Graph #2

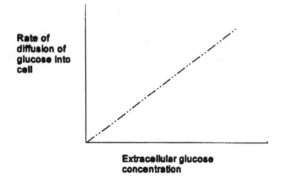

a. In what units might the rate of diffusion be measured? _____

b. Can you tell from these graphs whether glucose is moving into the artificial cell by diffusion or by protein-mediated transport? If it is mediated transport, is it active or passive?

c. Why does the line in graph #1 level off?

14. For each row in the table below, assume that the row represents two compartments separated by a membrane that is freely permeable to water but impermeable to solutes. Show which way water will move by osmosis for each row by drawing in an arrow in the direction of water movement in the column labeled "membrane." In the last column, tell the osmolarity of solution A relative to solution B.

SOLUTION A	MEMBRANE	SOLUTION B	OSMOLARITY OF A RELATIVE TO B
100 mM glucose		100 mM urea	
200 mM glucose		100 mM NaCl	
300 mOsM NaCl		300 mOsM glucose	
300 mM glucose		200 mM CaCl₂	

15. A one liter intravenous infusion of 300 mOsM NaCl is administered during a 15-minute period to a person whose starting osmolarity was 300 mOsM. What effect will this infusion have on the volume and osmolarity of the extracellular and intracellular fluid compartments? Explain your answer.

16. A red blood cell whose internal osmolarity is 300 osmol/L is placed into a solution that has a composition of 200 osmol/L NaCl and 100 osmol/L of urea.

    This solution is _____osmotic to the cell. This solution is _____tonic to the cell.

    Draw a fully labeled graph showing the volume change of the cell from the time it is placed into the solution (at the arrow) until it reaches equilibrium.

17. A man comes into the emergency room dehydrated after working out in the summer sun. His ECF volume is 13 liters and his ECF osmolarity is 340 mOsM. (Ignore ICF in this problem.)

    a. How many milliosmoles of solute are in his ECF when he comes in? _____

    b. An intravenous (IV) infusion of 1 liter of 160 mOsM NaCl is given. How many millimoles of solute are in the IV?

    c. Assuming that no water or solute move out of the ECF, calculate the new ECF volume and osmolarity after the man has been given the IV.

18. Assume that a membrane which is permeable to Na^+ but not to Cl^- separates two compart-ments. Two different sodium chloride solutions are placed in the compartments. The concentra-tion of NaCl on side 1 is 0.3 M, while the NaCl solution on side 2 is 2 M. In which direction will ion movement occur? Explain.
Will a membrane potential difference develop across the membrane? Explain.

19. If you are told that the extracellular K^+ concentration has increased from 2 mEq/L to 3 mEq/L, does this also mean that the extracellular fluid has become more positive? What happens to the resting membrane potential of a liver cell when the extracellular K^+ concentration increas-es? Explain.

20. You are doing an experiment and have an intracellular recording electrode stuck into a liver cell. The resting membrane potential is -70 mV relative to the outside of the cell. At the arrow head you add ouabain to the extracellular fluid bathing the cell. On the graph below draw what will happen to the resting membrane potential of the cell over time. Write an explanation of the forces at work in this situation.

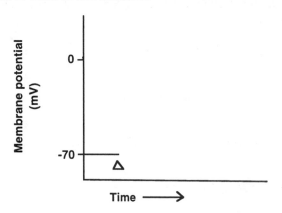

BEYOND THE PAGES

RUNNING PROBLEM: Cystic fibrosis

Cystic fibrosis: An antibiotic cure? (1996, 20 April) *Science News* 149:255. Abstract: Aminoglycoside antibiotics cause damaged CFTR genes bathed in the antibiotic to produce functioning protein in 5% of CF patients.

Cystic fibrosis controversy: A new theory hints that gene therapy in the womb can cure disease. (1997, 10 May) *Science News* 151:292. Abstract: J. Craig Cohen and his colleagues conducted research with mice that suggests that the disease may be preventable in humans. The researchers exposed mice fetuses with a mutant version of a gene, which is similar to the one that causes CF in people, to viruses carrying a working version of the gene.

Cystic fibrosis puzzle coming together. (1997, Feb 8) *Science News* 151:85. Abstract: Research indicates that cystic fibrosis results when a bacterium causes mucus to be produced in excess and a genetic defect causes the lungs' natural antibiotic, defensin, to be destroyed. The findings may lead to the development of more effective treatments.

READING

A is for ...: Caveola. (1995, May/June) *Science & Medicine*.

And still they are moving.... dynamic properties of caveolae. (1996, 24 June) *FEBS Letters* 389: 52-52.

Budding vesicles in living cells. (1996, March) *Scientific American*.

Caveolae and caveolins. (1996, August) *Current Opinion in Cell Biology* 8:542-548.

Caveolae caveat (secrets of cell caveolae explored). (1995, January) *Discover* 16:91.

Cellular mechanisms of phagocytosis (a series of six articles). (1995, 5 March) *Trends in Cell Biology* 5:89-119.

Cellular secretion: It's in the pits. (1997, March/April) *American Scientist* 85: 123-124.

Channelopathies: Their contribution to our knowledge about voltage-gated ion channels. (1997, June) *News in Physiological Sciences* 12: 105-112.

Epithelial secretion driven by anions other than chloride. (1993, April) *News in Physiological Sciences* 8: 91-93.

Epithelial sodium channels: Their role in disease. (1996, April) *News in Physiological Sciences* 11: 102-103.

Ion channels and human genetic diseases. (1996, February) *News in Physiological Sciences* 11: 36-42.

Liposomes revisited. (1995, 3 March) *Science* 267: 1275-1276.

The Mammalian Facultative Glucose Transporter Family. (1995, April) *News in Physiological Sciences* 10: 67-70.

Patch-clamp recording of ion channels. (1990, August) *News in Physiological Sciences* 5: 155-158.

Potocytosis: Sequestration and transport of small molecules by caveolae. (1992, 24 January) *Science* 255: 410-411.

Steps to the Na^+-K^+ pump and Na^+-K^+-ATPase (1939-1962). (1995, August) *News in Physiological Sciences* 10: 184-188.

What is new about the structure of the epithelial Na^+ channel? (1996, October) *News in Physiological Sciences* 11: 195-201.

CHAPTER 6
COMMUNICATION, INTEGRATION, AND HOMEOSTASIS

SUMMARY

The key ideas to take away from this chapter:

• What are the forms of communication in the body and when are they used?
• What important role do receptors play?
• What are the different signal transduction pathways?
• What is homeostasis and how is it maintained?
• What are the patterns in communication/response loops? (Make these your mental file folders.)

In order for a society of distinct individuals to exist, its members must learn to communicate effectively. The cells in a multicellular organism comprise such a society, and they have evolved effective and efficient means of communication so that each component can do its job correctly and respond to changes in the environment. This chapter introduces the concepts of cellular communication and homeostasis, and provides patterns by which you will be able to organize more detailed information that will be presented later for each specific physiological system. In other words, learn the patterns presented in this chapter, and use them to make files in your mental filing cabinet under which you can store specific information later.

Communication in the body is either electrical or chemical. Electrical communication is limited to nerves and muscles, while chemical communication is an activity in which all cells participate to some extent. Local communication is accomplished by paracrines, autocrines, and cytokines that work close to the cells that secrete them. Long distance communication is accomplished by any combination of the following: hormones from the endocrine system, cytokines from many cells, or electrical impulses in the nervous system. Chemical communication is limited by diffusion, as increasing distance increases the time required for action. Hormones (and sometimes cytokines) are secreted into the blood and carried to all parts of the body, but ultimately have to diffuse to their target cells. Therefore, electrical impulses are a faster form of long distance communication (milliseconds-minutes) than hormones (minutes-hours). However, electrical communication is short-lived, while chemical communication can have long lasting effects.

A cellular response to a signal depends more on the receptor than on the signal. Receptors turn on enzymatic machinery that ultimately directs the cellular response. If the receptor is on the cell membrane, it turns on signal transduction machinery to amplify the signal and relay it to the appropriate organelles. Signal ligands are called first messengers, and products of the transduction machinery are called second messengers. If the receptor is in the cytoplasm, then no signal transduction machinery is employed, but enzymes are still activated to direct the cellular response. Different cells can have different responses to the same signal depending on the receptor structure/function, and receptors can be down-regulated or up-regulated depending on a given situation.

Homeostasis means maintaining a "similar condition" despite changes in homeostatic parameters. Walter B. Cannon outlined four postulates regarding the properties of homeostasis (p. 156). The body uses a system of local control and reflex control pathways to maintain homeostasis. Local control depends on paracrines, autocrines, and cytokines. Reflex control pathways handle widespread or systemic challenges, and these reflex pathways are composed of nervous, endocrine, and cytokine components. (Is this starting to look familiar?)

Reflex pathways are further broken down into response loops and feedback loops. Response loops contain the following steps: sensor/receptor, efferent pathway, integrating center, afferent pathway, effector, response (cellular and systemic). This is an important pattern to learn, and the various endocrine and/or nervous system combinations involved serve as good file headings for future information. Feedback loops can be either negative or positive. Negative feedback loops turn off the response that created them, and are therefore homeostatic. Positive feedback loops continue the response that created them, are therefore not homeostatic, and must be controlled from outside the response loop. Feedforward control responses can anticipate changes.

TEACH YOURSELF THE BASICS

CELL-TO-CELL COMMUNICATION

• List the three basic methods of cell-to-cell communication.

_____

Gap Junctions Transfer Chemical and Electrical Signals Directly between Cells

• What kinds of signals pass through gap junctions (Fig. 6-1a)? [∫ p. 54]

_____

• Where do you find gap junctions?
_____

Paracrines and Autocrines Are Chemical Signals Distributed by Diffusion

• How do chemical signals secreted by cells spread to adjacent cells? _____

• This process limits communication to (long or short?) distances [∫ p. 118] _____

• Distinguish between paracrines and autocrines. (Fig. 6-1b)

_____

• Neuromodulators are paracrines secreted by _____.

Long-Distance Communication Is Carried Out by Electrical Signals, Hormones, and Neurohormones

• The chemical signals of the endocrine system are called _____.

• How do hormones reach their target cells? _____

• Why don't all cells react to all hormones? _____

_____ (Fig. 6-1c)

• In the nervous system, what kinds of signals are used to transmit information?

_____

• What is the difference between a neurotransmitter and a neurohormone?

_____

Cytokines Act as Both Local and Long-Distance Signals

• Cytokines are involved in (local communication / long distance communication / both?).

• How are cytokines different from peptide hormones?

_____

• What kinds of cell functions do cytokines control?

_____

RECEPTORS AND SIGNAL TRANSDUCTION

• Why do some cells respond to a chemical signal while other cells ignore it? (Fig. 6-2)

_____

Receptors as Well as Signal Molecules Determine the Cellular Response

• How do chemical signals interact with receptors? _____

• In what part(s) of a cell are receptors found? _____

_____

• How can a chemical signal have one effect in one tissue but a different effect in another tissue?

_____

Agonists and antagonists
• (Agonists/Antagonists?) are
      1. Molecules or drugs that mimic normal ligand
      2. Bind to and activate receptors
• (Agonists/Antagonists?) are
      1. Molecules or drugs that mimic normal ligand
      2. Bind and block ligand   Receptor never turned on

Up- and down-regulation of receptors enables cells to modulate cellular activity
• What causes down-regulation of a cell's receptors?

_____

- In down-regulation, the number of receptors (increases/decreases?) or their binding affinity (increases/decreases?).

- When a cell down-regulates its receptors, what happens to its responsiveness to the ligand that binds to that receptor?

_____

- Up-regulation of receptors (increases/decreases?) the number of receptors and (enhances/lessens?) the cell's response.

Receptors and disease
- Give an example of a disease involving abnormal receptors.

_____

Signal Molecules Are the First Messengers in a Series of Events

- Define a first messenger. _____

Lipophilic messengers enter the cytoplasm and combine with receptors inside the cell
- Where are the receptors for lipophilic first messengers found? (Fig. 6-3)

_____

- The general cellular response to a lipophilic first messenger is _____

_____

- Give an example of a lipophilic messenger. _____

Lipophobic messengers have surface membrane receptors
- What kinds of biomolecules are lipophobic messengers? _____

- Where are the receptors for lipophobic messengers found? Why in that location? [∫ p. 109]

_____

- Explain signal transduction. (Fig. 6-4)

_____

Signal Transduction Pathways Depend on Membrane Proteins

- Explain the role of second messengers in signal transduction pathways.

_____

Signal cascades and amplification
- What is a ligand? _____

• Explain the process of signal amplification (Fig. 6-6):

_____

_____

Membrane transducers
• Name two common membrane transducers and explain what happens when each is activated. (Fig. 6-8)

_____

_____

Second messengers
• Name one ionic and four chemical second messengers.

_____

Protein kinases
• What activates protein kinases? _____

• What do protein kinases do? _____

• List four common targets of phosphorylated proteins. _____

_____

HOMEOSTASIS

• Define homeostasis. [∫ p. 6] _____

_____

The Development of the Concept of Homeostasis

• What role did Claude Bernard play in the development of the concept of homeostasis? _____

_____

• What is a parameter? _____

• List Walter B. Cannon's four postulates.

1. _____

2. _____

3. _____

4. _____

The Failure of Homeostasis Results in Disease

• Name some common causes of pathological conditions. _____

_____

CONTROL PATHWAYS: RESPONSE AND FEEDBACK LOOPS

• Paracrines are most likely to influence (reflex/local?) responses.

Local control
> A. Involves paracrines or autocrines
> B. Response is restricted to where change took place (Fig. 6-12)

Reflex control
• List the three main components of a response loop. _____

_____

• Fill in the missing terms: Stimulus → _____ → _____ pathway →

_____ center → _____ pathway → _____ → response

Sensory Receptors
• Give two different meanings for the word "receptor." (Fig. 6-13)

_____

• What is the threshold of a receptor? _____

Afferent Pathway
• An afferent pathway always carries information away from a/an _____ to a/an

_____ .

Integrating center
• What is the role of an integrating center? _____

• Where are the integrating centers for nervous reflexes? _____

• Where are the integrating centers for endocrine reflexes? _____

Efferent Pathway
• An efferent pathway always carries information away from a/an _____ to a/an

_____ .

• In the nervous system, the efferent pathway is a/an _____ signal.

• In endocrine reflexes, the efferent pathway is a/an _____ signal.

• How are efferent signals named or described in nervous and endocrine pathways?

_____

Effectors
• Define and give some examples of effectors. _____

_____

Responses
• Distinguish between and give examples of cellular and systemic responses. _____

_____

Response Loops Begin with a Stimulus and End with a Response

• Name the seven steps of a response loop. (Fig. 6-14) _____

_____

• Distinguish between acclimatization and acclimation. _____

_____

Feedback Loops Modulate the Response Loop
• What is the purpose of a negative feedback loop? _____

_____

Negative feedback loops are homeostatic
• Negative feedback (Fig. 6-15) (opposes/reinforces?) the original stimulus. Most feedback loops are negative.

Positive feedback loops are not homeostatic
• Positive feedback loops (Fig. 6-16, 6-17) (oppose/reinforce?) the original stimulus.

• Give an example of a positive feedback loop. _____

_____

• What shuts off a positive feedback loop? _____

Feedforward Control Allows the Body to Anticipate Change and Maintain Stability

• Give an example of feedforward control. _____

• Feedforward control (opposes/enhances?) homeostasis.

Biological Rhythms Result from Changes in the Setpoint

• What is a circadian rhythm? (Fig. 6-18) _____

• Name some body functions that exhibit circadian rhythms. _____

Control Systems Vary in Their Speed and Specificity

• What is a neuroendocrine reflex? (Fig. 6-19, 6-20) _____

_____

Specificity
• Compare the specificity of nervous and endocrine control. How is the specificity of the reflex determined in each type of reflex?

_____

_____

Nature of the signal
• Electrical signals travel _____ distances; chemical signals travel _____ distances.

• The nervous system uses (chemical / electrical / both chemical and electrical?) signals.

• The endocrine system uses (chemical / electrical / both chemical and electrical?) signals. (Fig. 6-20, 6-6)

Speed
• Nervous reflexes are much (faster / slower?) than endocrine reflexes. [Review: diffusion principles, p. 118]

Duration of action
• Nervous control is of much (shorter/longer?) duration than endocrine control.

• Short-term functions tend to be under (nervous/endocrine?) control, while long-term functions tend to be under (nervous/endocrine?) control.

Coding for stimulus intensity
• Compare how stimulus intensity is coded in the nervous and endocrine systems.

_____

Pathways for Nervous, Endocrine, and Neuroendocrine Reflexes May Be Complex with Several Integrating Centers (Fig. 6-20)

• In endocrine reflexes (pattern 3), the integrating center is _____

and the efferent pathway is the _____.

• In nervous reflexes (pattern 1), the integrating center is _____ and the

efferent pathway is the _____.

• In neuroendocrine reflexes (patterns 2, 3, 4, 5), the afferent neuron leads to a/an

_____. The efferent pathway(s) is/are _____.

☛ *Take some time to draw out the different reflexes for yourself. Color-code the components and learn the distinctions between the reflex types.*

TALK THE TALK

acclimation	homeostasis
acclimatization	hormone
adenosine 3', 5'-cyclic monophosphate (cAMP)	inositol triphosphate (IP$_3$)
adenylate cyclase	integrating center
afferent pathway	local control
agonist	negative feedback
alpha (α) adrenergic receptor	neurocrine
amplifier enzyme	neuroendocrine reflex
autocrine	neurohormone
Bernard, Claude	neuromodulator
beta (β) adrenergic receptor	neuron
biorhythm	neurotransmitter
Cannon, Walter B.	nitric oxide (NO)
cascade	paracrine
central receptor	parameter
circadian rhythm	pathophysiology
connexin	pepsin
cyclic AMP (cAMP)	peripheral receptor
cyclic GMP(cGMP)	phospholipase C
cytokine	positive feedback loop
diacylglycerol (DAG)	protein kinase
down-regulation	receptor
drug tolerance	reflex control pathway
effector	response loop
efferent pathway	second messenger
epinephrine (adrenaline)	sensitivity
feedback loop	signal amplification
feedforward control	signal transduction
first messenger	stimulus
G protein	syncytium
gap junction	target cell
guanosine diphosphate (GDP)	threshold
guanosine triphosphate (GTP)	tonic control
guanosine 3', 5'-cyclic monophosphate (cGMP)	tyrosine kinase
guanylate cyclase	up-regulation
histamine	

ERRATA

p. 168, right column, discussion of pattern 2: The baby's mouth stimulates sensory receptors *linked to sensory neurons* that travel to the brain.

RUNNING PROBLEM - Diabetes mellitus

RUNNING PROBLEM - Diabetes mellitus

Diabetes mellitus is a family of related diseases. Most forms of non-insulin-dependent diabetes mellitus (NIDDM) seem to be associated with changes in the signal transduction pathway, and insulin levels in these patients are often normal to elevated until late in the disease process. Diabetes is an ideal disease for demonstrating the integration of physiological function, and various aspects of diabetes will be mentioned throughout the book. A detailed discussion of diabetes can be found in Chapter 21.

PRACTICE MAKES PERFECT

1. Describe the basic ways cells communicate with each other. Which modes are faster?

2. Why are there so many different types of receptors? Is the number of receptors on a cell constant?

3. What is the advantage of having a cascade of events as opposed to having just a single event drive a response?

4. Would you expect the secretion of growth hormone to be regulated through negative feedback? Explain.

MAPS

1. Use the following terms to design a map or flow chart describing the events that are associated with signal transduction and the activation of second messenger systems. You may add any terms that you deem necessary. You may draw a cell and associate the terms with different components of the cell (a structure/function map).

adenylate cyclase	GDP	membrane receptor
amplifier enzyme	GTP	phospholipase C
cAMP	guanylate cyclase	protein kinase
cGMP	hormone	second messenger
DAG	IP$_3$	target cell
first messenger		tyrosine kinase

2. Use the following terms to design a map or flow chart describing the events that are associated with reflex control pathways. You may add any terms you feel are necessary to complete the map.

afferent pathway   negative feedback   stimulus
effector   receptor   threshold
efferent pathway   response   tonic control
integrating center   sensory receptor

READING

A is for ...: Adenyl cyclase. (1995, November/December) *Science & Medicine*.

A is for...: Biomimetics. (1996, January/February) *Science & Medicine*.

Biologic roles of nitric oxide. (1992, May) *Scientific American*.

The biological switchboard: Nobels for finding how cells communicate. (1994 , October 24) *Newsweek*, pp. 65-66.

Boyd, C.A.R. and Noble, D. (eds.) (1993), The Logic of Life: the Challenge of Integrative Physiology. Oxford University Press.

Cannon, Walter B. (1929)Organization for physiological homeostasis. *Physiological Reviews* 9: 399-431.

The cycling of calcium as an intracellular messenger. (1989, October) *Scientific American*.

Cytokines: Molecular keys to homeostasis, development, and pathophysiology. (1993, December) *Journal of Cellular Biochemistry* 53 (4): 277-279.

Diseases caused by impaired communication among cells. (1980, March) *Scientific American*.

Drugs by design. (1993, December) *Scientific American*.

G proteins. (1992, July) *Scientific American*.

Hereditary and acquired defects in signaling through the hormone-receptor-G protein complex. (1994, February) *American Journal of Physiology* 266 (2 Pt 2): F163-174.

How are control systems controlled? (1994, January/February) *American Scientist* 82: 38-44.

How cells respond to stress. (1993, May) *Scientific American*.

How cells maintain stability. (1990, December) *Scientific American*.

Intercellular waves of communication. (1996, December) *News in Physiological Sciences* 11: 262-269.

Internal timekeeping. (1996, May/June) *Science & Medicine*.

Membrane-delimited cell signaling complexes: direct ion channel regulation by G proteins. (1993) *Journal of Membrane Biology* 131: 93-104.

The molecular basis of communication between cells. (1985, October) *Scientific American*.

Molecular machines that control genes. (1995, February) *Scientific American*.

Multidrug resistance in cancer. (1989, March) *Scientific American*.

New molecular targets for cancer therapy. (1996, September) *Scientific American*.

New clues found in circadian clocks. (1997, May 16) *Science* 276: 1030-1031.

Perspectives in imaging of second messengers. (1996, December) *News in Physiological Sciences* 11: 281-287. Use of fluorescent indicator dyes to track second messenger activity.

Protein kinases, phosphatases, and the control of cell volume. (1992, October) *News in Physiological Sciences* 7: 232-236.

Rasmussen, H. (ed) (1991) Cell communication in health and disease: Readings from Scientific American. W.H. Freeman and Co., New York.

The role of p53 in cancer development. (1994, September/October) *Science & Medicine*.

Signal transduction pathways as drug targets. (1995, November/December) *Science & Medicine*.

Special issue: Signal transduction. (1995, April 14) *Science*, Vol. 268.

Timing is everything. (biorhythms) (1994, July/August) *The Sciences*.

Venom peptides as human pharmaceuticals. (1997, September/October) *Science & Medicine*.

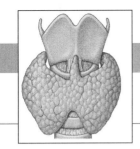

CHAPTER 7

INTRODUCTION TO THE
ENDOCRINE SYSTEM

SUMMARY

Themes to look for in this chapter:

• Compare peptide, steroid, and amine hormones with regard to synthesis, storage, release, transport, and cellular mechanism of action.
• How are the anterior and posterior pituitary lobes different? What hormones does each secrete?
• How are endocrine control pathways organized? What role does negative feedback play?
• What are the different types of hormone interactions?
• What factors are important for diagnosing endocrine pathologies?

This chapter focuses on the basic physiology of the endocrine system and its hormones. A hormone is defined here as a chemical, secreted by a cell or group of cells into the blood, that acts on a distant target and is effective at very low concentrations. Chemically, hormones fall into one of three classes: peptide, steroid, or amine. Generally, the three groups can be distinguished by how they are synthesized, stored, and released; how they are transported in the blood; and the mechanisms by which they cause a cellular response. Hormones can interact with other hormones, and in doing so, alter cellular response. Three types of hormone interaction are discussed in this chapter: synergism, permissiveness, and antagonism. You should familiarize yourself with these distinctions and, as you did with the information in Chapter 6, use them to make mental file folders under which you will store more specific information later.

The release of hormones can be under multiple levels of control: classic hormones are under direct control of the parameter they regulate, but other hormones can have a multi-level control pathway. Those hormones that fall into the latter category represent the more complex response loops. (Those mental file folders from Chapter 6 are already coming in handy; review Fig. 6-20.)

If a hormone causes the release of another hormone, that first hormone is called a trophic hormone. Most of the six hormones of the anterior pituitary are trophic hormones, and are in turn controlled by trophic neurosecretory hormones of the hypothalamus. The hypothalamus and the pituitary communicate via the hypothalamic-hypophyseal portal system Conversely, the posterior pituitary, which is derived of neural tissue, only releases two hormones that are actually neurohormones synthesized in the hypothalamus. There are clear distinctions between the anterior and posterior pituitary; know them, love them, recite them daily!

You should also familiarize yourself with the concepts of control pathways (more mental file folders). Specifically, learn how negative feedback affects hormone release in a control pathway and how feedback factors into diagnosing and understanding an endocrine pathology (primary or secondary?). While studying pathologies, you will also need to grasp the concepts of hypersecretion, hyposecretion, up- and down-regulation, and abnormal tissue responsiveness.

TEACH YOURSELF THE BASICS

HORMONES

• Hormones control:

1. _____

2. _____

3. _____

4. _____

• By doing what to their target cells?

1. _____

2. _____

3. _____

• Hormone action depends on specific _____ and hormone properties. [∫ p. 120]

• The distinction between nervous and endocrine system is no longer clear cut. What kinds of cells/tissues can secrete hormones? _____

Hormones Have Been Known Since Ancient Times

• List the four classic steps for identifying an endocrine gland.

1. _____

2. _____

3. _____

4._____

What Makes a Chemical a Hormone?

• Define a hormone.

_____

_____

• What is meant by "the cellular mechanism of action" of a hormone?

_____

• What is the half-life of a hormone?

_____

THE CLASSIFICATION OF HORMONES
☛ REVIEW organelle function. Can you describe all the steps of protein synthesis, starting with transcription? [∫ pp. 98-102]

• List the three chemical classes of hormones (Table 7-1).

_____

Most Hormones in the Body Are Peptides or Proteins

• Peptide hormones are composed of _____.

• Exclusion rule: If a hormone isn't a steroid or an amine, then it must be a peptide.

Peptide hormone synthesis, storage, and release (Fig. 7-3, 7-4)
• Peptide hormones are synthesized where in the cell? _____

• Explain the difference between the preprohormone, the prohormone, and the hormone. Where is each made? [∫ p. 102]

_____

_____

• How are peptide hormones released from endocrine cells? _____

Transport in the blood and half-life of peptide hormones
• Peptide hormones are (lipophobic/lipophilic?) and therefore (will/ will not?) dissolve in plasma.

• Describe the half-life of most peptide hormones. _____

Cellular mechanism of action of peptide hormones
• Peptide hormones are (lipophobic/lipophilic?), so they have (intracellular/membrane?) receptors. [∫ p. 151]

• What is the target cell's response to hormone-receptor binding? (Fig. 7-5) _____

_____

_____

• The target response time is (quick/slow?). Explain. _____

_____

Steroid Hormones Are Derived from Cholesterol

• List the tissues/organs from which steroid hormones are secreted.

_____

Steroid synthesis and release
• Steroid hormones are synthesized where in the cell? _____

• Can steroid hormones be stored? Explain. _____

• How are steroid hormones released from endocrine cells? _____

Transport in the blood and half-life of steroid hormones
• Steroid hormones are (lipophobic/lipophilic?) and therefore (will/ will not?) dissolve in plasma.

• Describe the half-life of steroid hormones. _____

Cellular mechanism of action of steroid hormones
• Steroid hormones are (lipophobic/lipophilic?), so they have (intracellular/membrane?) receptors.

• What is the target cell's response to hormone-receptor binding?_____

_____

_____

• The target response time is (quick/slow?). Explain. _____

_____

Amine Hormones Are Derived from Single Amino Acids

• List three different groups of amine hormones and the amino acids from which each group is derived.

_____

_____

☛ Be sure that you can list the tissues/glands that secrete steroid and amine hormones. If a hormone doesn't come from these tissues, then it must be a peptide!

CONTROL OF HORMONE RELEASE
☛ Review local and reflex control [∫ p. 158] and reflex pathways: input → integration → output. [∫ p. 167]

The Target of Most Trophic Hormones Is Another Endocrine Gland or Cell

• What is a trophic hormone?

_____

• In the names of trophic hormones, the suffix is usually –_____ and the root denotes _____

_____.

Negative Feedback Turns Off Hormone Reflexes

• Describe the negative feedback involved in the blood glucose/insulin reflex. _____

_____

☞ In complex endocrine reflexes, hormones can act as negative feedback signals. (Fig. 7-9)

Hormones Can Be Classified by Endocrine Reflex Pathways
☞ Classify hormones and control pathways by type of endocrine reflex pathway. [∫ p. 168]

Classic hormones
• In the simplest endocrine pathways, what is the receptor that senses the stimulus?

_____

Classic hormones with multiple controls
• Name two different ways that insulin release may be stimulated.

_____

Neurohormones
• Name three specific tissues that synthesize and secrete neurohormones. _____

_____

• What is the signal for release of neurohormones? [∫ Fig. 6-20, pattern 2]

_____

The Pituitary Gland
• Compare the tissue types that make up the anterior and posterior pituitary.

_____

Neurohormones of the Posterior Pituitary
• The posterior pituitary stores and releases what two hormones?_____

• Where are these hormones made and how do they get to the posterior pituitary? (Fig. 7-13) [∫ axonal transport, Ch. 8, Fig. 8-4]

_____

Hormones under trophic hormone control

• What is the function of hypothalamic trophic hormones? _____

_____

Hypothalamic Trophic Hormones and the Hypothalamic-Hypophyseal System

• How do hypothalamic hormones reach the anterior pituitary? (Fig. 7-14)

_____

• Describe a portal system. _____

_____

• What is the primary advantage of a portal system? _____

_____

Hormones of the Anterior Pituitary (Fig. 7-16)

• List (spell out!) the six hormones synthesized by endocrine cells of the anterior pituitary.

     1. _____

     2. _____

     3. _____

     4. _____

     5. _____

     6. _____

Put a star next to the hormone(s) that is/are not trophic hormones.

• Explain how negative feedback works in the hypothalamic-pituitary system. [∫ negative feedback, p. 162]

_____

• Distinguish between long-loop negative feedback and short-loop negative feedback.

_____

Multiple Hormones Can Affect a Single Target Simultaneously

• Name three types of hormone interactions.

_____

Synergism

• Explain a synergistic relationship between hormones A and B. _____

_____

Permissiveness

• Explain the relationship if hormone A is permissive for hormone B. _____

_____

Antagonism

• Explain the relationship if hormone A is antagonistic to hormone B. [∫ p. 150]

_____

ENDOCRINE PATHOLOGIES

• Cortisol secretion by the adrenal cortex is used as an example (Fig. 7-19). Fill in the hormones
of this pathway: Hypothalamus secretes _____→ anterior pituitary
secretes _____→ adrenal cortex secretes _____

Hypersecretion of a Hormone Exaggerates Its Effects

☛ Excess hormone secretion may be due to benign (adenomas) or malignant tumors, iatrogenic
causes.

The Effects of a Hormone Are Diminished or Absent with Hyposecretion

☛ Deficient hormone secretion may be due to atrophy of the endocrine gland, lack of vitamin's,
mineral's, etc.

Abnormal Tissue Responsiveness Can Be Due to Problems with Hormone Receptors or Second Messenger Pathways

• In abnormal tissue responsiveness, hormone levels are (normal/abnormal?).

• What are some reasons for abnormal tissue responsiveness? _____

_____

Up- and down-regulation

• (High/Low?) hormone levels will cause down-regulation of receptors. [ʃ p. 150]

Receptor and signal transduction abnormalities

• Mutations cause defects in receptor structure or signal transduction. [ʃ pp. 152-155]

    A. Receptor abnormalities: Lack of receptors or abnormal ones

       1. Example: Testicular feminizing syndrome

    B. Signal transduction abnormalities: Missing or abnormal transduction

       1. Example: Pseudohypoparathyroidism: Inherited defect in _____ protein

Diagnosis of Endocrine Pathologies Depends on the Complexity of the Reflex

• What is the difference between a primary endocrine pathology and a secondary endocrine pathology?

_____

_____

• Explain why the concentrations of trophic hormones change with primary and secondary pathologies.

_____

_____

• The questions below apply to the hypersecretion of cortisol (Fig. 7-21).

    1. High cortisol, high ACTH, low CRH: (primary /secondary?) pathology as a result of a defect in the (anterior pituitary / adrenal cortex?).

    2. High cortisol, low ACTH, low CRH: (primary /secondary?) pathology as a result of a defect in the (anterior pituitary / adrenal cortex?).

HORMONE EVOLUTION

• Why does insulin from cows, pigs, and sheep work in humans?

_____

• Why doesn't growth hormone from cows, pigs, and sheep work in humans?

_____

• What is a vestigial structure? _____

ORGAN BOX - The pineal gland

• Once thought to have no function, the pineal gland is now known to secrete _____.

• What is the only verified function of melatonin in humans? _____

TALK THE TALK

abnormal tissue responsiveness
adenohypophysis
adenoma
adrenocorticotrophic hormone
    (ACTH)
amine hormone
antagonism
anterior pituitary
antidiuretic hormone
atrophy
calcitonin
cAMP second messenger system
candidate hormone
castration
catecholamine
cellular mechanism of action
corticotropin
cortisol
co-secretion
cytokine
diabetes mellitus
down-regulation
endocrine gland
endocrinology
endogenous
erythropoietin
etiology
exogenous
follicle-stimulating hormone
goiter

gonadotropin
growth factor
growth hormone
half-life
hormone
hormone deficiency
hormone replacement therapy
hyperinsulinemia
hypersecretion
hyposecretion
hypothalamic-hypophyseal axis
hypothalamus
iatrogenic
immunocytochemistry
insulin
long-loop negative feedback
luteinizing hormone (LH)
melanocyte-stimulating hormone (MSH)
melatonin
metabolite
negative feedback
neurohypophysis
organotherapy
oxytocin
peptide hormone
permissiveness

pineal gland
pituitary gland
placebo effect
portal system
posterior pituitary gland
preprohormone
primary endocrine pathology
prohormone
prolactin
pseudohypoparathyroidism
secondary endocrine pathology
short-loop negative feedback
somatostatin
somatotropin
specificity
steroid hormone
synergism
tamoxifen
testes
testicular feminizing syndrome
thyroid hormone
thyroid-stimulating hormone
    (TSH)
thyrotropin
transcription factor
trophic hormone
vasopressin
vestigial

ERRATA

p. 184, second line of left column: should read "tissues" instead if "issues"

p. 189, Running Problem: TRH is thyrotropin-releasing hormone, not thyroid-releasing hormone.

p. 199, question 28: Should read "What characteristics do <u>receptors</u> (not hormones)..." and "Compare these molecules with the structure of <u>hormones</u> (not receptors)."

RUNNING PROBLEM - Graves' Disease

Graves' disease, an autoimmune condition in which the body produces antibodies to the TSH receptor on the thyroid cells, is the most common form of hyperthyroidism. It is more common in women than men (8 to 1), with typical onset between the ages of 20 and 40. The cause of antibody production is not clear. The most common treatments are antithyroid drugs and radioactive iodine (^{131}I) synthetic. The latter concentrates in the thyroid gland and destroys the cells. If too much tissue us destroyed, hypothyroidism may result, but this is easily treated with thyroid hormone pills. Surgery is a less common treatment, but it also reduces the amount of active thyroid tissue.

PRACTICE MAKES PERFECT

1. Fill in the chart below for the two primary types of hormones.

	Peptide	Steroid
Transport in plasma		
Synthesis site in endocrine cell		
Method of release from endocrine cell		
General response of endocrine cell to hormone stimulation (i.e., mechanism of action)		

2. Write "A" next to the statements that apply to the anterior pituitary, and "P" next to the statements that apply to the posterior pituitary.

_____ Connected to hypothalamus by nerve fibers

_____ Connected to hypothalamus by blood vessels

_____ Secretes hormones produced by hypothalamus

_____ Controlled by releasing hormones from hypothalamus

_____ Secretes peptide hormones

3. Which of the following seems to be a major function of the pineal gland?
    a) sense of appetite
    b) sense of thirst
    c) regulation of behavioral responses
    d) regulation of temperature responses
    e) acts as a biological clock

4. Match the items below to the questions. Answers may be used once, more than once, or not at all.

a) cortisol	b) aldosterone	c) growth hormone	d) adrenaline
e) vasopressin	f) prolactin	g) parathyroid hormone	h) thyroxine

1) Which hormone(s) is/are produced by the adrenal cortex _____

2) Which hormone(s) is/are produced by the anterior pituitary _____

3) Which hormone is the fight-or-flight hormone? _____

4) Name the two hormones most important for normal growth and development. _____

5) Name a hormone whose target is the kidney. _____

5. True/False? Defend your answer:

The adrenocorticosteroid hormones such as cortisol are useful in fight-or-flight situations such as being chased by a bear.

6. True or false? Defend your answer.

a) Endocrine gland cells that synthesize steroid hormones have lots of hormone stored in vesicles in the cytoplasm.

b) These same endocrine cells have lots of smooth endoplasmic reticulum and Golgi body.

7. Hormone XTC causes decreased neural transmission of pain, inducing a pleasant but misleading lack of pain. Another hormone, Y-ME, acts on XTC-producing cells to inhibit transcription of XTC-mRNA. The ultimate result of combining XTC and Y-ME is normal pain sensation.

(a) Would the interaction between XTC and Y-ME be best described as:

   1. synergism   2. antagonism   3. permissiveness   4. none of the above

(b) From the information given, would you say that hormone XTC is probably:

   1. a peptide hormone   2. a steroid hormone   3. Insufficient information to decide

8.You have been doing research on the pancreatic endocrine cells that secrete insulin and on the adrenal cortex cells that secrete corticosteroids. You prepared tissue samples for examination under the electron microscope, but the labels fell off the jars when the fixative dissolved the glue. You sent the tissue samples off anyway and got back the following description for one of them. Which tissue is being described? Defend your answer.
"...cells are close to blood capillaries. Numerous dense, membrane-bounded granules throughout the cytoplasm with reduced rough endoplasmic reticulum and free ribosomes. Cells with fewer secretory granules show an increase in rough ER and ribosomes."

9. A woman has secondary hypocortisolism due to a pituitary problem. If you give her ACTH, what happens to her cortisol secretion? Draw the complete control pathway, including the glands and hormones as part of your answer.

10. Graves' disease is caused by the production of auto-antibodies to the TSH receptor. These antibodies interact with the TSH receptor to stimulate the thyroid gland in a manner similar to TSH. The antibodies are not subject to negative feedback. Which of the sets of lab values below would indicate Graves' disease? BRIEFLY defend your choice. (HINT: On which tissue would you find TSH receptors?)

	Serum thyroxine	Serum TSH
Patient A	6 μg/100 mL	1.5 μIU*/mL
Patient B	16 μg/100 mL	0.75 μIU/mL
Patient C	2.5 μg/100 mL	20 μIU/mL
Patient D	12 μg/100 mL	10 μIU/mL
Normal	4-11 μg/100 mL	1.5-6 μIU/mL

*IU = international units, a standard way of quantifying TSH amounts.

BEYOND THE PAGES

GRAPHS

Graph the data for the Graves' disease patients in problem #10 above. What kind of graph is most appropriate: a scatter plot? a bar graph? a line graph?

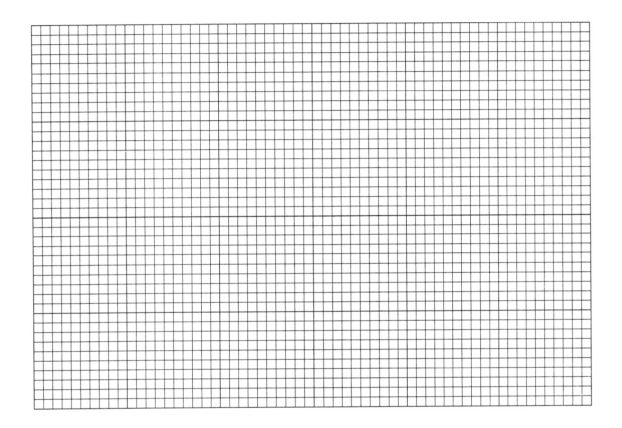

READING

A bright future for the sunshine hormone. (melatonin). (1995, March/April) *Science & Medicine.*

The anterior pituitary gland — its cells and hormones. (1979, July) *Bioscience* 29 (7): 408-413.

Arendt, Josephine. (1995) *Melatonin and the mammalian pineal gland.* Chapman & Hall, London.

Cell volume: A second message in regulation of cellular function. (1995, February) *News in Physiological Sciences* 10: 18-21. Some peptide hormones activate their targets by altering cell volume.

Gastrin processing: From biochemical obscurity to unique physiological actions. (1997, February) *News in Physiological Sciences* 12: 9-15.

Human exposure to putative pheromones and changes in aspects of social behavior. (1991) *Journal of Steroid Biochemistry & Molecular Biology* 39 (4B): 647-659.

In search of human skin pheromones. (1994) *Archives of Dermatology* 130 (8): 1048-1051.

Menstrual synchrony: Agenda for future research. (1995) *Psychoneuroendocrinology* 20(4): 377-83.

MHC-dependent mate preferences in humans. Proceedings of the Royal Society of London, Series B. (1995, 22 June) *Biological Sciences* 260 (1359): 245-249. Female students were asked to rate the odors of t-shirts worn by male students.

Molecular pathways of steroid receptor action. (1992) *Biology of Reproduction* 46: 163-167.

Nerve cells that double as endocrine cells. (1981, June) *Bioscience* 31 (6): 445-448.

New aspects of cytosolic calcium signaling. (1996, February) *News in Physiological Sciences* 11: 13-16.

Non-genomic mechanisms of action of steroid hormones. (1995) *Ciba Foundation Symposium* 191: 24-37.

Nontranscriptional effects of steroid hormones. (1993) *Receptor* 3 (4): 277-291.

Steroid receptors and molecular chaperones. (1995, July/August) *Science & Medicine.*

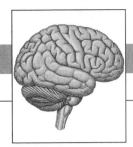

SUMMARY

This chapter contains information that forms the foundation of our understanding of cell excitability and nerve cell function. Be warned! It is very heavy with terminology and contains some of the most difficult information you will need to master in physiology. If you did not understand the concept of electrochemical gradients and the movement of ions along those gradients, go back and learn this <u>before</u> you begin this chapter [∫ pp. 136-141].

So what do you need to take from this chapter? Here are some key points:

- Be able to map the divisions of the nervous system.
- Learn the different types of neural tissue and their functions.
- How does electrical signaling work? (Spend most of your time on this!)
  Include in your study: What roles do voltage-gated ion channels play? What are the differences between graded potentials and action potentials? How are action potentials conducted and what factors affect conduction of action potentials?
- Study how things work at the synapse.

The nervous system is a complex arrangement of branching divisions, each with specific functions. Take some time to map out the divisions of the nervous system and become familiar with their duties. The nervous system is composed of two basic cell types: neurons and glial cells. The neurons are the functional units of the system, producing electrical and chemical signals to communicate with other cells. The glial cells are the supporting cells, providing support or insulation for the neurons.

A neuron has three parts: the cell body, the dendrites, and the axon. Axons lack organelles for protein synthesis, so the cell body ships materials to the axon terminals via axonal transport. Neurons synapse on their target cells. Information passes from the presynaptic cell, across the synapse, to receptors on the postsynaptic target cell(s).

Electrical signaling in neurons is a result of ion movement across the membrane. Compare creation of membrane potential with electrical signaling. What specific cellular components make electrical excitability possible? It is very important that you take the time to learn the specifics of how ion movement creates graded potentials and action potentials (APs). Graded potentials are summed, and if they surpass the threshold, an AP is created. APs cannot be summed: they are all or none. Conduction mechanisms allow APs to travel the axon at high speeds and at full strength. Conduction is affected by axon diameter and resistance. Saltatory conduction takes advantage of myelin sheaths and nodes of Ranvier to give high speed conduction at biologically reasonable axon diameters. Take time to learn how conduction works and how it is altered.

Cells still communicate by electrical or chemical means (memories of Chapter 6 . . .). At electrical synapses, electrical signals pass directly from cell to cell. At chemical synapses, neurotransmitters are released onto the postsynaptic cell. Neurotransmitters are made in the cell body or axon terminal and stored in synaptic vesicles until signaled for release. When signaled, synaptic vesicles are released by calcium-dependent exocytosis. (If you don't remember this story, review this chapter and Chapter 6 in the text.) Chemical signals can work directly on ion channels or by second messenger systems.

TEACH YOURSELF THE BASICS

ORGANIZATION OF THE NERVOUS SYSTEM
(Try putting this section into a concept map or flow chart)
• Information flow in the nervous system follows a reflex pathway. List all six parts of a reflex. [∫ p. 158]

_____

• The nervous system divides into what two major divisions? _____

• The peripheral nervous system (PNS) can be divided into what two divisions?

_____

• Afferent neurons link sensory receptors to _____.
  (see Ch. 10)

• Efferent neurons carry information (to/from?) the CNS. Efferent neurons can be classified into
  two groups:

_____

• Somatic motor neurons (Ch. 12) control _____.

• Autonomic neurons (visceral nervous system) control _____

_____.

• Autonomic neurons can be divided into _____.

• What is the enteric nervous system? _____

_____.

CELLS OF THE NERVOUS SYSTEM

• Name the two primary cell types found in the nervous system. _____

Neurons Are Excitable Cells That Generate and Carry Electrical Signals

• Name the three main parts of a typical neuron. _____

• How do interneurons, (afferent) sensory neurons, and efferent neurons differ in their function?

_____

_____

_____

The cell body is the control center of the neuron
- List the organelles typically found in the neuron cell body. _____

_____

Dendrites receive incoming signals
- Describe the structure and function of dendrites._____

_____

- What takes place at the integrating region of a neuron? _____

_____

Axons carry outgoing signals to the target
- Describe the structure and function of an axon. _____

_____

- What is a nerve and how is it different from a neuron? _____

_____

- At the axon terminal, a/an (chemical/electrical?) signal is usually translated into a/an (chemical/electrical?)

 signal called _____.

- Axons may divide, forming branches known as _____.

- What membrane-bound organelles are found in the axon terminal? _____

_____

- What is a neurocrine?
_____

- Explain the relationship between the following terms: axon terminal, presynaptic cell, postsynaptic cell, synapse, synaptic cleft, target cell.

_____

_____

_____

- What is the purpose of axonal transport? _____

_____

• Describe the mechanism by which vesicles move during fast axonal transport. _____

_____

• Forward or _____ axonal transport moves _____

   from the _____ to the _____.

• Backward or _____ axonal transport moves _____

   from the _____ to the _____.

Glial Cells Are the Support Cells of the Nervous System

• Are there more neurons or glial cells in the nervous system?_____

• Do glial cells communicate with other cells? How? _____

• What is the function of glial cells? _____

• Describe the extracellular matrix of neural tissue. _____

• What is the function of Schwann cells, oligodendrocytes, astrocytes, and microglia? Where are
   they found?

_____

_____

_____

_____

• Describe the structure and function of myelin. (Fig. 8-6c) _____

_____

• One (Schwann cell/oligodendrocyte?) forms myelin around portions of several axons in the
   (CNS/PNS?) but in the (CNS/PNS?), each (Schwann cell/oligodendrocyte?) associates with
   only one axon. (Fig. 8-6b)

• What are the nodes of Ranvier? _____

ELECTRICAL SIGNALS IN NEURONS

• Neurons and muscle cells are considered excitable tissues because they do what? _____

_____

• Name and describe the two basic types of electrical signals.

_____

_____

• How do cells create electrical signals? _____

_____

Changes in the Membrane Potential Create Electrical Signals

• (True/false?) All cells have resting membrane potentials.

• Explain the resting membrane potential difference of the cell. [∫ pp. 136-141]

_____

_____

_____

• What two factors influence membrane potential? [These two factors are related by Goldmann equation, Appendix A, p. 738]

_____

• Which ion is more concentrated in the ECF: Na^+ or K^+ ? Which is more concentrated in the ICF? _____

• To which ion is the resting cell membrane most permeable? Na^+ or K^+

• What is the role of the Na^+/K^+-ATPase found in cell membranes? _____

_____

• _____ is the major ion contributing to resting cell membrane potential.

• What is an average value for resting cell membrane potential? (Include units!) _____

• Describe how you might measure a cell's membrane potential in the laboratory. (Fig. 8-7)

_____

_____

• Explain the following terms associated with membrane potentials.

Depolarize: _____

Hyperpolarization: _____

Repolarize: _____

Overshoot: _____

• If the membrane potential of a cell decreases, does the cell become more positive or negative? _____

Ion Movement across the Cell Membrane Creates Electrical Signals

• Which ions are most important in the electrical signaling of excitable tissues? _____

• A sudden (increase/decrease?) in Na^+ permeability allows Na^+ to (enter/leave?) the cell. When Na^+ moves, it is moving (down/against?) its concentration gradient and (down/against?) its electrical gradient.

• The (influx/efflux?) of Na^+ ions (depolarizes/hyperpolarizes/repolarizes?) the membrane potential, creating an electrical signal.

• Describe two examples of how ion movement can hyperpolarize a cell. _____

_____

_____

_____

Gated Ion Channels Control the Ion Permeability of the Neuron

• How do cells alter their permeability to ions? _____

_____

• How are ion channels classified? _____

• Name four ions that move through membrane channels: _____

• Mechanically gated ion channels open in response to _____.

• Chemically gated ion channels open in response to _____.

• Voltage-gated ion channels open in response to _____.

• In the axon, _____ gated K$^+$ channels open in response to _____

   and K$^+$ flows from (cytoplasm/ECF?) to (cytoplasm/ECF?).

• In the axon terminal, _____gated Ca^{2+} channels open in response to _____

   and Ca^{2+} moves from (cytoplasm/ECF?) to (cytoplasm/ECF?).

• The movement of Ca^{2+} is a signal that initiates what event? _____

• When chemically gated Cl$^-$ channels on a neuron open, Cl$^-$ moves (out of / into?) the cell. What
   does Cl$^-$ movement do to the membrane potential?

_____

Graded Potentials Reflect the Strength of the Stimulus That Initiates Them

• What is a graded potential? _____

_____

• What is local current flow ? _____

_____

• What determines the strength of the initial depolarization? _____

_____

• Opening K$^+$ or Cl$^-$ channels will cause (depolarizing/hyperpolarizing?) graded potentials.

• Graded potentials travel through neurons until reaching the _____ _____.

• Where are the trigger zones in sensory neurons? _____

   In efferent neurons? _____

• If a graded potential reaches threshold at the trigger zone, what happens? _____

_____

• If a graded potential does not reach threshold at the trigger zone, what happens?

_____

• Depolarizing graded potentials are also called _____. (EPSP)

• Hyperpolarizing graded potentials are also called _____ _____ _____. (IPSP)

• What is an average threshold for a mammalian neuron? (include units!) _____

Multiple Graded Potentials Are Integrated in the Trigger Zone

• What happens when several graded potentials reach the trigger zone at the same time?

_____

• Several sub-threshold EPSPs can sum to create a/an _____ potential at the trigger zone. (Fig. 8-10a)

• A hyperpolarizing IPSP can offset _____ and prevent threshold from being reached. (Fig. 8-10b)

• Distinguish between spatial and temporal summation. (Fig. 8-12, 8-13)

_____

_____

_____

• What is a grand postsynaptic potential (GPSP)?

_____

• Why is summation a useful process for a neuron? _____

_____

Action Potentials Travel Long Distances without Losing Strength

• Define an action potential (AP). _____

_____

• How is an action potential similar to a graded potential? _____

_____

• How do APs differ from graded potentials? (Table 8-2) _____

_____

_____

• Because APs either don't occur or occur at maximal level, they are sometimes called

_____-_____-_____ phenomena.

Action potentials represent movement of Na$^+$ and K$^+$ across the membrane
• In one sentence, explain how action potentials are generated. (Fig. 8-14) _____

_____

• List the three phases of the action potential. Beneath each phase, tell what is happening to ion permeability and which ions are moving, in what direction (into or out of the cell). (Fig. 8-14)

1. _____ phase

Voltage-gated _____ channels (open / close?) → permeability (increases / decreases?)

and _____ flows into (ICF / ECF?) . The membrane potential (hyperpolarizes / depolarizes?).

When the membrane potential reaches 0 mV, there is no longer a(n) _____

gradient moving the _____ ions, but the _____ gradient keeps

the ions moving into the cell. As the _____ ions continue in, membrane potential approaches

the _____ potential (E$_{ion}$), the point at which electrical and chemical gradi-

ents exactly _____ each other. However, the AP peaks at +30 mV because the ion

channels _____.

2. _____ phase of AP is due to a/an (decrease/ increase?)

in _____ permeability. _____ gated ion channels open in a delayed

response to depolarization and _____ moves from (ECF/ ICF?) to (ECF/ ICF?), down an

electrochemical gradient.

Ion (influx / efflux?) makes membrane potential difference more (positive / negative?).

3. _____ phase is due to movement of what ion(s) in what

direction(s)?_____

Na⁺ *channels in the axon have two gates*

• Name the two gates of the voltage-gated Na^+ channels.

_____

• In the resting neuron, the activation gate is (open/closed?) and the inactivation gate is (open/closed?). (Fig. 8-15a)

• In this configuration, what can you say about Na^+ movement through the channel? _____

• Depolarization opens the _____ gate. (Fig. 8-15b)

• Na^+ moving into the ICF down its electrochemical gradient (Fig. 8-12c) causes (more /less?)

depolarization and creates a (negative / positive / feedforward?) feedback loop. (Fig. 8-16)

• The inactivation gate closes in delayed response to _____and

this closure stops Na^+ _____.

• What must happen to the gates before the next action potential can take place?

_____

Action Potentials Will Not Fire during the Refractory Period

• Define refractory period. _____

_____

• How do the absolute and relative refractory periods differ?

_____

_____

• Why can a greater-than-normal stimulus trigger an AP during the relative refractory period but not during the absolute refractory period?

_____

_____

• Refractory periods limit the _____ of AP transmission.

Stimulus Intensity Is Coded by the Frequency of Action Potentials

• Explain how stimulus intensity can be coded by action potentials if all action potentials are identical.

_____

_____

• How does one neuron transmit information about stimulus intensity to the next neuron?

_____

The Na$^+$/K$^+$ -Pump Plays No Direct Role in the Action Potential

• (True / false?) Following an action potential, the Na$^+$ concentration gradient reverses.

• (True / false?) If the Na$^+$/K$^+$-ATPase is poisoned, the neuron will immediately become unable to fire action potentials.

Action Potentials Are Conducted from the Trigger Zone to the Axon Terminal

• Explain what is meant by the conduction of action potentials. _____

_____

• Conduction ensures that electrical energy is replenished so that the electrical signal does not lose

_____ over distance like _____ potentials.

• Cellular mechanisms of conduction are (similar to / different from?) those initiating the AP.

☛ _There is no one action potential. An AP is simply a representation of membrane potential in a membrane segment at given period in time. A series of recording electrodes along an axon will measure a series of identical APs in different stages, just like falling dominoes frozen in different positions. (Fig. 8-19b)_

• When Na$^+$ channels in the middle of an axon open, depolarizing local current flow will spread in both directions along the axon. Why then don't action potentials reverse and move back toward the cell body?

_____

_____

The Diameter and Resistance of the Neuron Influence the Speed of Conduction

• List two factors that affect the speed of AP conduction. (These are called cable properties).

_____

• The larger the diameter of the axon, the (faster / slower?) an action potential will move through it.

• Why don't all animals have giant axons?

_____

_____

• What is meant by the resistance of a membrane? _____

_____

• What mechanism allows some animals to have high-resistance, small diameter axons that rapidly transmit signals?

_____

• The nodes of Ranvier are gaps in PNS axons between Schwann cells that have (high / low?) concentrations of voltage-gated Na^+ and K^+ channels in nodes.

• Explain saltatory conduction. _____

_____

_____

• What happens to conduction through axons that have lost their myelin? _____

Electrical Activity in the Nervous System Can Be Altered by a Variety of Chemical Factors

• What happens to AP conduction in a neuron whose Na^+ channels have been blocked? (Table 8-4)

_____

• What happens to the likelihood of firing an action potential when the extracellular K^+ increases?

_____

• Explain how hypokalemia decreases neuronal excitability. What happens to the membrane potential difference in this instance? (Fig. 8-23d)

_____

_____

☛ *Because of the importance of K^+ in nervous system function, the body regulates K^+ levels tightly.*

CELL-TO-CELL COMMUNICATION IN THE NERVOUS SYSTEM

Information Passes from Cell to Cell at the Synapse

• Define a synapse and give its three components.

_____

_____

• Name two kinds of synapses. _____

Electrical synapses
• Explain how information is transmitted at electrical synapses. _____

_____

• What is the main advantage of an electrical synapse? _____

Chemical synapses
• Explain how information is transmitted at chemical synapses. _____

_____

_____

The Nervous System Uses a Variety of Neurotransmitters

• Name five classes of neurotransmitters. _____

_____

• Acetylcholine (ACh) is made from _____, found in membrane phospholipids, and

_____, a metabolic intermediate linking glycolysis and the citric acid cycle.
[∫ p. 91]

• Where is ACh synthesized (Fig. 8-25)? _____

• What happens to ACh once released into the synaptic cleft?

_____

• Some amino acid, amine, and purine neurotransmitters are synthesized in the axon terminal using

enzymes brought from the _____ _____ via _____ axonal transport.

• How are these neurotransmitters deactivated? _____

• Where are polypeptide neurotransmitters made? _____

• Nitric oxide (NO) is an unstable gas synthesized from _____ and _____.

• Give an example of NO function. _____

• What division of the nervous system has the largest variety of neurotransmitters ? _____

• Name the three primary neurotransmitters of the peripheral nervous system.

_____

Calcium Is the Signal for Neurotransmitter Release

• Describe the steps of neurotransmitter release, beginning with the depolarization that reaches the axon terminal.

_____

_____

• Describe what happens to neurotransmitters once they are released into the synapse.

_____

_____

Not All Postsynaptic Responses Are Rapid and of Short Duration

• In fast postsynaptic responses, what does the neurotransmitter do to the postsynaptic cell?

_____

• How do slow postsynaptic responses differ from fast responses?

_____

_____

Some Signals Move from the Postsynaptic to the Presynaptic Cell

• How can postsynaptic cells communicate with their presynaptic cells?

_____

Disorders of Synaptic Transmission Are Responsible for Many Diseases

• Some nervous system disorders are caused by synaptic transmission abnormalities. Give some examples of these disorders.

_____

• List some pharmacological agents that can alter synaptic transmission.

_____

Development of the Nervous System Depends on Chemical Signals

• How do embryonic nerve cells find their target cells?

_____

_____

• Synapse formation and maintenance require that what take place?

_____

When Neurons Are Injured, Segments Separated from the Cell Body Die

• If a neuron is damaged at the cell body, what happens to the neuron? _____

• Describe what happens to a neuron if the cell body survives but the axon is severed.

_____

_____

_____

TALK THE TALK

absolute refractory period
acetylcholine (ACh)
acetylcholinesterase
action potential
activation gate
afferent neuron
after-hyperpolarization
all-or-none
astrocyte
autonomic neuron
axon
axon hillock
axon terminal
axonal transport
axoplasmic flow
brain
calcium channel
cell body (soma)
cell processes
central nervous system, CNS
chemical synapse
choline
chloride channel
collateral
conduction
demyelinating diseases
dendrite
depolarization
depression
efferent neuron
electrical synapse
end plate potential
enteric nervous system
epinephrine
excitable tissue
excitatory post-synaptic potential
   (EPSP)

fast axonal transport
fast synaptic potential
generator potential
glia
graded potential
hyperkalemia
hyperpolarization
hypokalemia
inactivation gate
inhibitory post-synaptic
potential
   (IPSP)
initial segment
innervated
interneuron
ion channel
kinesin
local anesthetic
local current flow
membrane resistance
microglia
multiple sclerosis
myasthenia gravis
myelin
Na^+/K^+-ATPase, role in
   action potentials
nerve
nerve growth factor
nervous system
neuroglia
neuromodulator
neuromuscular junction
neuron
neurotoxin
neurotransmitter
neurotrophic factor
nitric oxide

nodes of Ranvier
norepinephrine
oligodendrocyte
overshoot
parasympathetic neuron
Parkinson's disease
peripheral nervous system
plasticity of synapses
positive feedback loop
postsynaptic response
postsynaptic cell
potassium channel
presynaptic cell
receptor potential
relative refractory period
repolarization
resting membrane potential
   difference
saltatory conduction
schizophrenia
Schwann cell
sensory neuron
slow axonal transport
slow synaptic potential
sodium channel
sodium permeability
somatic motor neuron
spatial summation
spinal cord
sympathetic neuron
synapse
synaptic cleft
synaptic vesicle
temporal summation
threshold
trigger zone
Wobbler mice

ERRATA

p. 224, Running Problem. The question was omitted but is included in the problem conclusion on p. 234. The question is "In Guillain-Barré syndrome, what would you expect the results of a nerve conduction test to be?"

PRACTICE MAKES PERFECT

1. What is probably the biggest advantage of using the nervous system for a homeostatic response rather than the endocrine system?

2. On the figure below, label the boxes with either K^+ or Na^+ to show the ion concentration in the two body compartments.

3. Anatomically trace a signal from one end of a neuron to the other, using the correct sequence. Assuming that there is a chemical synapse, describe how a signal passes from one neuron to the next.

4. In the year 2045 you are part of a scientific expedition to investigate newly discovered life forms on the planet Zwxik in another solar system. You are assigned to study a single-cell organism found in the aqueous swamps of Zwxik. In one of your first studies (see Chapter 5), using a radioisotope of calcium, you observed that calcium moves freely in and out of the cell during daylight hours but was unable to get into or out of the cell in the dark. [question omitted here; parts re-labeled]

You follow up this study with an electrophysiology study. You use an intracellular electrode to measure the electrical charge inside the cell and find that it has a resting membrane potential of +60 mV when the outside fluid is arbitrarily set at 0 mV. From additional studies you find that the ion concentrations in the cell and the surrounding swamp water are as follows:

	Cell	Swamp
K+	50	50
Na+	175	15
Cl-	200	90

All concentrations in millimoles per liter.

You now use contemporary molecular biology techniques to insert some protein channels into the cell membrane. These channels will allow both Na^+ and K^+ to pass; no other molecules can go through them.

a) Predict which ion(s) will move. Tell what direction it/they will move and what force(s) is acting on it/them.

b) The Nernst potential for an ion describes two things: 1) the resting membrane potential if it were determined by only one permeable ion, and 2) the point at which ion movement across a membrane ceases because electrical and osmotic work directly oppose each other. Using the Nernst equation [Appendix A, p. 738], determine the Nernst potential for Na+ in the above situation. You have measured the temperature in the swamp to be 110° F., R and F are the same values as given in the Appendix, z = 1.

_____ 5. During repolarization of a nerve fiber:
      a. potassium ion leaves the cell
      b. potassium ion enters the cell
      c. neither occurs, as the fiber is in its refractory period

_____ 6. Treatment of a nerve cell with cyanide, an inhibitor of ATP synthesis, will: (circle all that are correct)
      a. immediately reduce the resting membrane potential to zero
      b. cause a slow increase in the intracellular sodium concentration
      c. immediately prevent the cell from propagating action potentials
      d. inhibit the movement of potassium ions across the membrane

_____ 7. The conduction of an action potential along a nerve axon: (circle all that are correct)
      a. is faster for a strong stimulus than a weak one
      b. occurs at a constant velocity
      c. is faster in unmyelinated nerve fibers than in myelinated nerve fibers
      d. decreases in magnitude as it is propagated along the axon
      e. is slower in a long nerve than in a short nerve of the same diameter

8. Fill in the following table that asks you to list different types of voltage- and chemically gated ion channels in neurons. Tell which ion(s) moves through the channel. For all channels, fill in the blocks showing the type of gating (chemical or electrical) and the physiological process that is linked to the ion movement. For action potentials, state which phase the ion movement is linked to.

	Location of channel	Ion(s) that moves	Chem or volt gating?	Physiological process in which ion participates
NEURONS				
	dendrite			
	axon			
	axon hillock			
	axon terminal			

9. What is the function of the ground electrode when recording an action potential in a nerve?

10. What would you guess is the reason(s) that stretching the nerve disrupts its functions?

11. Explain to your parents or your unscientific roommate what an action potential graph looks like, what it represents, and how it is generated.

MAPS

1. Draw a map showing the relationship of the central and peripheral nervous system, in as much detail as you can. Label the parts with the corresponding parts of a nervous reflex: AP for afferent pathway, IC for integrating center, and EP for efferent path.

BEYOND THE PAGES
READING

Action potentials in dendrites: Do they convey a message? (1996, April) *News in Physiological Sciences* 11:101-102.

Calcium in synaptic transmission. (1982, October) *Scientific American.*

How does an ion channel sense voltage? (1997, October) *News in Physiological Sciences* 12: 203-210.

Neurodegenerative prion diseases. (1996, September/October) *Science & Medicine.*

Progenitor cells of the central nervous system: A boon for clinical neuroscience. (1997, December) *Journal of NIH Research* 9:31-36.

Recent breakthroughs in neurotransmitter release: Paradigm for regulated exocytosis? (1995, February) *News in Physiological Sciences* 10: 42-45.

The release of acetylcholine. (1985, April) *Scientific American.*

Synapse formation in the developing brain. (1989, December) *Scientific American.*

RESOURCES

The Jackson Laboratory in Bar Harbor, Maine is a non-profit research institution that emphasizes genetic research. They develop and keep track of the mutant mouse and rat strains being developed that have yielded so much valuable information for medical research. Each summer they have outstanding undergraduate students come to the lab to conduct biomedical research under the guidance of staff scientists. To learn more about the Jackson Laboratory and its programs, check out its web site at www.jax.org

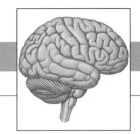

CHAPTER 9

THE CENTRAL NERVOUS SYSTEM

SUMMARY

This entire chapter should fit into your mental file folder for the central nervous system division of the nervous system. Details, details, details . . .

Key learning tasks for this chapter include:

• Be able to describe the gross anatomy and cellular anatomy of the central nervous system.
• Know how information is transmitted through the spinal cord to the brain.
• Know the major divisions of the brain and their functions.
• Learn the organization of the cerebral cortex.
• Understand how we learn information and store it into memory.

The central nervous system (CNS) is made up of the brain and the spinal cord. Both are encased in membrane and protected by bone. Cerebrospinal fluid (CSF), secreted by the choroid plexus, cushions the CNS and provides a chemically controlled environment. Impermeable brain capillaries create a blood-brain barrier to protect the brain from a possibly toxic blood chemical. Gray matter consists of nerve cell bodies, dendrites, and axon terminals. The cell bodies either form layers or cluster into nuclei. White matter, on the other hand, is made primarily of myelinated axons.

The spinal cord is the main pathway between the brain and effectors of the body. The spinal cord is protected by the vertebrae. Severing the spinal cord leads to loss of sensation or paralysis. In the spinal cord, each spinal nerve has a dorsal and ventral root. Dorsal roots carry incoming sensory information while ventral roots carry information from the CNS to effectors.

The brain is divided into the brain stem, the cerebellum, and the cerebrum. The brain stem plays a major role in the unconscious functions of the body, and it is where many neurons cross sides. The cerebellum processes sensory information and coordinates body movements. The cerebrum, consisting of two hemispheres connected by the corpus callosum, accomplishes the higher order brain functions like memory and perception. The cerebrum houses the basal ganglia, participating in movement control, and the limbic system, linking higher brain function and emotion.

The diencephalon has the thalamus and hypothalamus. The thalamus is an integration and relay center for sensory information; the hypothalamus plays a key role in homeostasis, autonomic function, and endocrine function.

The cerebral cortex houses our reasoning ability. Neurons here are highly ordered into vertical columns and horizontal layers. The cortex has sensory areas, motor areas, and association areas. The sensory areas are specialized for specific sensory information: visual cortex, auditory cortex, and olfactory cortex. Motor areas direct skeletal movement. Association areas integrate sensory information and create our perception of the sensory world. For example, spoken language is processed in Wernicke's area and Broca's area. Each cerebral hemisphere has become lateralized, not sharing certain functions with the other hemisphere. CNS neurons exhibit plasticity.

The CNS has a variety of neurotransmitters (NT) and neuromodulators (NM) that create complex pathways for information storage and transmission. Diffuse modulatory systems influence a wide range of body functions. Sleep is a reversible state of inactivity. Sleep consists of REM and slow-wave (non-REM) sleep.

is our acquisition of knowledge. Associative learning occurs when we learn to associate two stimuli. Nonassociative learning includes imitative behaviors like language. Habituation is showing a decreased response to a repeated stimulus, and sensitization is just the opposite. Memory has multiple storage levels: short-term and long-term. Short-term memories can be consolidated into long-term memories.

This chapter presents many details with which you are simply going to have to get familiar. Look at it as your opportunity to practice your pure memorization skills. HINT: Break this information up in a way that makes the most sense to you, then memorize the details. For example, if it makes more sense for you to divide by function, go that way. If you prefer to move around by anatomy, do that. You'll probably have to change game plans later in the chapter, but do what works best for you. If you memorize things quickly, then this chapter won't seem that hard. If you have trouble memorizing, start making flash cards, maps, charts, or whatever helps you learn the best.

TEACH YOURSELF THE BASICS

EVOLUTION OF NERVOUS SYSTEMS

• What property of neural networks is the most difficult to duplicate in an artificial intelligence system?

_____

• What is a ganglion? _____

• What is a spinal reflex? _____

• Which division of the brain is most developed in sophisticated animals? _____

ANATOMY OF THE CENTRAL NERVOUS SYSTEM

The Skull and Vertebral Column Protect the Central Nervous System

• Why do the brain and spinal cord require protection? _____

• List the three types of support structures that protect the CNS. (Fig. 9-1b)

_____

• Name the bones that protect the brain and spinal cord. _____

• The stacked vertebrae are separated by tissue disks. What type of tissue makes up these disks? What is the function of these disks?

_____

• What are meninges? _____

• List the three layers of meninges, moving from bone to tissue. (Fig. 9-1b)

_____

Cerebrospinal fluid and the ventricular system

• Fluid within the skull is divided into what two distinct extracellular compartments?

_____

• What is the choroid plexus and what is its function? (Fig. 9-3c)

_____

• How do ependymal cells secrete fluid into the ventricles?

_____

• Describe the anatomical route followed by cerebrospinal fluid from its point of secretion to its point of reabsorption back into the blood.

_____

• Name the two functions of the cerebrospinal fluid. _____

• Describe the composition of the cerebrospinal fluid. How does CSF compare with other extracellular fluids?

_____

_____

• When physicians need a sample of CSF, how do they acquire it?

_____

The Blood-Brain Barrier Protects the Brain from Harmful Substances in the Blood

• What is the only fuel source for neurons under normal metabolic circumstances? _____

• Neurons have a high rate of oxygen consumption so that they can produce ATP for what purpose? [∫ p. 121]

_____

• Why is a large percentage of the blood pumped by the heart directed to the brain?

_____

• What is the blood-brain barrier and what is its function?

_____

_____

• What kinds of molecules can cross the blood-brain barrier? _____

• Why are brain capillaries less leaky than other capillaries? _____

• What cells induce formation of tight junctions in brain capillaries and how do they do this?

_____

• Name two brain areas that lack a blood-brain barrier. What are the functions of these areas that require their contact with blood?

_____

_____

Neurons of the Central Nervous System Are Grouped into Nuclei and Tracts
☛ *Before you continue, see if you can map the divisions of the nervous system from memory . . .*

• Name the two cell types found in the CNS._____

• What is the difference between gray matter (Fig. 9-1e,i) and white matter?

_____

• Describe two ways neuron cell bodies are organized in the brain and spinal cord.

_____

_____

• Where are tracts found and what do they do? _____

• What are the differences between ascending, descending, and propriospinal tracts?

_____

_____

• Tracts in the CNS are equivalent to _____ in the PNS.

THE SPINAL CORD

• The spinal cord is divided into what four regions? (Fig. 9-1a)

_____

• How are the bilateral pairs of nerves that exit the spinal cord named? _____

• Explain the relationships between the following terms: axons, brain, cell bodies, columns, dorsal horns, dorsal roots, dorsal root ganglia, efferent signals, gray matter, nuclei, sensory information, spinal reflexes, tracts, ventral horns, ventral roots, white matter. (Fig. 9-5)

_____

_____

_____

_____

_____

_____

_____

THE BRAIN

• What is meant by compartmentalization in the brain?

_____

• Define parallel processing. _____

_____

• Define plasticity. _____

• List the three main areas of the adult brain. (Fig. 9-1g,h)

_____

• What are ventricles? _____

The Brain Stem

• List the three sections of the brain stem. (Fig. 9-6 b) _____

• What are cranial nerves? How many pairs are there, and, in general terms, what do they innervate?

_____

_____

• The medulla oblongata ("medulla") contains the _____ fiber tracts

that convey information between which two areas? _____

• These fibers cross to the other side of the body in the region called the _____.

• Where is the pons located and what is its primary function?

_____

• Where is the midbrain located and what is its primary function?

_____

• What is the reticular formation and what are its best known functions?

_____

The Cerebellum ("Little brain")

• The cerebellum receives sensory input from what sensory receptors?

_____

The Diencephalon ("Between-brain")

• Name the two main sections of the diencephalon. _____

• What is the pineal gland? [∫ p. 196] _____

• What is the primary function of the thalamus? _____

• Sensory information destined for which part of the brain travels through the thalamus?

_____

• Can the thalamus integrate information? _____

• Where is the hypothalamus located? _____

• Describe the relationship between the pituitary and the hypothalamus. [∫ p. 185]

_____

• Name eight major functions of the hypothalamus.

_____

_____

_____

☛ *REVIEW: Hypothalamic control of anterior pituitary hormone release [∫ p. 184]*

• The hypothalamus receives input from _____.

• What is the function of the suprachiasmatic nucleus? [∫ p. 164-5] _____

The Cerebrum (Figure 9-8)

• The two hemispheres of the cerebrum are connected at the _____ _____. (Fig. 9-12)

• Name the four lobes found in each hemisphere. (Fig. 9-8)

_____

• What is the adaptive significance of the intricate folding of the surface of the cerebrum?

_____

• Name the three major clusters of nuclei in the interior of the cerebrum and give their functions.

_____

_____

_____

Organization of the cerebral cortex
• What are some higher brain functions that arise in the cerebral cortex?

_____

• Describe the anatomical and functional arrangement of the cortical neurons. (Fig. 9-1i)

_____

• Name the following functionally specialized areas (not necessarily correlated with anatomical lobes).

     1. _____ direct perception

     2. _____ direct movement

     3. _____ integrate information, direct voluntary behaviors (Fig. 9-10)

• The primary somatic sensory cortex in the _____ lobe receives information

  from_____.

• Damage to this region results in _____.

• The visual cortex in the _____ lobe receives information from _____.

• The auditory cortex in the _____ lobe receives information from _____.

• The olfactory cortex, a small region in the _____ lobe, receives information from _____.

• The primary motor cortices process information about _____.

• Damage to the right motor cortex is exhibited as problems with movement on the

_____ side of the body.

• Define perception. _____

• What brain areas integrate sensory information into perception? _____

• Describe cerebral lateralization. _____

_____

• The right side of the brain is associated with what functional skills? _____

• The left side of the brain is associated with what functional skills? _____

BRAIN FUNCTION

• Name two non-invasive techniques for studying brain function. _____

• To what degree is behavior genetically determined? _____

Neurotransmitters and Neuromodulators Influence Communication in the Central Nervous System

☞ *Review synaptic transmission of impulses [∫ p. 224]*

• Signal meaning in the nervous system depends on what factors?

_____

• What is the difference between neurotransmitters and neuromodulators?

_____

• Name at least seven common neurotransmitters in the CNS. (Table 9-3)

_____

• Describe diffuse modulatory systems and list some actions they influence. (Table 9-4)

_____

States of Arousal and the Reticular Formation

• Define arousal. _____

• What distinguishes arousal from sleep? sleep from coma? coma from death?

_____

_____

• How do brain wave patterns differ with different states of arousal? _____

_____

Sleep
• Describe circadian rhythms. _____

• In mammals, the internal clock appears to be in the _____ nucleus
  of the hypothalamus.

• Name the two major phases of sleep and describe them.

_____

_____

• Why do we sleep?

_____

The Hypothalamus Is the Primary Integrating Center for Many Homeostatic Reflexes

• What is the fight-or-flight response? Where in the brain is it mediated? By what chemical(s)?

_____

_____

• What are some physical manifestations of the fight-or-flight response?

_____

Emotion and Motivation Are Complex Neural Pathways

• What is motivation? _____

• What is the amygdala a center for? _____

• Name the brain regions involved in emotional neural pathways. (Fig. 9-15)

_____

Learning and Memory Change Synaptic Connections in the Brain

Learning
• What is learning?_____

• Distinguish between associative and nonassociative learning.

_____

_____

• Distinguish between habituation and sensitization.

_____

_____

Memory
• What brain region is important for learning and memory? _____

• Explain anterograde amnesia. _____

• Describe the steps of memory processing and consolidation.

_____

_____

• Distinguish between reflexive and declarative memory.

_____

_____

• What is a memory trace? _____

Plasticity and long-term potentiation
• Define long-term potentiation. _____

_____

• What events may take place at the cellular level to explain LTP? _____

_____

Language Is the Most Elaborate Cognitive Behavior

• Why is language considered a complex behavior? _____

_____

• Language ability is found primarily in the _____ cerebral hemisphere, even in most left-handed or ambidextrous people.

• Name the two processes required for communication. _____

• Damage to Wernicke's area results in _____ aphasia, which is

_____

• Damage to Broca's area results in _____ aphasia, characterized by

_____

Personality and Individuality Are a Combination of Experience and Inheritance

• If we all have similar brain structure, what makes us different?

_____

TALK THE TALK

affective behavior
alpha wave
amygdala
anterograde amnesia
arachnoid membrane
artificial intelligence
association areas
associative learning
auditory cortex
basal ganglion
blood-brain barrier
brainstem
Broca's area
central nervous system (CNS)
cerebellum
cerebral cortex
cerebral lateralization
cerebrospinal fluid
choroid plexus
circadian rhythm
cognitive behavior
column
consciousness
consolidation
cranium
declarative (explicit) memory
deep wave sleep
delta wave
diencephalon
diffuse modulatory system
dopamine
dorsal horn
dorsal root

dorsal root ganglion
dura mater
electroencephalography
emotion
ependyma
expressive aphasia
fight-or-flight response
gamma-aminobutyric acid
    (GABA)
ganglia
glutamate
glycine
gray matter
habituation
hippocampus
hypothalamus
L-DOPA
learning
limbic system
long term memory
long term potentiation
medulla oblongata
memory trace
meninges
mesencephalon
midbrain
motivation
nerve fiber
NMDA receptor
nonassociative learning
nuclei
olfactory cortex
parallel processing

Parkinson's disease
perception
pia mater
plasticity
pons
primary motor cortex
primary somatic sensory cortex
propriospinal tracts
pyramids
receptive aphasia
reflexive (implicit) memory
REM (rapid eye movement)
sleep
reticular formation
satiety
sensitization
short-term memory
skull
sleep
sleepwalking
spinal reflex
suprachiasmatic nucleus
thalamus
tract
ventral horn
ventral root
ventricle
vertebral column
visual cortex
visual imaging
Wernicke's area
white matter
working memory

ERRATA

Fig. 9-3, p. 240: The part labels for (b) and (c) are reversed.

Fig. 9-14, p. 252: Stage 4 sleep should also be labeled as "slow wave" sleep, to correspond with the discussion in the text.

RUNNING PROBLEM: TIAs

The opening section has a misleading statement. The vast majority of patients who suffer a TIA do <u>not</u> lose consciousness. They may experience loss of sensation, loss of ability to speak or move, but generally do not lose touch with the world.

PRACTICE MAKES PERFECT

1. You are walking to class, pondering the intricacies of physiology, when you suddenly trip over an uneven place in the sidewalk. Unhurt but embarrassed and angry, you jump up and glance around to see if anyone is watching. From your knowledge of neuroanatomy and function, BRIEFLY explain how the following areas of the brain might be involved in this scenario.

    a) cerebrum
    b) cerebellum

2. What does an electroencephalogram tell us?

3. Give the function of the following parts of the brain:

    a. cerebrum _____

    b. hypothalamus_____

    c. brain stem _____

    d. cerebellum _____

_____ 4. Meninges are:

a. bacterial infections of the brain or spinal cord

b. connective tissue coverings around the central nervous system

c. synapses between the meningeal nerve fibers and the postsynaptic membranes of other neurons

d. non-neuronal cells in the brain and spinal cord that help regulate the ionic concentrations of the extracellular space

_____ 5. The cerebellum:

a. if destroyed would result in the loss of all voluntary skeletal muscle activity

b. initiates voluntary muscle movement

c. is essential for the performance of smoothly coordinated muscular activity

d. contains the center responsible for the regulation of body temperature

e. is of no importance in the control of posture and balance

_____ 6. Which of the following is true of REM sleep?

a. The average level of activity of brain cells decreases

b. Students sleeping peacefully through a physiology lecture are awakened and report that they had no dreams

c. This type of sleep occurs periodically through the night and accounts for 25% of total sleeping time

d. The blood flow to the brain decreases markedly

e. REM sleep develops immediately after a person has fallen asleep

7. You are visiting Jurassic Park when you come upon a *T. rex* that has escaped. Describe the steps of the fight-or-flight reflex that takes place in your body, beginning with the sight of the *T. rex*. Trace the neural pathways through different parts of the CNS to the various effectors that carry out the fight or flight response.

8. Create a map of the central nervous system using the following terms:

amygdala	dorsal horns	pons
arachnoid membrane	dorsal root ganglia	primary motor cortex
association areas	dorsal roots	primary somatic sensory cortex
auditory cortex	dura mater	propriospinal tracts
axons	efferent signals	pyramids
basal ganglion	ependyma	reticular formation
blood-brain barrier	ganglia	sensory information
brain	gray matter	skull
brainstem	hippocampus	spinal reflex
cell bodies	hypothalamus	suprachiasmatic nucleus
central nervous system	limbic system	thalamus
cerebellum	medulla oblongata	tract
cerebral cortex	meninges	ventral horn
cerebrospinal fluid	mesencephalon	ventral root
choroid plexus	midbrain	ventricle
columns	nuclei	vertebral column
cranium	olfactory cortex	visual cortex
diencephalon	pia mater	white matter

BEYOND THE PAGES
READING

Adult cortical plasticity and reorganization. (1997, January/February) *Science & Medicine.*

Amyloid protein and Alzheimer's disease. (1991, November) *Scientific American.*

Anesthesiologists wake up to the biochemical mechanisms of their tools. (1997, December) *Journal of NIH Research* 9: 37-41.

Anesthesiology. (1985, April) *Scientific American.*

Apolipoprotein E and Alzheimer's disease. (1995, September/October) *Science & Medicine.*

Artificial intelligence: A debate. (1990, January) *Scientific American.*

Astrocytes. (1989, April) *Scientific American.*

Barriers in the developing brain. (1997, February) *News in Physiological Sciences* 12: 21-31.

The biology of sleep apnea. (1996, September/October) *Science & Medicine.*

The blood-brain barrier. (1986, September) *Scientific American.*

Brain damage caused by prenatal alcohol exposure. (1996, July/August) *Science & Medicine.*

The brain's immune system. (1995, November) *Scientific American.*

Breaching the blood-brain barrier. (1993, February) *Scientific American.*

Can science explain consciousness? (1994, August) *Scientific American.*

Chemical signaling in the brain. (1993, November) *Scientific American.*

Dopamine receptors and psychosis. (1995, September/October) *Science & Medicine.*

Dyslexia. (1996, November) *Scientific American.*

Emotion, memory, and the brain. (1994, June) *Scientific American.*

GABAergic neurons. (1988, February) *Scientific American.*

The gene that rewards alcoholism. (1996, March/April) *Science & Medicine.*

The machinery of thought. (1997, August) *Scientific American.*

The mammalian choroid plexus. (1989, November) *Scientific American.*

Memory storage and neural systems. (1989, July) *Scientific American.*

Molecular basis for functional differences on AMPA-subtype glutamate receptors. (1996, April) *News in Physiological Sciences* 11: 77-82.

Molecular mechanisms of antiepileptic drugs. (1997, July/August) *Science & Medicine.*

Neural transplants work. (1996, December) *News in Physiological Sciences* 11: 255-261. Dopaminergic neurons transplanted to treat Parkinson's disease.

The neurobiology of fear. (1993, May) *Scientific American.*

Neurological effects of serotonin. (1995, July/August) *Science & Medicine.*

Obsessive-compulsive disorder. (1997, March/April) *Science & Medicine.*

Pathophysiology of the migraine aura. (1996, July/August) *Science & Medicine.*

Perceiving shape from shading. (1988, August) *Scientific American.*

The physiology of perception. (1991, February) *Scientific American.*

Plasticity in brain development. (1988, December) *Scientific American.*

Polymer-based drug delivery to the brain. (1996, July/August) *Science & Medicine.*

Prostanoids: Intrinsic modulators of cerebral circulation. (1997, April) *News in Physiological Sciences* 12: 72-77.

The puzzle of conscious experience. (1995, December) *Scientific American.*

Rx for addiction. (1991, March) *Scientific American.*

Scaling of the mammalian brain: The maternal energy hypothesis. (1996, August) *News in Physiological Sciences* 11: 149-156.

Sleep and energy conservation. (1993, December) *News in Physiological Sciences* 8: 276-281.

Understanding Parkinson's disease. (1997, January) *Scientific American.*

Visualizing the mind. (1994, April) *Scientific American.*

Vomiting — Its ins and outs. (1994, June) *News in Physiological Sciences* 9: 142-147

A window on the sleeping brain. (1983, April) *Scientific American.*

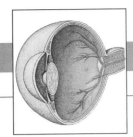

SUMMARY

This reflex pattern should look familiar: stimulus → sensory receptor → sensory neurons CNS → integration, perception. It is the same reflex pattern you first learned in Chapter 6 [∫ p. 159]. The sensory neurons work like the typical neurons you studied in Chapter 8. As you learned in Chapter 9, the CNS integrates sensory information. This chapter presents you with more details, but you already know the basic patterns — retrieve them from the mental file folders you've been making.

Key points to learn in this chapter:

• Know the similarities and differences of somatic senses and special senses: where the receptors are, how the receptors work, where those senses are integrated.
• Learn how lateral inhibition and receptive fields influence our perception of stimuli.
• Understand how the nervous system can tell how strong a stimulus is and where it is coming from.
• Know the difference between tonic and phasic receptors.

The somatic senses are touch-pressure, temperature, pain, and proprioception. The five special senses are vision, hearing, taste, smell, and equilibrium. Sensory stimuli are converted into electrical potentials by specialized receptors. Receptors have an adequate stimulus, a particular form of energy to which they are most responsive. There are five types of sensory receptors: chemoreceptors, mechanoreceptors, thermoreceptors, photoreceptors, and nociceptors. If the stimulus depolarizes the receptor membrane potential past threshold, then an action potential results and sensory information is sent to the CNS. (If this is unfamiliar, go back to Chapter 8 and review neuron function.) The CNS integrates the information and creates our perception of the stimulus.

Receptive fields are created when primary neurons converge on secondary neurons. Lateral inhibition enhances contrast within the receptive field so that sensation is more easily located (except in hearing and smell). Auditory and olfactory localization is accomplished by interpreting the timing of stimuli. Stimulus intensity is coded by the number of activated receptors and by frequency of the action potentials transmitted. Tonic receptors send information for the duration of a stimulus; phasic receptors respond only to changes in stimulus intensity.

The facts above summarize the themes in sensory physiology. The rest of the chapter provides details about the special senses and somatic senses. Study the chapter one sense at a time and become comfortable with how each sensory organ/receptor transduces energy, how that information is sent to the CNS, where the information is integrated in the CNS, and how the information becomes perception. Then think about what responses the body might exhibit upon perceiving specific sensory information.

TEACH YOURSELF THE BASICS

GENERAL PROPERTIES OF SENSORY SYSTEMS

• Using the terms associated with a reflex, describe how a stimulus is converted into perception.

• If a stimulus is below threshold, will the receptor show a change in its membrane potential?

_____

Receptors Are Transducers That Convert Stimuli into Electrical Signals

• Name and describe the simplest sensory receptors.

_____

• Sensory receptors are (circle all that are correct) specialized neurons / specialized nonneural cells.

• Name the five major types of receptors and describe the stimuli that activate each receptor type. (Table 10-2)

_____

_____

_____

_____

• Explain the law of specific nerve energies. _____

_____

Sensory Pathways Carry Information to the Central Nervous System Integrating Centers

• What is the relationship between primary, secondary, and tertiary sensory neurons?

_____

_____

_____

• Name the parts of the brain that process the following types of sensory information. (Fig. 10-2)

Visual information = _____ Sound = _____

Somatic senses = _____ Smell = _____

Equilibrium = _____ Taste = _____

• All sensory information except smell is processed through what part of the brain? _____

• What happens to the perceptual threshold for a stimulus when we "tune it out"? Explain the mechanism that allows us to do this.

_____

_____

Receptive fields
• What is the receptive field of a sensory neuron? _____

• Explain how receptive fields and convergence create some areas of the body that can sense two pins separated by only a few millimeters, while in other areas two pins 30 mm apart are sensed as separate stimuli.

_____

_____

_____

Sensory Transduction Converts Chemical and Mechanical Stimuli into Graded Potentials

• How is sensory transduction similar to signal transduction which you studied in Chapter 6? [∫ p. 151]

_____

_____

• Explain the concept of adequate stimulus for a receptor. _____

_____

• Define threshold. What happens in a sensory neuron if a stimulus is above threshold?

_____

_____

• How do stimuli create electrical signals in sensory receptors? _____

☛ *Review: electrical signals in neurons and the role of ions, beginning on p. 208.*

• Define receptor (or generator) potential. Are receptor potentials more like action potentials or graded potentials? [∫ p. 210]

_____

Stimulus Coding and Processing Is Used to Determine Location, Intensity, Duration, and Nature of a Stimulus

• Name four attributes of stimuli that must be preserved during nervous system processing.

_____

Sensory modality
• Explain what is meant by labeled line coding. _____

_____

Location of the stimulus
• How can the brain tell which part of the body is sending sensory information? _____

_____

• Give an example of the topographical organization of the cerebral cortex.

_____

• How does the brain determine where sound and smell stimuli originate?

_____

• Explain how lateral inhibition enhances contrast, allowing better localization of stimuli. (Fig. 10-6)

_____

_____

Intensity and duration of the stimulus
• Name two ways in which stimulus intensity is coded.

_____

• The amplitude of a receptor potential increases in proportion to stimulus intensity [∫ Fig 8-18, p. 219], but all action potentials are identical in amplitude. How then can the primary sensory neuron send information about the strength of a stimulus?

_____

_____

• Compare the response of tonic receptors and phasic receptors to a constant stimulus. (Fig. 10-8)

_____

• What happens during adaptation of a receptor?

_____

• What is the advantage of phasic receptors? Give some examples.

_____

SOMATIC SENSES

• Name the four somatosensory modalities.

_____

Pathways for Somatic Perception Project to the Somatosensory Cortex and Cerebellum

• Describe the sensory pathways for the following stimuli. (Table 10-3)

Nociception, temperature, some touch stimuli: _____

Most touch and proprioception: _____

• Sensations from the left side of the body are processed in the (left / right?) hemisphere of the

brain because _____.

Touch-pressure receptors
• Where are touch-pressure receptors found? _____

• Describe the structure and function of the Pacinian corpuscle. (Fig. 10-1b)

_____

_____

• Are Pacinian corpuscles tonic or phasic? _____

Nociceptors and pain
• What stimuli activate nociceptors? _____

• Why is it inaccurate to call nociceptors "pain receptors"? _____

• What are the differences between slow pain and fast pain fibers?

_____

☞ *Review: How do axon diameter and myelination affect conduction? See Ch. 8, p. 220.*

• List some chemicals that activate or sensitize nociceptors.

_____

• Draw the complete reflex pathway for the withdrawal reflex when a frog's leg is placed in hot water. (Stimulus, receptor, afferent path, etc.)

• Where do ascending pain pathways terminate in the brain?

_____

• Explain how pain perception can be modulated by the brain.

_____

• Explain or diagram the gating theory of pain modulation. (Fig. 10-11)

• What is referred pain and why does it occur? (Fig. 10-12b)

_____

Temperature receptors
• Describe the anatomy of temperature receptors. _____

• Do temperature receptors adapt? Explain. _____

_____

CHEMORECEPTION: SMELL AND TASTE

• List the five special senses. _____

• What are the technical terms for smell and taste? _____

Olfaction

• Where do primary olfactory neurons terminate? (Fig. 10-13a) _____

• Describe the olfactory epithelium. (Fig. 10-13b) _____

_____

• What role do olfactory binding proteins play in olfaction? _____

• What are pheromones? _____

Taste Is a Combination of Five Basic Sensations

• List and briefly describe the five taste sensations. _____

_____

_____

• Where are taste receptors (buds) located? (Fig. 10-14a) _____

• What roles do water and taste binding proteins play in taste? _____

_____

• Which taste sensations activate the taste buds using receptors and second messenger systems?
  Name the receptors and second messenger systems for each sensation. (Fig. 10-15) [∫ p. 152-155]

_____

_____

• Which taste sensations activate taste buds by altering ion channels? Explain what happens to the ion channels and to membrane potentials when the taste ligands bind.

_____

_____

• What is specific hunger? Give an example. _____

_____

THE EAR: HEARING

• The ear contains sensory receptors for what two functions? _____

Sound Waves Vary in Their Pitch and Loudness

• Define hearing. _____

• Name two ways sound waves can be characterized. _____

Transduction of Sound Is a Multistep Process

• Trace the anatomical path followed by sound wave energy as it moves from air through the inner ear. (Fig. 10-17)

_____

_____

• List the steps through which the energy of sound waves in air is converted into action potentials in the sensory neuron.

_____

_____

The Middle Ear Transfers Sound from the Eardrum to the Cochlea

• Describe the anatomy and function of the middle ear.

_____

_____

• Name the three bones of the middle ear in the order in which a sound wave would reach them. (Figs. 10-17 and 10-19a):

_____

• What functions do these little bones serve in the conduction of sound?

_____

The Cochlea of the Inner Ear Is Filled with Fluid

• Describe the structure of the cochlea, naming all fluids, windows, and ducts. (Fig. 10-17)

_____

_____

• Compare the composition of perilymph and endolymph.

_____

• Describe the location and structure of the organ of Corti.

_____

Sound Transduction through the Cochlea Depends on Movement of Hair Cell Stereocilia

☛ *Error in text: p. 284, right column: Sound waves reach the cochlea through the* <u>*oval window,*</u> *not the round window. All figures and other mention of this are correct in the text.*

• Explain how hair cells convert fluid waves into action potentials. Include a description of stereocilia, ion channels, protein bridges, and neurotransmitters. (Fig. 10-19)

_____

_____

_____

Sounds Are Processed First in the Cochlea

• List the four properties of sound waves used for sound discrimination. In what part(s) of the auditory system are these properties processed?

_____

_____

• What role does the basilar membrane play in sound processing? (Fig. 10-21)

_____

• How is loudness coded by the auditory system? _____

Auditory Pathways Project to the Auditory Cortex

• Trace the anatomical path that action potentials follow from the auditory sensory neurons to their final destination in the brain.

_____

• How does the brain localize sound? _____

Hearing Loss May Result from Mechanical or Neural Damage

• List and explain the three different forms of hearing loss.

_____

_____

THE EAR: EQUILIBRIUM

• Define equilibrium. _____

• What receptors in the body provide sensory information about body position?

_____

• How are hair cells of the inner ear similar to hair cells of the cochlea? _____

_____

The Vestibular Apparatus Is Filled with Endolymph

• Explain the anatomy of the vestibular apparatus. (Fig. 10-22)

_____

_____

_____

• What fills the lumen of the inner ear? _____

The Vestibular Apparatus Provides Information about Movement and Position in Space

• Compare the functions of the semicircular canals and otolith organs. _____

_____

• Why is the presence of endolymph in the inner ear key to the body's ability to sense rotation? (Fig. 10-23)

_____

• What force acts on the otolith organs to alert them to changes in head position?_____

Equilibrium Pathways Project Primarily to the Cerebellum

• Trace the anatomical pathway that action potentials follow from the hair cells of the inner ear to their final destination in the brain.

_____

• Describe the descending (efferent) pathways from the brain to effectors in an equilibrium reflex.

_____

THE EYE AND VISION

• Define vision. _____

• How are light waves like sound waves? _____

_____

☛ *Error in Fig. 10-26, p. 290: The color bar is upside down. Violet should be at the top and orange at the bottom.*

• What is the frequency and wavelength range of visible light? _____

The Optic Tract Extends from the Eye to the Visual Cortex

• Describe the function(s) of each of the following components of the eye (Fig. 10-27):

Lens _____

Retina _____

Cornea _____

Pupil _____

Aqueous and vitreous humors _____

Photoreceptors _____

Optic disk

_____

• Trace the anatomical route action potentials follow from the photoreceptors to their final destination in the brain.

_____

• What do the extrinsic eye muscles, eyelids, lacrimal apparatus do? _____

_____

The Lens Focuses Light on the Retina

• How does the eye control the amount of light hitting the retina? _____

• How does the eye focus light onto the retina? _____

Image focusing and the lens
• What happens to a beam of light when it passes from air into a medium of different density such

as the cornea? _____

• Compare the focal point and the focal distance. _____

_____

• List two ways to change the focal distance for an object. _____

_____

• Define accommodation. (Fig. 10-29c) _____

_____

• Ciliary muscles are attached to the lens by _____. (Fig. 10-30)

• To make the lens more round, the ciliary muscles (contract / relax?) which (increases / decreases?) tension on the zonulas.

• When the lens is more rounded, the focal distance becomes (shorter / longer?). (Fig. 10-29)

• Explain the following vision problems and tell what shape lens would correct for each. (Fig. 10-31)

Presbyopia _____

Myopia _____

Hyperopia _____

• What is astigmatism? _____

Light entering the eye and the pupil
• Explain the pupillary reflex. _____

_____

• Explain "shallow depth of field" and tell what change in the pupil would lengthen the depth of field.

_____

Phototransduction Occurs at the Retina

• Describe the layers of the retina. (Fig. 10-32) _____

_____

• What is phototransduction? _____

• What is the optic disk and why is it also called the blind spot? _____

_____

• What is the function of melanin in the pigment epithelium? _____

• How are the photoreceptors in the fovea different? (Fig. 10-33) _____

_____

• The image projected onto the retina is upside down. Why then do we see things in the correct orientation?

_____

Photoreceptors: rods and cones
• Distinguish between rods and cones. _____

_____

• Describe the three-segment structure of rods and cones and tell what process(es) occurs in each segment. (Fig. 10-34)

_____

_____

• What are visual pigments? How many are there? _____

_____

• The perceived color of an object depends on the color(s) of light that the object (reflects / absorbs?).

Phototransduction
• Describe the rhodopsin molecule and explain how it changes when activated by light.

_____

_____

• Describe the state of each of the following in a rod in the dark. (Fig. 10-36)

Rhodopsin molecule _____

Transducin _____ cGMP levels _____

Na^+ channels _____ K^+ channels _____

Membrane potential _____ Neurotransmitter release _____

• Describe the state of each of the following in a rod exposed to light. (Fig. 10-36)

Rhodopsin molecule _____

Transducin _____ cGMP levels _____

Na^+ channels _____ K^+ channels _____

Membrane potential _____ Neurotransmitter release _____

• Why do our eyes require some time to adjust to changes in light intensity? _____

_____

Signal Processing in the Retina Occurs When Light Strikes Visual Fields

• What is convergence? _____

• Which neurotransmitter is released from photoreceptors? _____

• This neurotransmitter excites some bipolar neurons but inhibits others. Explain how one signal

molecule can have opposing effects. _____

• What is the relationship between photoreceptors, bipolar neurons, ganglion cells; and visual
  fields? (Fig. 10-38)

_____

• Describe the two ganglion cell types. _____

_____

• Visual acuity is greatest when a ganglion cell has (many / few?) photoreceptors in its visual field.

• Vision is better when light entering the eye (is high intensity / is high contrast?).

Visual Processing in the Central Nervous System Takes Place in the Visual Cortex

• Optic nerves enter the brain at the optic _____. (Fig. 10-39)

• In this region, (all / some?) fibers from the right field of vision cross to the left side of the brain.

• Fibers from the optic chiasm to the midbrain serve what purpose? _____

_____

• Fibers from the optic chiasm to the lateral geniculate body of thalamus serve what purpose?
  (Fig. 10-27a)

_____

• Explain how we see things in three dimensions with binocular vision. _____

_____

TALK THE TALK

accommodation
acuity
adaptation of receptors
adequate stimulus
amplification
ampulla
analgesic drug
aqueous humour
astigmatism
auditory cortex
basilar membrane
bipolar neurons
bitter
bleaching
blind spot
central hearing loss
cerumen
chemoreceptor
ciliary muscle
cochlea
cochlear duct (scala media)
cochlear implant
coding sound for pitch
cold receptor
color-blindness
conductive hearing loss
cone
convergence
cornea
crista
cupula
decibel (dB)
depth of field
electromagnetic wave
endolymph
equilibrium
eustachian tube
fast pain
focal distance
focal point
fovea
ganglion cells
gating theory
G_{olf}
gustation

gustducin
hair cell
helicotrema
hertz (Hz)
hyperopia
incus
ischemia
kinocilium
labeled line coding
labyrinth
lacrimal apparatus
lateral geniculate nucleus
lateral inhibition
law of specific nerve energies
lens
M cell
macula
malleus
mechanoreceptor
melanin
Meniere's disease
modality
monellin
myopia
nociceptor
olfaction
olfactory binding protein
olfactory bulb
olfactory neuron
ophthalmoscope
opsin
optic chiasm
optic disk
optic nerve
optics
organ of Corti
otitis media
otolith
otolith membrane
otolith organ
oval window
P cell
Pacinian corpuscle
perception
perceptual threshold

perilymph
phantom limb pain
phasic receptor
pheromone
photoreceptor
phototransduction
pinna
pitch
presbyopia
primary sensory neuron
proprioception
pupillary reflex
receptive field
receptor (generator) potential
referred pain
refraction
retina
retinal
rhodopsin
rods
round window
saccule
salt appetite
salty
secondary sensory neuron
semicircular canal
sense organ
sensorineural hearing loss
slow pain
somatic senses
somatosensory cortex
somatosensory receptor
sound wave
sour
special senses
specific hunger
stapes
stereocilia
substance P
sweet
taste bud
tectorial membrane
thaumatin
thermoreceptor
threshold

tonic receptor

topographical organization in the
  visual cortex

transducin

transduction

two-point discrimination test

tympanic duct (scala tympani)

tympanic membrane

umami

utricle

vestibular apparatus

vestibular duct (scala
vestibuli)

visible light

visual field

visual pigment

vitreous chamber

vomeronasal organ

warm receptor

zonula

ERRATA

p. 284, right column: Sound waves reach the cochlea through the <u>oval window</u>, not the round window. All figures and other mention of this are correct in the text.

p. 291, Fig. 10-26: The color bar of the spectrum is upside down. Blue-violet should be at the top, at 380 nm, and red-orange at the bottom, at 750 nm.

RUNNING PROBLEM - Meniere's Disease

This condition was first described in 1861 by Prosper Meniere. It has an occurrence rate of 1-2:10,000 people and is relatively rare in people under 30. Men are 2-3 times more likely to develop Meniere's disease than women.

Meniere's disease: differential diagnosis and treatment. [Review] (1997, March) *American Family Physician* 55(4):1185-90, 1193-4.

PRACTICE MAKES PERFECT

1. Why aren't you constantly aware of your clothing touching your body?

2. If a salty solution is placed at the rear of the tongue, it cannot be tasted. Why not?

3. TRUE / FALSE? Explain your reasoning:
    Light travels through nerves and blood vessels before striking the rods and cones.

4. Chemoreceptors
    a. are involved in the perception of taste
    b. are found in the olfactory mucosa
    c. cover the retina except at the optic discs
    d. are found in the cochlea
    e. can be described by both a and b

5. How does astigmatism differ from nearsightedness?

6. A sensory nerve is termed (circle the best one):
    a. efferent
    b. afferent
    c. monoefferent
    d. none of the above

7. If a person with a hearing impairment has good bone conduction and no air conduction, where would you predict the problem to be?

8. Stare hard at a bright light, then shift your gaze to a white sheet of paper. What do you see on the paper? Can you explain this in terms of what you know about rhodopsin and bleaching?

9. You are sitting up straight on a stool and are spinning to your left. Suddenly, you are stopped and you remain sitting up straight. Which one of the three semicircular canals is involved in this sensation?

10. Which type of visual receptor cell would you expect to find more of if you dissected the retina of a nocturnal (i.e., active at night) animal? Why?

11. The organ of Corti sends electrical information about sound to the brain. The pitch of sound is determined by (circle the best answer):
    a. the location of the activated hair cells on the basilar membrane
    b. the frequency of the action potentials received by the brain
    c. the amplitude of the action potentials received by the brain
    d. a and b

MAPS

1. Create a map of the inner ear using the following terms and any functions/actions that you wish to add.

ampulla	inner ear	saccule
crista	macula	semicircular canals
cupula	otolith membrane	utricle
endolymph	otolith organs	vestibular apparatus
hair cells	otoliths	vestibular nerve

2. Using the vocabulary list from Talk the Talk as a starting point, select related terms and create similar maps for vision, hearing, taste, and smell.

BEYOND THE PAGES

TRY IT - Here are some fun demonstrations of sensory physiology.

✎ Taste and Smell

The relationship between taste and smell can be shown with the following experiment. Cut a small slice of potato and a small slice of apple. Remove the skin or peel. Close your eyes and pinch your nose shut. Have a friend put either a piece of apple or potato in your mouth. Chew it with your nose pinched shut. Can you tell whether it is apple or potato? Repeat the experiment with your nose open so that you can smell what you are eating.

✎ Mapping Taste Receptors

You can map the location of taste receptors on the tongue. Assemble the following materials: cotton swabs, 1 tsp. sugar dissolved in 1 T. water, 1 tsp. salt dissolved in 1T. water, vinegar diluted 1:1 with water, monosodium glutamate (Accent®) dissolved in a little water. Dry the tongue with a tissue. Dip a cotton swab in one of the solutions. Touch the swab to the tip, sides, and back of the tongue, and notice where you taste the dissolved solute. On the drawing of the tongue below, record the area where you noticed the taste. Rinse your mouth with water, dry the tongue with a tissue, and repeat using each of the different solutions.

✎ The Blind Spot

In each field of vision we have a **blind spot** at the **optic disc**, the point on the retina where the fibers from the rods and cones converge to form the **optic nerve**. Normally we are not aware of this blind spot, but the following test will allow you to find it. Hold the X in the figure below directly in front of your right eye and about 20 inches from your face. Close your left eye and focus your right eye on the X. You should see both the X and the black dot. Keeping your left eye closed and focusing on the X, bring the page closer to your face until the dot disappears from your field of vision. With a ruler, have a friend measure the distance from the page to your eye at which the dot disappears.

X   ●

Focal distance for the blind spot: _____ cm = _____ mm

You can now calculate the distance from the blind spot to the fovea in the eye, using the following formula:

$$A/B = a/b \qquad \text{where}$$

A= the distance between the center of the X and the center of the dot above (in mm)
B= the the focal distance for the blind spot (in mm)
a = the distance between the optic disk and the fovea (unknown)
b = the distance from the lens of the eye to the retina (assume 20 mm)

What was the distance between the optic disk and the fovea? _____

✎ Negative after-images
Stare at a bright light like a penlight for about 30 seconds. Rapidly shift your gaze to a plain white surface. You should see a dark reverse image of the light. This after-image occurs because the cones have become fatigued (adapted) to the bright light.

✎ Complementary after-images
You can also get color after-images. Place a red or green cardboard square under a bright light and stare at it intently for about 30 seconds. Then transfer your gaze to a piece of white paper. The colors will change and you will see the complement of the color at which you were staring. The **complement** appears opposite the color on a traditional color wheel. The three main pairs of complementary colors are red-green, purple-yellow, and blue-orange. When you stare at one color, certain cones become fatigued, leaving the other cones to react to the light. If you looked at a red square, you'll see green on the paper, and vice versa.

✎ Eye dominance
With both eyes open, hold a cardboard cone or tube with both hands at arm's length directly in front of your nose. Center an object in the opening. Without moving the tube, close your right eye. Does the object move out of the opening? Now repeat with your left eye. If the object stayed in the opening when your right eye was open, then you are right-eye dominant. If it stayed in the opening with your left eye open, then you are left-eye dominant. Eye dominance usually matches hand dominance, that is, if you are right-handed, you will be "right-eyed."

✎ Depth perception
Because the images seen with each eye are not identical, the brain can interpret the two views into a three-dimensional representation of the image. With one eye closed, the field of vision becomes two-dimensional, and the relative distances between near and far objects become much harder to gauge. Try the following to demonstrate depth perception.

1. Extend your arms to the side, palms facing front and index fingers extended. Close one eye. Now try to touch the tips of the index fingers to each other about one foot in front of your face. Without input from both eyes, this can be more difficult than it would seem.

2. Close one eye and pick up a very large needle and thread. Keeping one eye closed, try to thread the needle. Repeat with both eyes open. Which was easier?

✎ Binaural localization of sound

Our ability to pinpoint the location of the source of sounds depends on the reception of sound waves by both ears. Our brain uses two parameters in processing the information received: the difference in loudness between the two ears and the difference in the time the sound reaches each ear.

Have the subject close his or her eyes. The tester should move around the subject with a "cricket clicker," metronome, or other noisemaker, turning it on periodically and asking the subject to identify the location of the sound source. You should experiment with the distance the noisemaker should be placed from the subject's head, as this will vary with the loudness of other masking noises in the room.

Repeat at the same distances, but this time have the subject block one ear with a finger. With one ear blocked, is the subject as accurate in pinpointing the location of the sound source?

✎ Two-point discrimination

Two pins touched to the skin close together will sometimes be perceived as a single point if only a single secondary receptive field is stimulated (see Fig. 10-4, p. 267). In order for two points to be perceived, two different receptive fields must be stimulated. The distance by which the points must be separated in order for them to be felt as distinct points is known as the **two-point threshold**.

You need a small ruler and a compass with two points or two <u>dull</u> pins.

1. Have the subject close his or her eyes. Use a single point of the compass or set the points 5, 15, 25, and 50 mm apart. In random order, touch the compass to the skin. Try to keep the pressure equal each time. **Do not push so hard that you break the skin!**

2. Note the distance between pins when the subject first feels two distinct points. Vary the distance between the points randomly. Repeat several times for accuracy. Record the results below.

Site	Minimum distance for two-point discrimination (mm)
fingertip	
back of neck	
cheek	
back of calf	
lateral surface of forearm	

How do these distances relate to the functions of these body parts?

THERMORECEPTORS

Our perception of temperature is related to stimulation of cold receptors, warmth receptors, and pain receptors. Most parts of the body have many more cold receptors than warmth receptors. All of the thermoreceptors **adapt** very rapidly, i.e., the intensity of the signal decreases with time so that the perception of the stimulus decreases. This means that our temperature receptors are much more sensitive to *changes in temperature* than to constant temperatures. (Are thermoreceptors tonic or phasic?) The following experiment shows adaptation of temperature receptors.

✎ Adaptation of receptors
1. Place one hand in 40° C. water and the other hand simultaneously into a large tub of ice water. Leave them there for one minute.

2. Now place both hands into a tub of room temperature water. Do they feel the same? Can you explain what you do feel based on the past thermal history for each hand?

✎ Referred pain
In certain cases when a pain receptor fires, it will stimulate other neurons that run in the same nerve. This can lead to sensations of pain far from the actual site of the stimulus. One example of referred pain is irritation of the abdominal side of the diaphragm that is sensed as a pain in the shoulder. Another example can be demonstrated by placing your bent elbow into ice water. Leave it there for a minute or until you cannot stand the feeling any more. Was the sensation in your elbow or elsewhere?

VISUAL REFLEXES

✎ Pupil dilation reflex
This reflex helps regulate the amount of light that enters the eye and strikes the retina.

With a penlight, shine light into one eye of the subject. What happens to the pupil? Repeat, this time watching both pupils. Do they respond simultaneously?

Have the subject sit facing the window or a lighted area of the room. Note the pupil size. Keep watching while the subject places a hand over one eye. What happens to the pupil of the uncovered eye? This is known as the **consensual reflex**.

✎ Accommodation reflex
The ability of the eye to focus at different distances depends on the accommodation reflex. Smooth muscle fibers (**ciliary muscle**) attach to ligaments around the edge of the lens of the eye. In normal distance vision, the ciliary muscle is relaxed and objects that are about twenty feet away are in focus. Because light rays from objects that are closer than twenty feet are not parallel, these rays are refracted to a principal focus point that is behind the retina. As a result, objects closer than twenty feet will be blurred.

In the accommodation reflex, as the ciliary muscles contact, the lens becomes more spherical. The **focal distance** shortens so that it strikes the retina and the close objects come into focus. As we age, the lens loses its elasticity and we lose our ability to accommodate for near and far vision. This is the reason why around age 40 we often start to need to use reading glasses or bifocals.

1. Have subject look at a distant object. Note the size of the pupils.

2. As you continue to watch, ask the subject to shift focus to a nearer object on the same line of sight. What happens to the pupils? The nervous pathways that cause the lens to accommodate also cause a change in pupillary size at the same time.

✎ Near point of accommodation

1. Lay a meter or yard stick flat on the table, perpendicular to the edge. In one hand take a 3x5 in. card with a letter "e" from the newspaper pasted on it. Close your right eye. Place your nose at the end of the meter stick.

2. Starting at arm's length, move the card in toward your face. Note the distance at which you can no longer see the "e" clearly. This distance is known as the **near point of accommodation**.

As discussed previously, this distance increases with age as the lens loses elasticity. Average distances for the near point of accommodation are 10 cm at age 20 and 13 cm at age 30. By age 70, the near point averages 100 cm.

Find the near point of accommodation for each eye. Left _____ cm Right _____cm

✎ Ciliospinal reflex

While watching the subject's pupils, suddenly pinch the skin on at the nape of his or her neck. Watch the pupil of the eye on the side you pinched. (This is called the **ipsilateral** side.) What happens? Sudden pain stimulates the sympathetic nerves and causes a pupillary response.

READING

Acupuncture:
Acupuncture is moving out of the realm of "alternative medicine" and into traditional medicine. It was recently approved by the Federal Drug Administration as a legitimate form of therapy.

Alterations in electrical pain thresholds by use of acupuncture-like transcutaneous electrical nerve stimulation in pain-free subjects. (1992, September) *Physical Therapy*. 72(9):658-667.

Electroacupuncture suppresses a nociceptive reflex: Naltrexone prevents but does not reverse this effect. (1988, June 14) *Brain Research*. 452(1-2):227-31.

Electroacupuncture suppression of a nociceptive reflex is potentiated by two repeated electroacupuncture treatments: The first opioid effect potentiates a second non-opioid effect. (1988, June 14) *Brain Research*. 452(1-2):232-236.

Conquer chronic pain and more with acupuncture. (The Best of Alternative Medicine). (1994, December) *Prevention*. 46(12): 76-80.

Nod to an ancient art; the FDA has OK'd acupuncture needles - and they could help you. (1996, May 13) *U.S. News & World Report.* 120(19): 78(2).

Umami

Basic properties of umami and effects on humans. (1991, May) *Physiology & Behavior* 49(5):833-841.

Cephalic-phase insulin release induced by taste stimulus of monosodium glutamate (umami taste). (1990, December) *Physiology & Behavior* 48(6):905-908.

A comparison of English and Japanese taste languages: Taste descriptive methodology, codability and the umami taste. (1986, May) *British Journal of Psychology* 77(Pt 2):161-174.

Kawamura, Y. and Kare, M.R. (eds.)(1987) *Umami: A Basic Taste.* Marcel Dekker, Inc, New York.

Pruning of rat cortical taste neurons by an artificial neural network model. (1995, September) *Journal of Neurophysiology.* 74(3):1010-1019.

Responses of neurons in the primate taste cortex to glutamate. (1991, May) *Physiology & Behavior* 49(5):973-979.

Umami taste of monosodium glutamate enhances the thermic effect of food and affects the respiratory quotient in the rat. (1992, November) *Physiology & Behavior* 52(5):879-884.

Taste

Sweet and salty: Transduction in taste. (1995, August) *News in Physiological Sciences* 10: 166-170.

Olfaction

The functionality of the human vomeronasal organ (VNO): Evidence for steroid receptors. (1996, June) *Journal of Steroid Biochemistry & Molecular Biology* 58(3):259-265.

How we smell: The molecular and cellular bases of olfaction. (1998, February) *News in Physiological Sciences* 13.

The molecular logic of smell. (1995, October) *Scientific American.*

The noses have it. (human pheromone studies) (1995, November) *Scientific American.*

A novel family of putative pheromone receptors in mammals with a topographically organized and sexually dimorphic distribution. (1997, August 22) *Cell* 90(4):763-773.

Olfactory receptors, vomeronasal receptors, and the organization of olfactory information. (1997, 22 August) *Cell* 90(4):585-587.

Pheromone transduction in the vomeronasal organ. (1996, August) *Current Opinion in Neurobiology*. 6(4):487-493.

A scent circuit. (electronic noses) (1996, January/February) *American Scientist*.

Unconscious odors. (1997, October 3) *Science*. 278(5335):79.

Pain/perception

Biosensors. (1991, August) *Scientific American*.

Memories of pain. (1996, November/December) *Science & Medicine*.

New developments in microneurography of human C fibers. (1996, August) *News in Physiological Sciences* 11: 170-174.

Phantom limbs. (1992, April) *Scientific American*.

Hearing

Afferent and efferent synaptic transmission in hair cells. (1996, August) *News in Physiological Sciences* 11: 161-165.

The functional replacement of the ear. (1985, February) *Scientific American*.

The hair cells of the inner ear. (1983, January) *Scientific American*.

Listening with two ears. (1993, April) *Scientific American*.

The middle ear muscles. (1989, August) *Scientific American*.

Recent advances in cochlear neurotransmission: Physiology and pathophysiology. (1995, August) *News in Physiological Sciences* 10: 178-183.

Vision

Art, illusion, and the visual system. (1988, January) *Scientific American*.

Blind spots. (1992, May) *Scientific American*.

The discovery of the visual cortex. (1988, September) *Scientific American*.

The excimer laser in ophthalmology. (1997, January/February) *Science & Medicine*.

The genes for color vision. (1989, February) *Scientific American*.

How photoreceptor cells respond to light. (1987, April) *Scientific American*.

How the human eye focuses. (1988, July) *Scientific American*.

The molecules of visual excitation. (1987, July) *Scientific American.*

The silicon retina. (1991, May) *Scientific American.*

What the brain tells the eye. (1990, April) *Scientific American.*

CHAPTER 11

EFFERENT PERIPHERAL NERVOUS SYSTEM: THE AUTONOMIC AND SOMATIC MOTOR DIVISIONS

SUMMARY

So, what should you take from this chapter?

• Be able to compare and contrast the autonomic and somatic motor divisions.
• Be able to do the same for the sympathetic and parasympathetic divisions. Where are the ganglia? Where are the preganglionic, postganglionic neurons? What neurotransmitters are secreted from each?
• Learn the autonomic receptor types and their affinities. When and where are they used?
• Be familiar with how things work at neuroeffector and neuromuscular junctions.

This chapter starts filling in the details of the peripheral nervous system – mainly the autonomic nervous system. Learn the details so that you can apply them later as you learn specific examples of efferent function. Break out the colored pens and the paper! Design color-coded pictures and diagrams to help organize the material in this chapter. Recreate figures in the book and change them to fit your learning style.

Remember from Chapter 8 that the peripheral nervous system is divided into the somatic motor division and the autonomic division. The autonomic division is further divided into the sympathetic and parasympathetic divisions. Sympathetic and parasympathetic divisions are often antagonistic, allowing tight control over homeostasis. Review Cannon's postulates on homeostasis [∫ Ch. 6, p. 156] and see how the two autonomic divisions fulfill all four postulates. Somatic motor neurons always innervate skeletal muscle; autonomic neurons innervate smooth muscle, cardiac muscle, glands, and some adipose tissue.

An autonomic pathway consists of a preganglionic neuron from the CNS to an autonomic ganglion, and a postganglionic neuron from the ganglion to a target. Preganglionic divergence onto multiple synapses allows rapid control over many targets. Sympathetic pathways have ganglia near the spinal cord, while parasympathetic pathways have ganglia near their target. The synapse between a postganglionic neuron and its target is called a neuroeffector junction. Autonomic axons have varicosities from which neurotransmitter is released. CNS control over the autonomic division is linked to centers in the hypothalamus, pons, and medulla. Autonomic responses can be spinal reflexes, and the cerebral cortex and limbic system can also influence autonomic output.

Here are some points to remember: All preganglionic neurons secrete acetylcholine (ACh). Sympathetic postganglionic neurons secrete norepinephrine; parasympathetic postganglionic neurons secrete ACh. The type and concentration of neurotransmitter plus the receptor type at the target determine autonomic response. Cholinergic receptors respond to ACh and come in two types: nicotinic at the ganglia and muscarinic at parasympathetic neuroeffector junctions. Adrenergic receptors respond to epinephrine and norepinephrine and come in three types: α on most sympathetic tissue, β_1 on heart muscle and kidney, and β_2 on tissues not innervated by sympathetic neurons.

Somatic motor pathways have only one neuron that originates in the CNS and projects to a skeletal muscle. Somatic motor neurons always excite muscles to contract, and muscle contraction can only be inhibited by inhibiting the somatic motor neuron. One motor neuron may branch to control many muscle fibers. The synapse between the neuron and the skeletal muscle is called the neuromuscular junction. ACh is the neurotransmitter at the neuromuscular junction, and the motor end plates of the muscle cell have a high concentration of nicotinic ACh receptors.

TEACH YOURSELF THE BASICS

THE AUTONOMIC DIVISION

• What is a mixed nerve? _____

• Why is the autonomic division also called the visceral nervous system?

_____

• Characterize and compare the parasympathetic and sympathetic divisions.

_____

_____

The Autonomic Division Consists of Two Efferent Neurons in Series

• In the space below, draw and label an autonomic pathway. (Fig. 11-2) Include the following terms: ganglion, CNS, preganglionic neuron, and postganglionic neuron.

• List the targets of autonomic neurons. _____

• The synapse between an autonomic neuron and its target cell is called the _____ junction. [∫ p. 225]

• Compare the sites of neurotransmitter release in autonomic and somatic motor neurons. (Fig. 11-3)

_____

_____

Sympathetic pathways

• Sympathetic preganglionic neurons originate from which regions of the spinal cord? (Fig. 11-4)

_____

• Sympathetic ganglia are found close to the (spinal cord / target?) and along the descending aorta.

• Sympathetic postganglionic neurons project (long / short?) axons to their targets.

Parasympathetic pathways

• Long parasympathetic preganglionic axons leave the brain stem via the_____ nerve

  (IX) or the _____nerve (X). [∫ p. 245]

• Other parasympathetic preganglionic neurons originate from which regions of the spinal cord? (Fig. 11-4)

_____

• Parasympathetic ganglia are located close to the (spinal cord / target?).

• Parasympathetic postganglionic neurons project (long / short?) axons to their targets.

The Adrenal Medulla Secretes Catecholamines

• Where are the adrenal glands located? (Fig. 11-6a) _____

• The adrenal medulla is embryonically related to what tissue? _____

• What are the two catecholamines secreted by this gland? _____

• The adrenal cortex is embryonically related to what tissue? _____

• What kind of hormone is secreted by the adrenal cortex? [∫ p. 60] _____

• Explain why it is correct to call the adrenal medulla a modified sympathetic ganglion. (Fig. 11-6c)

_____

_____

Acetylcholine and Norepinephrine Are the Primary Autonomic Neurotransmitters

• Cholinergic neurons secrete the neurotransmitter _____.

• Adrenergic neurons secrete the neurotransmitter _____.

Chemical classification of autonomic neurons
• Match the following:

    1. Adrenergic neurons   _____ autonomic preganglionic neurons

    2. Cholinergic neurons   _____ sympathetic postganglionic neurons

                                        _____ parasympathetic postganglionic neurons

                                        _____ sympathetic neurons innervating sweat glands

• Name some neurotransmitters used by non-adrenergic, non-cholinergic neurons. _____

Neurotransmitter release
☛ Review: neurotransmitter synthesis, storage, and release [Ch 8, p. 225]

• In the autonomic division, where in the neuron are neurotransmitters made and packaged into vesicles? (Fig. 11-8)

• What is the signal for neurotransmitter release? _____

• Does modulation take place on the (preganglionic neuron / postganglionic cell / or both?).

Termination of neurotransmitter activity
• List three ways neurotransmitter activity is terminated.

_____

• What is monoamine oxidase? _____

Autonomic Neurotransmitter Receptors
☛ Multiple receptor forms allow variable responses to each neurotransmitter. [∫ p. 149]

Cholinergic receptors
• List the two subtypes of cholinergic receptors, name the tissues on which they are found, and explain what happens to the target cell when these receptors are activated.

_____

_____

Adrenergic receptors
• Explain what happens to the target cell when an adrenergic receptor is activated.

_____

• Name the two major subtypes of adrenergic receptors. (Table 11-1) _____

• What happens to intracellular Ca^{2+} concentrations when α receptors are activated?_____

• Compare the locations of β_1 and β_2 receptors. _____

• Compare the affinity of α, β_1, and β_2 receptors for epinephrine and norepinephrine.

_____

_____

• Which receptor is usually associated with smooth muscle contraction? _____ relaxation? _____

Agonists and antagonists
• List some agonists and antagonists for autonomic receptors.

_____

The Sympathetic and Parasympathetic Branches Maintain Homeostasis

• Use the example of heartbeat regulation by the autonomic divisions to explain Cannon's postulates of tonic control and antagonistic control.

_____

_____

_____

• Explain why epinephrine will cause some blood vessels to constrict but other vessels to dilate.

_____

The Autonomic Division Is Regulated by the Brain

• What regions of the brain contain most of the control centers for autonomically regulated functions? (Fig. 11-9)

_____

• List some autonomic functions that do not require any input from the brain.

_____

Disorders of the Autonomic Nervous System Are Relatively Uncommon

• In some autonomic pathologies, diminished sympathetic input over a period of time will cause

the number of neurotransmitter receptors in the target tissue to (up / down?)-regulate. [∫ p. 150]

THE SOMATIC MOTOR DIVISION

☛ *Practice: Recreate Table 11-4 to help solidify the comparison of the somatic motor and autonomic divisions.*

• Somatic motor pathways have (one / two?) neuron(s) from the CNS to the target tissue, which is

always_____ _____ muscle. Somatic motor input is (excitatory / inhibitory / may be

either excitatory or inhibitory?).

Anatomy of the Somatic Division

• Where in the CNS are the cell bodies of somatic motor neurons found? _____

• The synapse of a somatic motor neuron and its target is known as the _____
junction. (Fig. 11-11)

• Describe the structure and function of the motor end plate. (Fig. 11-12) _____

_____

The Neuromuscular Junction

☛ *Review: what happens as an action potential reaches the axon terminal. [Ch. 8, p. 226]*

• Describe what happens to the postsynaptic cell when acetylcholine binds to ACh receptors on
the motor end plate.

_____

_____

TALK THE TALK

acetylcholinesterase
adrenal medulla
adrenalin
adrenergic neuron
adrenergic receptor
alpha receptor
anticholinesterase
autonomic ganglion
autonomic neuron
beta$_1$ receptor
beta$_2$ receptor
bouton

bungarotoxin
catecholamine
cholinergic neuron
denervation hypersensitivity
intrinsic neuron
monoamine oxidase (MAO)
motor end plate
muscarine
muscarinic receptor
myasthenia gravis
neuroeffector junction
neuromuscular junction

nicotinic receptor
non-adrenergic, non-cholinergic
   neuron
parasympathetic branch
postganglionic neuron
preganglionic neuron
somatic motor neuron
sympathetic branch
sympathetic cholinergic neuron
vagus nerve
varicosity

RUNNING PROBLEM: Nicotine addiction

• The science of smoking (how nicotine creates addiction). (1996, May 11) *Economist* 339, (7965): p22.
• Pharmacology of nicotine: addiction and therapeutics. [Review] (1996) *Annual Review of Pharmacology & Toxicology* 36:597-613.
• Nicotine addiction and treatment. [Review] (1996) *Annual Review of Medicine* 47:493-507.
• Nicotine and addiction. The Brown and Williamson documents. (1995, July 19) *JAMA.* 274(3):225-233.
• Reasons for tobacco use and symptoms of nicotine withdrawal among adolescent and young adult tobacco users—United States, 1993. (1994, Oct 21) *Morbidity & Mortality Weekly Report.* 43(41):745-750.

PRACTICE MAKES PERFECT

_____ 1. The parasympathetic nervous system is characterized by:

a. long preganglionic and short postganglionic nerve fibers
b. short preganglionic and long postganglionic nerve fibers
c. direct connections (no synapses) between the central nervous system and the innervated organs
d. mediation by adrenaline (epinephrine)
e. the axons coming into immediate contact with the effector organs lying inside the central nervous system

2. Review agonists and antagonists. In the following scenarios, would the experimental drug be an agonist or antagonist?

a) You have just discovered a new vertebrate species while doing fieldwork in the rain forest. After obtaining proper permission, you begin studying the effects of different known neurotransmitters on the heart rate of this organism. You inject ACh alone and observe that it decreases heart rate. After letting the heart rate return to resting values, you inject GABA. You see that GABA increases the heart rate. After letting the creature rest again, you inject ACh + GABA and notice that the heart rate does not change. Without further experiments you can't be certain, but at this point, do you think that GABA is an agonist or antagonist to ACh? Explain your reasoning.

b) You're having a very successful expedition, and the next day you find another new vertebrate species that has some very interesting properties. The creature has bright red skin early in the morning, but by midday, the creature has dull brown skin. From other observations, you suspect that this process is under the control of the autonomic nervous system. Therefore, you decide to test this idea by injecting the animal with various chemicals. The results of your experiments are in the table.

	Initial skin color	Inject	Resulting skin color
1	fading from red to brown	norepinephrine	color returns to bright red
2	dull brown, beginning to change to red	ACh	skin returns to the dull brown
3	beginning to turn from red to brown	chemical that binds ACh and prevents it from acting for several hours	skin stays red
4	red	chemical named Compactine	skin changes to dull brown

From these experiments, do you think Compactine is acting as an agonist or antagonist to ACh?

3. In what way(s) is autonomic neurotransmitter action different from that of somatic motor neuron neurotransmitter action?

4. Certain drugs inhibit the activity of monoamine oxidase. In general terms, predict what these drugs would do to autonomic activity.

MAPS

1. Create a map that shows all efferent divisions of the nervous system. Show all neurons, all receptor types on their respective tissues, and all neurotransmitters.

READING

The adrenal chromaffin cell. (1985, August) *Scientific American.*

Transmitter release at adrenergic nerve endings: Total exocytosis or fractional release? (1997, February) *News in Physiological Sciences* 12: 32-36.

Using antibodies to unwire the sympathetic nervous system. (1995, June) *News in Physiological Sciences* 10: 101-106.

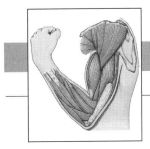

SUMMARY

Here's what you should know after reading this chapter:

• What are the structural components of skeletal muscle fibers? Know your sarcomere.
• What causes skeletal muscle contraction? Describe the molecular events.
• Discuss the differences in skeletal muscle fibers. Where do you find them?
• How do muscles meet their energy requirements?
• Now answer those same questions for smooth muscle.
• Be able to compare the three muscle types.

More details! Break this chapter into sections and spend some time learning all you can about muscles. Concentrate on skeletal muscles, breaking the information into structure, then function. Draw pictures and make maps to help you really grasp this material.

There are three types of muscles: skeletal, smooth, and cardiac. Remember from the last chapter that skeletal muscle contracts in response to somatic motor neurons. Our skeletal muscles move us around. Smooth and cardiac muscles respond to nervous and chemical signals. They move things through us and help maintain homeostasis.

Take some time to learn the basic anatomy of skeletal muscle – both macro and micro. Be able to rattle off information about all the components of a sarcomere. According to the sliding filament theory, contraction boils down to stationary myosin fibers (thick filaments) pushing along mobile actin fibers (thin filaments), driven by the power of ATP consumption. Here are the molecular events in a nutshell (you should be able to elaborate on these): an action potential created by opening an acetylcholine-gated ion channel depolarizes the muscle fiber. Depolarization spreads across the muscle membrane and t-tubule system, opening Ca^{2+} channels in the sarcoplasmic reticulum. Ca^{2+} flows into the cytoplasm, where it binds to troponin. Ca^{2+}-binding moves tropomyosin, which uncovers actin-binding sites on myosin. Myosin can now bind fully to actin and complete the power stroke. For relaxation, Ca^{2+} is pumped back into the sarcoplasmic reticulum, decreasing intracellular Ca^{2+} concentrations. Ca^{2+} unbinds from troponin, actin-myosin binding is partially blocked again, and the muscle fiber relaxes.

Skeletal muscle fibers are classified by their speed of contraction and their resistance to fatigue. Learn the difference between fast-twitch glycolytic fibers, fast-twitch oxidative fibers, and slow-twitch oxidative fibers. Where would you expect to find each? Fatigue is the condition of not being able to generate or sustain muscle power. It is affected by a variety of factors.

The tension possible in a single muscle fiber is determined by the length of its sarcomeres before contraction. Force in a muscle can be increased by increasing the stimulus, up to a point of maximal contraction, or tetanus. Isotonic contraction creates force and moves loads, while isometric contraction creates force without movement. A motor unit is a group of muscle fibers and the somatic motor neuron that controls them. Recruitment is a means of creating more force within a skeletal muscle. When done asynchronously, recruitment helps prevent fatigue.

Smooth muscle is much more fatigue-resistant due to its design. The same intracellular fibers create contraction in both skeletal and smooth muscle, but the molecular events are different. In contrast to skeletal muscle, smooth muscle depolarization is due to Ca^{2+} influx, and contraction is graded according to this influx. Smooth muscle molecular events in a nutshell: Ca^{2+} binds to calmodulin, and the Ca^{2+}-calmodulin complex activates myosin light chain kinase (MLCK). Active MLCK phosphorylates myosin, which in turn activates myosin ATPase. This results in a power stroke and contraction as actin and myosin slide past each other. For relaxation, Ca^{2+} is pumped out of the cytoplasm, and phosphatase dephosphorylates myosin. Smooth muscle can have unstable membrane potentials that produce action potentials if the cell depolarizes past threshold. If threshold isn't reached, then only slow wave potentials are generated. If threshold is reached, the resulting action potential is called a pacemaker potential.

Cardiac muscle has qualities of both skeletal and smooth muscle. It is striated like skeletal muscle, but is under autonomic and endocrine control like smooth muscle. Cardiac muscle has intercalated disks, gap junctions that allow action potentials to spread throughout the tissue.

TEACH YOURSELF THE BASICS

• List the three types of muscles. (Fig. 12-1) _____

SKELETAL MUSCLE

• How do skeletal muscles attach to bones? [∫ p. 63] _____

• Define the following terms:

Origin _____

Insertion _____

Joint _____

Flexor _____

Extensor _____

Antagonistic muscle groups _____

Skeletal Muscles Are Composed of Muscle Fibers

• What is the difference between a muscle and a muscle fiber? (Fig. 12-3) _____

_____

• What do you find in skeletal muscle besides muscle fibers? _____

• Explain the following terms: (Table 12-1)

Sarcolemma _____

Sarcoplasm _____

Myofibril _____

Sarcoplasmic reticulum (SR) (Fig. 12-4) _____

Transverse tubules (t-tubules) _____

• Why does muscle tissue have many glycogen granules in the cytoplasm? _____

Myofibrils are the contractile structures of a muscle fiber
• Name the two contractile proteins of the myofibril. _____

• Name the two regulatory proteins. _____

• Describe the structure of myosin. (Fig. 12-3e) _____

_____

• Describe the structure of actin. (Fig. 12-3f) _____

_____

• What is the difference between G-actin and F-actin? _____

_____

• Describe the relationship between actin and myosin in a myofibril. _____

_____

• What is a crossbridge? _____

• What is a sarcomere? (Fig. 12-3c, 12-5) _____

• Explain what creates the following bands of a sarcomere. How many of each are in one sarcomere?

A band _____

H zone _____

I band _____

M line _____

Z disk _____

• What is the function of titin? _____

• What is the function of nebulin? _____

Muscles Shorten When They Contract

• What is muscle tension? _____

• Briefly explain the sliding filament theory of contraction. _____

_____

_____

• According to this theory, the amount of tension developed by a muscle fiber depends on _____

_____.

Sliding filament theory of contraction

• In the sliding filament theory, which filaments move and which remain stationary?

_____

• Explain why the A bands of the sarcomere do not shorten during contraction.

_____

• Explain why the I band and H zone almost disappear during contraction.

_____

• What is a motor protein? _____

• Describe the power stroke in terms of actin and myosin. _____

_____

• Explain the molecular events of contraction, beginning with the rigor state. (Fig. 12-8)

_____

_____

_____

_____

• Why do muscles freeze in the state known as rigor mortis after death? _____

_____

Contraction Is Regulated by Troponin and Tropomyosin

• Describe the regulatory role of troponin, tropomyosin, and calcium in muscle contraction and relaxation. (Fig. 12-9)

_____

_____

_____

Acetylcholine from Somatic Motor Neurons Initiates Excitation-Contraction Coupling

• What is excitation-contraction coupling? (Table 12-2) _____

_____

• To initiate contraction, (nicotinic / muscarinic?) (adrenergic / cholinergic?) receptors on the motor end plate combine with (norepinephrine / ACh?) and open (Na^+ / K^+ / nonspecific monovalent cation?) channels. Net (Na^+ entry / Na^+ efflux / K^+ entry / K^+ efflux?) depolarizes the cell, creating an end-plate potential (EPP). The EPP results in a/an (graded potential / action potential?) that spreads across the (cytoplasm / sarcolemma?). The electrical signal then spreads into the _____ , where _____ receptors open (Na^+ / K^+ / Ca^{2+} ?) channels in the _____. (Fig. 12-10)

• What is the immediate signal for contraction: ACh, an action potential, Na^+, or Ca^{2+}?

• A single contraction-relaxation cycle is called a _____.

• What is the latent period and what creates it? _____

_____

Skeletal Muscle Contraction Depends on a Steady Supply of ATP

• Where does the ATP needed for muscle contraction come from? _____

• What is the role of phosphocreatine in muscle?

_____

• Compare the ATP yield for aerobic and anaerobic metabolism of one glucose. [∫ p. 90]

_____

• True or false? Explain your answer. Muscles can only use glucose for energy.

_____

Muscle Fatigue Has Multiple Causes

• What is muscle fatigue? _____

• What factors are believed to contribute to muscle fatigue?

_____

_____

• Why do H^+ and inorganic phosphate accumulate during exercise?

_____

• K^+ (enters / leaves?) the muscle fiber with each action potential (AP). With repeated APs, what effect will this have on the fiber's membrane potential? _____

What effect would this change in membrane potential have on Ca^{2+} release from the sarcoplasmic reticulum?

_____

• What is central fatigue and why is it felt to be a protective mechanism?

_____

Skeletal Muscle Fibers Are Classified by Speed of Contraction and Resistance to Fatigue

• List the three classifications of muscle fibers and describe their speed of contraction and resistance to fatigue. (Table 12-3)

_____

_____

_____

• What factor determines speed of tension development? _____

• What factor determines duration of contraction? _____

• Which fiber type has the longer duration of contraction? _____

• Compare the types of movements for which the three fiber types are best suited.

_____

_____

• What contributes to fatigue in fast twitch fibers? _____

• What is the role of myoglobin in skeletal muscle? Which fiber type has the most myoglobin?

_____

• What are the differences in the two types of fast-twitch muscle?

_____

_____

Tension Developed by Individual Muscle Fibers Is a Function of Fiber Length

• What determines muscle tension at the molecular level? (Fig. 12-14)

_____

• For a single muscle fiber, explain why long and short sarcomeres develop less tension than sarcomeres at optimal length. (Fig. 12-14)

_____

_____

Force of Contraction Increases with Summation of Muscle Twitches

• Explain the process of summation in a muscle fiber. [∫ neuronal summation, p. 211] (Fig. 12-15)

_____

_____

• Would you consider summation in a muscle to be an example of spatial or temporal summation?

_____

• Describe the difference between unfused tetanus and fused (complete) tetanus. (Fig. 12-15)

_____

_____

One Somatic Motor Neuron and the Muscle Fibers It Innervates Form a Motor Unit

• What is a motor unit? _____

• A muscle responsible for fine muscle movements will have (more /fewer?) muscle fibers in its motor units.

• Fibers in motor unit are of the (same / different?) fiber type.

• During development, what determines what kind of muscle fibers will be in each motor unit?

_____

• Do muscle fibers exhibit plasticity, the ability to alter metabolic characteristics with use? _____

Contraction in Intact Muscles Depends on the Types and Numbers of Motor Units in the Muscle

• Explain the two ways that the body can vary force of contraction in a muscle. (Fig. 12-17)

_____

• What is recruitment and how does it take place?

_____

• What is asynchronous recruitment and why is it helpful for avoiding fatigue during a sustained contraction?

_____

_____

MECHANICS OF BODY MOVEMENT

• What does the term mechanics mean when applied to muscle physiology?

_____

Isotonic Contractions Move Loads but Isometric Contractions Create Force without Movement

• Compare and contrast isotonic and isometric contractions. (Fig. 12-18) _____

_____

• What are the series elastic elements (Fig. 12-19a) and what role do they play in isometric contractions?

_____

• What is an eccentric contraction? _____

Bones and Muscles around Joints Form Levers and Fulcrums

• The body uses lever and fulcrum systems to move loads. What are the levers of the body and

what are the fulcrums? _____

• What advantage does the body gain by using lever and fulcrum systems?

_____

• The work done by muscle = Force X _____ (Fig. 12-20)

• For a given muscle, what is the relationship between the speed of contraction and the load being

moved by the muscle? _____

Muscle Disorders Have Multiple Causes

• List three major causes of skeletal muscle pathologies.

_____

SMOOTH MUSCLE

• Where in the body are smooth muscles found? _____

• Compare the speed and duration of smooth and skeletal muscle contraction. (Fig. 12-22)

_____

• Why are some smooth muscles in the body tonically contracted? Give an example.

_____

• What is muscle tone? _____

Smooth Muscle Fibers Are Much Smaller Than Skeletal Muscle Fibers

• Compare the arrangement of cytoplasmic fibers in smooth muscle to that of skeletal muscle.

_____

• Compare the shape and size of a smooth muscle fiber to that of a skeletal muscle fiber.

_____

• What is the function of the dense bodies in smooth muscle?

_____

• Describe smooth muscle myosin and its ATPase activity.

_____

• What are caveolae and what is their function in smooth muscle?

_____

• What is the difference between single-unit and multi-unit smooth muscle? Give an example of where each type is found.

_____

_____

Smooth Muscle Can Vary Its Force of Contraction

• Why can't single-unit smooth muscle vary its force of contraction the same way that skeletal

and multi-unit smooth muscle can? _____

• How does single-unit smooth muscle vary its force of contraction?

_____

Phosphorylation of Proteins Plays a Key Role in Smooth Muscle Contraction

• Where does the Ca^{2+} for contraction in smooth muscle come from? (Fig. 12-27, Table 12-4)

_____

• In skeletal muscle, Ca^{2+} entering the cytoplasm binds to _____. In

smooth muscle, Ca^{2+} entering the cytoplasm binds to _____.

• In skeletal muscle, Ca^{2+} binding to troponin does what? _____

In smooth muscle, C binding to calmodulin does what? _____

• Looking at its name, can you tell what myosin light chain kinase (MLCK) does? Explain.

_____

• Phosphorylation of myosin light chains (inhibits / enhances ?) the ATPase activity of myosin

and (stimulates / inhibits?) crossbridge formation and (contraction / relaxation ?). (Fig. 12-28a)

• What two events are necessary for relaxation in smooth muscle?

_____

• What is a latch state and what are its advantages? _____

_____

• In some smooth muscle, actin can be regulated by a protein called _____.

• Actin regulation also requires phosphorylation of a protein. Explain the steps for activation of actin. (Fig. 12-28b)

_____

Some Smooth Muscles Have Unstable Membrane Potentials

• What are the differences between a resting membrane potential, slow wave potentials, and pace-maker potentials? (Fig. 12-29)

_____

_____

_____

Calcium Entry Is the Signal for Smooth Muscle Contraction

• Not all Ca^{2+} influx into smooth muscle is through voltage-gated ion channels. What other kinds

   of gated channels are there? [∫ p. 113] _____

• Explain the term Ca^{2+}-induced Ca^{2+} release. _____

_____

• How can a single smooth muscle fiber grade (vary) the force of its contractions?

_____

• Explain how stretching a smooth muscle could cause it to contract. What is the term used to describe this type of contraction?

_____

• Ca^{2+} entry into the cytoplasm of a cell will (depolarize / hyperpolarize?) the cell's membrane potential.

• In pharmacomechanical coupling, chemically gated Ca^{2+} channels open to allow Ca^{2+} entry and contraction, but the muscle fiber's membrane potential does not change. Why not?

_____

Smooth Muscle Contraction Is Regulated by Chemical Signals

• Hyperpolarization of a smooth muscle cell will usually (enhance / inhibit ?) contraction.

• List four neurotransmitters that alter smooth muscle activity. _____

• What do each of the following chemicals do to smooth muscle contraction?

   nitric oxide? [∫ p. 225] _____ histamine? _____ epinephrine? _____

• A smooth muscle is stretched and initiates a myogenic contraction in response. Yet over time, the muscle begins to relax even though it remains stretched. Explain how this occurs.

_____

CARDIAC MUSCLE

• How is a cardiac muscle fiber similar to a skeletal muscle fiber?

_____

• How is a cardiac muscle fiber different from a skeletal muscle fiber?

_____

• How is a cardiac muscle fiber similar to a smooth muscle fiber?

_____

• How is a cardiac muscle fiber different from a smooth muscle fiber? (see Table 12-5)

_____

TALK THE TALK

A band
acidosis
actin
aerobic metabolism in muscle
anaerobic metabolism in muscle
antagonistic muscle groups
asynchronous recruitment
atherosclerosis
atrophy
beta-oxidation in muscle
botulinum toxin
Ca^{2+}-induced Ca^{2+} release
caldesmon
calmodulin
caveolae
central fatigue
contraction
creatine kinase (CK)
crossbridge
crossbridge tilting
dense bodies
dihydropyridine (DHP) receptor
eccentric action
end-plate potential (EPP)
endothelium-derived relaxing
    factor (EDRF)
excitation-contraction coupling
extensor
fast-twitch fiber
fast-glycolytic muscle fiber
fast-intermediate muscle fiber
fatigue, muscle
flexor
H zone
Huxley, Sir Andrew

I band
in vitro motility assay
insertion of a muscle
intercalated disk
isometric contraction
isotonic contraction
isozyme
joint
"knockout" mice
lactic acid, muscle
latch state
latent period
length-tension relationships,
    muscle
lever and fulcrum system
load
M line
motor unit
multi-unit smooth muscle
muscle
muscle cramp
muscle fiber
muscular dystrophy
myofibril
myogenic contraction
myoglobin
myosin
myosin light chain kinase
    (MLCK)
myosin light chain phosphatase
nebulin
Niedeigerke, Rolf
origin of a muscle
pacemaker potentials
pharmacomechanical coupling

phosphocreatine
power stroke
protein kinase
recruitment
relaxation
rigor mortis
rigor state
sarcolemma
sarcomere
sarcoplasm
sarcoplasmic reticulum
series elastic element
single-unit smooth muscle
skeletal muscle
sliding filament theory of
    contraction
slow wave potential
slow-twitch fiber
sphincter
striated muscle
summation in muscle fibers
tendon
tension
tetanus
thick filament
thin filament
titin
tone
transverse tubule (t-tubule)
tropomyosin
troponin
twitch
unfused tetanus
visceral smooth muscle
Z disk

PRACTICE MAKES PERFECT

1. Muscle contraction requires which ion to be present and uses what form of cellular energy?

2. Which neurotransmitter is secreted by somatic motor neurons? _____

3. TRUE or FALSE? A single motor neuron can have synapses on more than one muscle fiber.

4. Why does Na^+ entry exceed K^+ efflux when ACh-gated channels open at the motor end plate?

5. Which of the following characteristics are typical of vertebrate skeletal muscle fibers? (circle all that apply):
   a. When stimulated, they produce Ca^{2+}-dependent action potentials.
   b. Depolarization of the muscle fiber results in an influx of Ca^{2+} from ECF, which causes contraction.
   c. They twitch spontaneously.
   d. Each muscle fiber is a multi-nucleated single cell.
   e. None of the above

6. Why is muscle contraction faster than relaxation?

7. Describe and explain the difference between temporal summation in muscle and temporal summation in nerves.

8. Why does a fatigued muscle take longer to relax than a "fresh" muscle?

9. What would be the disadvantage of having gap junctions between skeletal muscle cells like there are between cardiac muscle cells?

10. TRUE or FALSE? Muscle length changes during an isotonic muscle contraction.

11. Which of the following statements applies to isometric muscle contraction? (circle all that apply)
   a. Thick and thin filaments slide past each other.
   b. An isometric twitch lasts longer than an isotonic twitch.
   c. Thick and thin filaments do not slide past each other.
   d. Maximum tension is always generated, irrespective of initial fiber length.

12. In order to move a load, isotonic muscle tension should be:
   a. greater than the load
   b. equal to the load
   c. less than the load
   d. equal to the square root of the load

13. Can you give some examples of motor proteins that you studied in the nervous system? In the chapter on cell organelles?

14. In the muscular disorder myasthenia gravis, an autoimmune response causes a reduction in the number of ACh receptor sites at the motor end plate. Predict the symptoms of this disease and explain them based on your knowledge of the neuromuscular junction.

15. You have isolated the leg muscle of a frog and the nerve that innervates the muscle. To obtain a muscle contraction, you can either electrically stimulate the muscle itself (direct stimulation) or you can stimulate the nerve (indirect stimulation), creating an action potential in the nerve. In both instances, your recording shows a latency period, a delay between the time of stimulation and the peak of the contraction. Do you expect the latency with direct stimulation to be greater or less than the latency with indirect stimulation? Explain briefly.

MAPS

1. Map the anatomical organization of skeletal muscle (see map in Anatomy Summary), then add functions where appropriate.

2. Use a map to relate the following words into a cohesive, orderly description of muscle contraction. You must use all the words. You may add other words to make the relationships clear.

acetylcholine	crossbridge	myosin ATPase	thick filament
actin	dihydropyridine receptor	power stroke	thin filament
action potential	end plate potential	sarcoplasmic reticulum	tropomyosin
binding site	ion channel	swing	troponin
calcium	myosin	t-tubule	

READING

The in vitro motility assay: A window into the myosin molecular motor. (1996, February) *News in Physiological Sciences* 11: 1-6.

Lessons from a new model: Intracellular Ca^{2+} mediates muscle plasticity. (1997, August) *News in Physiological Sciences* 12: 194-195.

Long- and short-term regulation of the Na^+-K^+ pump in skeletal muscle. (1996, February) *News in Physiological Sciences* 11: 24-30.

Motions of myosin heads that drive muscle contraction. (1997, December) *News in Physiological Sciences* 12: 249-254.

Mouse models of muscular dystrophy: Gene products and function. (1992, October) *News in Physiological Sciences* 7: 195-198.

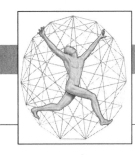

SUMMARY

What should you take from this chapter?

- A good review of nervous and autonomic reflexes, and a chance to review skeletal muscle contractions.
- A background in skeletal muscle reflexes. What are the structure and function of muscle spindles? Golgi tendon organs? Know the differences between alpha and gamma motor neurons.
- What are the differences among the three types of movement?

This chapter integrates all the information you've learned so far and then adds some. If the old material doesn't immediately click, now's the time to go back and review.

Remember the components of a nervous reflex? If not, look in this chapter and back to Chapter 6 to see where it all began. Nervous reflexes are classified in different ways. There are somatic reflexes and visceral reflexes, named for the neurons involved. There are spinal reflexes and cranial reflexes, named for the location where information is integrated. There are monosynaptic and polysynaptic reflexes, named for the number of neurons involved. Some nervous reflexes are innate, others are learned. [☛ Go back to Chapter 9, p. 253 and review the process of learning.] Neural activity can be altered by neuromodulators. Presynaptic modulation is more specific than postsynaptic modulation. Think about how modulation relates to divergence and convergence.

Some autonomic reflexes are spinal reflexes integrated in the spinal cord and modulated by the brain. Other reflexes are integrated in the brain itself, in the hypothalamus, thalamus, and brain stem. [☛ Go back to Chapter 9 and review CNS anatomy.] Many homeostatic reflexes are controlled by the brain. All autonomic reflexes are polysynaptic, and many exhibit tonic activity. [☛ Go back to Chapter 11 and review autonomic anatomy and physiology.]

Now for the new material. Remember that skeletal muscle contraction is controlled by somatic motor neurons. [☛ See Chapter 12, skeletal muscle contraction.] Contractile fibers are called extrafusal fibers, and they are controlled by alpha motor neurons. There are receptors within skeletal muscles that respond to stretch and length stimuli to protect the muscle from damage. Muscle spindles, made of intrafusal fibers, report information about length to the CNS. Sensory neurons wrap around the non-contractile centers of spindles, sending information to the CNS. Gamma motor neurons return commands to the contractile ends of the intrafusal fibers. Stretching the muscle fiber initiates a stretch reflex that creates contraction. Alpha-gamma coactivation ensures that the spindle stays active during contraction. Golgi tendon organs are found at the junction of muscle and tendon. They respond to both stretch and contraction, yielding a reflexive relaxation.

A myotatic unit is a set of synergistic and antagonistic muscles that control a single joint. Reciprocal inhibition is necessary to allow free movement by one member of the antagonistic pair. Flexion reflexes move a limb away from a harmful stimulus. Crossed extensor reflexes are postural reflexes that help maintain balance while one foot is off the ground. Central pattern generators, like those for breathing, are neural networks that spontaneously generate rhythmic muscle movements.

There are three categories of movement: reflex movements, voluntary movements, and rhythmic movements. Reflex movements are the least complex, and rhythmic movements are the most complex. The CNS coordinates and plans movements.

TEACH YOURSELF THE BASICS

NERVOUS REFLEXES

• List the steps in a nervous reflex. _____

_____

• Compare the roles of negative feedback [∫ p. 162] and feedforward responses [∫ p. 163] in the control of body movement.

_____

_____

Nervous Reflex Pathways Can Be Classified Different Ways (Table 13-1)

• What is the difference between somatic reflexes and autonomic (visceral) reflexes?

_____

• What is the difference between a spinal reflex and a cranial reflex?

_____

• What is the difference between an innate reflex and a learned reflex?

_____

• What is the difference between a monosynaptic reflex (Fig. 13-1a) and a polysynaptic reflex (Fig. 13-1b)?

_____

• Compare convergence and divergence. _____

_____

Modulation of Neuronal Activity

• What is the advantage of convergence in the CNS? _____

_____

• Compare the effects of excitatory postsynaptic potentials (EPSPs) and inhibitory postsynaptic

potentials (IPSPs) in the postsynaptic cell. _____

_____

• Explain why presynaptic modulation is more selective than postsynaptic modulation. (Fig. 13-3)

_____

_____

_____

AUTONOMIC REFLEXES

• Where are autonomic reflexes integrated? [∫ Fig. 11-9, p. 316]

_____

• Describe the influences that higher brain centers can have on autonomic reflexes.

_____

• Autonomic reflexes are (always monosynaptic / always polysynaptic / may be either?).

SKELETAL MUSCLE REFLEXES

• What is the only way to inhibit skeletal muscle contraction? _____

• How do alpha and gamma motor neurons differ? _____

_____

• List the three types of sensory receptors associated with skeletal muscles.

_____

Muscle Spindles Respond to Muscle Stretch

• What is the function of muscle spindles? _____

• What is the difference between intrafusal and extrafusal muscle fibers?

_____

• Describe the stretch reflex, using the standard steps of a reflex.

Stimulus_____Receptor_____

Afferent path_____Integrating center_____

Efferent path _____Effector _____

Tissue response _____ Systemic response _____

• Explain alpha-gamma coactivation. (Fig. 13-7) _____

_____

• What is the function of the muscle spindle reflex? _____

_____

Golgi Tendon Organs

• Describe the structure of the Golgi tendon organs. (Fig. 13-5c) _____

_____

• Describe the Golgi tendon reflex, using the standard steps of a reflex.

Stimulus_____Receptor_____

Afferent path_____Integrating center_____

Efferent path _____Effector _____

Tissue response _____Systemic response _____

• What is the function of this reflex? _____

Myotatic Reflexes and Reciprocal Inhibition

• What is a myotatic unit? _____

• Describe a monosynaptic stretch reflex, the knee jerk reflex, using the standard steps of a reflex. (Fig. 13-9)

Stimulus_____Receptor_____

Afferent path_____Integrating center_____

Efferent path _____Effector _____

Tissue response _____Systemic response _____

• Describe reciprocal inhibition in the knee jerk reflex, using the standard steps of a reflex. (Fig. 13-9)

Stimulus_____Receptor_____

Afferent path_____Integrating center_____

Efferent path _____Effector _____

Tissue response _____Systemic response _____

• How can a single stimulus, transmitted through a single sensory neuron, create two opposing responses?

_____

Flexion Reflexes and the Crossed Extensor Reflex

• Describe the flexion reflex (withdrawal reflex), using the standard steps of a reflex. There are two efferent pathways because this reflex involves reciprocal inhibition. (Fig. 13-10)

Stimulus_____Receptor_____

Afferent path_____Integrating center_____

Efferent path 1_____ Effector 1_____

Efferent path 2_____ Effector 2_____

Tissue response 1_____ Tissue response 2_____

Systemic response _____

• Why does this reflex take longer than the knee jerk reflex? _____

• Now describe the crossed extensor reflex that often accompanies the flexion reflex. (Fig. 13-10)

Stimulus_____Receptor_____

Afferent path_____Integrating center_____

Efferent path 1_____ Effector 1_____

Efferent path 2_____ Effector 2_____

Tissue response 1_____ Tissue response 2_____

Systemic response _____

• What is a central pattern generator?

_____

• What role do central pattern generators play in movement? _____

_____

THE INTEGRATED CONTROL OF BODY MOVEMENT

Types of Movement (Table 13-2)

• Name and describe the three basic types of movement.

_____

_____

_____

Integration of Movement within the Central Nervous System

• Movement is controlled at what three levels of the central nervous system? (Fig. 13-12, Table 13-3)

_____

• Describe the role of the following in planning and executing movement:

Thalamus _____

Cerebral cortex _____

Basal ganglia _____

Cerebellum _____

• Explain and give an example of a feedforward postural reflex. (Fig. 13-14)

_____

CONTROL OF MOVEMENT IN VISCERAL MUSCLES

• How does reflex control of visceral muscles differ from reflex control of skeletal muscles?

_____

TALK THE TALK

alpha motor neurons
alpha-gamma coactivation
autonomic reflexes or visceral reflexes
basal ganglia
central pattern generators
convergence
cranial reflexes
crossed extensor reflex
divergence
excitatory postsynaptic potentials (EPSPs)
extrafusal muscle fibers
feedforward postural reflexes
flexion reflexes
gamma motor neurons
Golgi tendon organ
inhibitory postsynaptic potentials (IPSP)
intrafusal fibers

joint capsule mechanoreceptors
monosynaptic reflex
muscle spindle
muscle tone
myotatic unit
nervous reflexes
neuromodulators
polysynaptic reflex
postsynaptic modulation
postural reflexes
presynaptic modulation
reciprocal inhibition
rhythmic movements
somatic reflexes
spinal reflexes
stretch reflex

PRACTICE MAKES PERFECT

1. In the left-hand column, arrange the words below in the proper order:

response, efferent path, afferent path, receptor, integrating center, effector

In the right-hand column, give the parts of any <u>biological</u> reflex discussed or demonstrated in the textbook except the knee jerk reflex.

Steps of a reflex	Example
Stimulus	

2. Proprioceptors monitor (circle the best one):
a. limb and muscle state
b. taste perception
c. blood pressure in the carotid artery
d. a and c

3. Look at the pathways diagramed in Fig. 13-10 (p. 373) and Fig. 13-11 (p. 375). Can you find examples of convergence, divergence, and presynaptic or postsynaptic inhibition in these figures?

BEYOND THE PAGES

TRY IT

PROPRIORECEPTION

Proprioreceptors are special receptors in the muscles and joints that send messages to the central nervous system about the position of the body parts relative to each other. Proprioreception normally works with visual cues. These exercises will show you the effectiveness of proprioreception alone.

Get a partner and try the following exercises with your eyes closed.

1. Have your partner time how long you can stand on one leg with your eyes closed and arms extended. Note the extent of swaying. Repeat with the eyes open.

2. Extend your arms to the side at shoulder level, palms down. With your eyes closed and without bending your elbows, try to bring your arms together in front of you, palms down, so that the index fingers meet side by side. Repeat with your eyes open.

3. Extend your arms to the side again, palms facing front. With eyes closed, bend your elbows and try to touch the tips of your index fingers together in front of you. Repeat with your eyes open.

In all three tests, which was easier? Why? What can you conclude from these exercises?

READING

Controlling computers with neural signals. (1996, October) *Scientific American.* Developing ways paralyzed people can use electrical impulses to command computers.

The mammalian muscle spindle. (1997, February) *News in Physiological Sciences* 12: 37-41.

Neural networks for vertebrate locomotion. (1996, January) *Scientific American.*

CHAPTER 14
CARDIOVASCULAR PHYSIOLOGY

SUMMARY

This is a chapter full of new concepts. Here are some key points to watch for while studying this chapter:

- How do pressure, volume, resistance, vessel length, and fluid viscosity relate to fluid flow?
- How are action potentials (APs) generated in cardiac contractile and autorhythmic cells?
- How does cardiac action potential generation differ from AP generation in other excitable tissues?
- What are the electrical events of the cardiac cycle?
- What are the mechanical events of the cardiac cycle? How do they relate to the electrical events?
- How do pressure and volume change during a cardiac cycle?
- How is heart rate generated? regulated?
- What factors affect stroke volume? cardiac output (CO)?

This chapter discusses how the heart creates blood flow and examines the variables that influence blood flow through the circulatory system. Blood flow is controlled by the natural laws of pressure, resistance, volume, vessel length, and the viscosity of the fluid. The heart consists of striated contractile muscle cells and specialized myocardial autorhythmic cells called pacemakers. The most important pacemaker is the sinoatrial (SA) node, and depolarization of this node transmits action potentials to other parts of the heart. Pacemaker activity is controlled by specialized ion channels and is influenced by the autonomic nervous system. Calcium, Na^+, and K^+ ions are involved in contraction of the myocardial contractile cells. The catecholamines, epinephrine and norepinephrine, affect the strength and duration of contraction by altering how much Ca^{2+} is available for binding to troponin. Cardiac output, the volume of blood pumped per ventricle per minute, is determined by heart rate and stroke volume.

TEACH YOURSELF THE BASICS

OVERVIEW OF THE CARDIOVASCULAR SYSTEM

- Why did organized cardiovascular (CV) systems become necessary as animals increased in complexity?

- Describe the basic structure of a cardiovascular system.

- Describe the general function of the cardiovascular system.

The Cardiovascular System Transports Material throughout the Body

• List at least five substances transported by the blood. _____

_____

The Cardiovascular System Consists of the Heart and Blood Vessels

• How do arteries differ from veins? _____

• What ensures one-way flow of blood through the system? _____

• Describe the structure of the heart. _____

_____

• Why do we say that the heart functions like two pumps pumping in series when, anatomically, the two sides of the heart are next to each other and look like they are pumping in parallel?

_____

_____

• Trace a drop of blood from the left ventricle to the stomach and back to the left ventricle. (Fig. 14-1)

_____

_____

_____

• Compare the pulmonary circulation with the systemic circulation. _____

_____

• What is a portal system? _____

_____

• Name the three portal systems of the body. [∫ p. 185]

_____

• Give the adjective for each of these organs: liver _____ lung _____

kidney _____ heart _____ brain_____

PRESSURE, VOLUME, FLOW, AND RESISTANCE

• Liquids and gases flow from areas of_____pressure to areas of_____ pressure.

• How does the cardiovascular system create a region of higher pressure? _____

• As blood moves away from the heart, what happens to the pressure?_____

• The highest pressure in the blood vessels is found in the _____ and the lowest

   pressure is found in the _____. (Fig. 14-2)

The Pressure of Fluid in Motion Decreases over Distance

• What is the difference between hydrostatic pressure (Fig. 14-3a) and hydraulic pressure (Fig. 14-3b)?

_____

_____

• What units are used to measure pressure in the cardiovascular system? _____

When Volume Decreases, Pressure Increases

• What happens to the pressure inside a water-filled balloon when you squeeze on it? _____

• What is driving pressure? _____

• What happens to pressure when the heart relaxes or the blood vessels dilate? _____

Blood Flows from an Area of Higher Pressure to One of Lower Pressure

• What is a pressure gradient? (Fig. 14-4) _____

• Fluid is flowing through two identical tubes. In tube A, the pressure at one end is 150 mm Hg
   and the pressure at the other end is 100 mm Hg. In tube B, the pressure at one end is 75 mm Hg
   and the pressure at the other end is 10 mm Hg. Which tube will have the greatest flow? (Answer
   in the appendix at the end of the workbook.)

Resistance Opposes Flow

• What two factors create friction in blood flowing through blood vessels?

_____

• Define resistance (R). _____

• When resistance increases, flow (increases / decreases?).

• Express this relationship in a mathematical equation: _____

• Name the three parameters that influence resistance for fluid flowing through a tube.

_____

• In humans, which of these factors are relatively constant and which play a significant role in determining resistance to blood flow?

_____

• The relationship between these factors and resistance was expressed mathematically by Jean Poiseuille. Write the equation known as Poiseuille's Law:

_____

• When the radius of a tube decreases, what happens to the resistance of that tube? _____

   What happens to flow through that tube? _____

• Write the equation that expresses the relationship between resistance and radius. _____

• If the radius of a tube doubles, what happens to the resistance? (Fig. 14-5) _____

• Define vasoconstriction and vasodilation in terms of diameter and resistance.

_____

• Write the equation that expresses the relationship between flow, resistance, and the pressure gradient.

_____

Velocity of Flow Depends on the Flow Rate and the Cross-Sectional Area

• What is the difference between flow (flow rate) and the velocity of flow? (Give units.) (Fig. 14-6)

_____

_____

• What factors have the biggest influence on the flow rate? _____

   On the velocity of flow? _____

• Fluid is flowing through a tube at a constant rate of flow. What happens to the velocity of flow

   if the tube suddenly narrows? _____

• Write the equation that expresses the relationship between the flow rate (Q), the velocity of flow
   (v), and the cross-sectional area (a) of a tube.

_____

☛ *Table 14-2 summarizes the rules for blood flow.*

CARDIAC MUSCLE AND THE HEART

The Heart Has Four Chambers

• The heart is a muscle that lies in the center of the _____ cavity, surrounded by

   the _____ membrane. (Fig. 14-7)

• True or false? The base of the heart is the pointed end that angles downward.

• The alternate term for cardiac muscle is _____.

• The heart has (how many?) _____ chambers, separated by a wall known as the _____.

The _____ are the lower chambers and the _____ are the upper

chambers. Which chambers have the thickest walls? _____

• Name the blood vessels that connect to each chamber and tell where they are bringing blood
from or taking blood to.

_____

_____

_____

• Name the vessels that supply blood to the heart muscle itself. _____

• Name two functions of the fibrous connective tissue rings that surround openings to major
   arteries and chambers (Fig. 14-7a).

_____

_____

One-way flow in the heart is ensured by the heart valves
• Name the valves of the heart and tell where they are located. (Fig. 14-9)

_____

_____

• What are the chordae tendineae and what is their function? (Fig. 14-9)

_____

• How are the atrioventricular (AV) and semilunar valves similar and how are they different?

_____

_____

Cardiac Muscle Cells Contract without Nervous Stimulation

• Why are some myocardial cells considered to be autorhythmic? _____

_____

• What is a pacemaker potential? [∫ p. 354] _____

_____

• What are intercalated disks? _____

• What role do desmosomes play in the heart? [∫ p. 54] _____

• What role do the autorhythmic cells play in the generation of force by the heart?

_____

• Compare the contractile cardiac muscle cells to skeletal muscle cells. _____

_____

_____

• Why do myocardial cells have a high rate of oxygen consumption?

_____

• What anatomical feature of myocardial cells allows coordinated contraction?

_____

Excitation-Contraction Coupling in Cardiac Muscle Is Similar to Skeletal Muscle

• The action and regulation of actin and myosin in contractile myocardium is most like that of smooth muscle or skeletal muscle?

• Explain how the events occurring between the action potential in the muscle fiber and the binding of Ca^{2+} to troponin are different in cardiac muscle.

_____

_____

_____

• How is Ca^{2+} removed from the cytoplasm of a cardiac muscle cell?

_____

Cardiac Muscle Contraction Can Be Graded

• How do cardiac muscle cells create graded contractions? _____

_____

• How do catecholamines enhance the force of cardiac muscle contraction? [∫ p. 181]

_____

• What kind of (cholinergic / adrenergic?) receptors are found on the myocardium? _____

• Signal transduction of epinephrine or norepinephrine uses what second messenger system? [∫ p. 314] (Fig. 14-12) _____

• When voltage-gated Ca^{2+} channels are phosphorylated, their probability of opening is (increased / decreased?) and (more / less?) Ca^{2+} enters the cell.

• What is phospholamban and what role does it play in altering cardiac muscle contraction?

_____

_____

When Cardiac Muscle Is Stretched, It Contracts More Forcefully

• True or false? The force of cardiac muscle contraction depends on the length of the muscle fiber when contraction begins. [∫ p. 340] (Fig. 14-13)

• In the intact heart, the length of the sarcomeres at the beginning of a contraction is a reflection

of _____

Action Potentials in Myocardial Cells Vary According to Cell Type

• What ion is important in cardiac muscle action potentials but plays no role in skeletal muscle or

neuronal action potentials? _____

Myocardial contractile cells

• Do myocardial contractile cells have a stable or unstable membrane potential? _____

• Compare the resting membrane potential of a myocardial contractile cell to that of a typical
neuron. [∫ p. 208]

_____

• In the myocardial contractile cell, the rapid depolarization phase is due to the entry of _____ .

How does this compare with a neuron? _____

• The repolarization phase is due to _____ (influx / efflux?). How does this compare with a
neuron?

_____

• Why is the shape of the myocardial contractile cell's action potential different from that of a
neuron?

_____

• In Fig. 14-14, p. 401, what causes the small fall in the membrane potential between points 1 and 2?

(Answer in the appendix at the end of the workbook.) _____

• Compare the duration of the contractile myocardial cell AP to that of a skeletal muscle.

_____

Myocardial autorhythmic cells

• Do myocardial autorhythmic cells have a stable or unstable membrane potential? _____

• What happens when the pacemaker potential reaches -40 mV? (Fig. 14-16)

_____

• Why is the membrane potential in these cells unstable?

_____

• The slow depolarization phase is due to the net movement of what ion(s) in which direction(s)?

_____

• The rapid depolarization phase is due to the entry of _____. How does this compare with a neuron?

_____

• The repolarization phase in autorhythmic cells is due to _____ (influx /efflux ?). How does

   this compare to a neuron? _____

• What effect do the catecholamines have on the rate of depolarization in pacemaker cells?

• To what kind of receptor are the catecholamines binding? _____

• What happens to heart rate if the depolarization rate of autorhythmic cells decreases? _____

• The combination of which neurotransmitter and which receptor slows heart rate? _____

☞ *Table 14-4 compares AP generation in cardiac and skeletal muscle.*

THE HEART AS A PUMP

• Define a cardiac cycle. _____

_____

Electrical Conduction in the Heart Coordinates Contraction

• Where do electrical signals in the heart originate? _____

• If you cut all nerves leading to the heart, will it continue to beat? Explain.

_____

• How do electrical signals spread throughout the heart? _____

• Starting at the sinoatrial (SA) node, describe or map the spread of electrical activity through the heart. Be sure to include all of the following terms. (Fig. 14-19)

atrial conducting system   bundle of His   right atrium
atrioventricular (AV) node   internodal pathway   septum
AV node delay   left atrium   ventricle
bundle branches   Purkinje fibers

• What is the purpose of AV node delay? _____

Pacemakers Set the Heart Rate

• If the SA node is damaged, will the heart continue to beat? at the same rate? Explain.

_____

_____

The Electrocardiogram Reflects the Electrical Activity of the Heart

• What is an electrocardiogram (ECG)? _____

_____

• What kind of information can one learn from an ECG? _____

_____

• Does the ECG show ventricular contraction? _____

• What is Einthoven's triangle? (Fig. 14-20) _____

_____

• How is an ECG different from the action potential of a myocardial contractile cell?

_____

_____

• Name the waves of the ECG, tell what electrical event they represent, and with what mechanical event each wave is associated. (Fig. 14-21, 14-22)

_____

_____

_____

• In some leads of an ECG, the R wave goes down instead of up. Does this mean that the ventricles are repolarizing instead of depolarizing?

_____

• What are the technical terms for rapid and slow heart rate? _____

• What is a normal range for heart rate? _____

• When looking at an ECG, how can you determine the heart rate? _____

• Why is it important to see that the P-R segment on an ECG has a constant duration?

_____

• Explain how you would identify an arrhythmia on an ECG.

_____

The Cardiac Cycle

• Define systole and diastole. _____

_____

• Is the heart in atrial and ventricular systole at the same time? Explain. _____

_____

The heart at rest: atrial and ventricular diastole (Fig. 14-25a)
• When both the atria and ventricles are relaxing:

Which valves are open? _____

Is blood flowing into the heart? Which chambers? _____

• What wave or wave segment on the ECG is associated with this phase? _____

Completion of ventricular filling: atrial systole (Fig. 14-25b)
• In a person at rest, how much of ventricular filling depends on atrial contraction? _____

• What wave or wave segment on the ECG is associated with this phase? _____

• Is there a valve between the atria and the veins emptying into them? What prevents blood from being pushed backward into the veins when the atria contract?

_____

Early ventricular contraction and the first heart sound (Fig. 14-25c)
• What electrical event precedes ventricular systole? _____

• As the ventricles contract, why do the AV valves close? _____

• What creates the heart sounds? _____

• Explain what is happening during isovolumic ventricular contraction. _____

_____

_____

• What happens to pressure within the ventricles during isovolumic ventricular contraction?_____

• What is happening to the atria during this phase of the cycle? _____

• What wave or wave segment on the ECG is associated with this phase? _____

The heart pumps: ventricular ejection (Fig. 14-25d)
• Why do the semilunar valves open, allowing blood to be ejected into the arteries?

_____

• Blood flows out of the ventricles, therefore you know that ventricular pressure must be

   (lower / higher ?) than arterial pressure.

• What wave or wave segment on the ECG is associated with this phase? _____

Ventricular relaxation and the second heart sound (Fig. 14-25e)
• As the ventricles relax, the ventricular pressure (increases / decreases ?).

• Why do the semilunar valves close? _____

• What creates the second heart sound? _____

• During this phase, are the AV valves open or closed? Explain. _____

_____

• When do the AV valves open? _____

• What wave or wave segment on the ECG is associated with this phase? _____

Pressure-Volume Curves Represent One Cardiac Cycle

Answer the following questions using the pressure-volume loop shown in Fig. 14-26, p. 413.

• At point A, is the atrium relaxed or contracting? _____

Is the ventricle relaxed or contracting? _____

• From point A to point B, the ventricular volume is increasing. Why doesn't ventricular pressure

also increase substantially? _____

• Define end-diastolic volume. _____

• What event begins at point B and continues until point D? _____

• From point B to point C, why does pressure increase without a change in ventricular volume?

_____

• From point C halfway to point D, why is volume decreasing as pressure continues to increase?

_____

• At point D, why does the aortic valve close? _____

• Define end-systolic volume. _____

• At what point(s) on the graph are both the aortic and mitral valves closed? _____

• Adaptively, why doesn't the ventricle empty itself with each contraction? _____

_____

☛ *The Wiggers diagram (Fig. 14-27) summarizes electrical and mechanical cardiac events.*

Stroke Volume Is the Volume of Blood Pumped by One Ventricle in One Contraction

• Define stroke volume. (Give units!) _____

• How do you calculate stroke volume? _____

• If the end-diastolic volume increases and the end-systolic volume decreases, has the heart

  pumped more or less blood? Make up an example. _____

  _____

Cardiac Output Is a Measure of Cardiac Performance

• Define cardiac output (CO). (Give units!)

  _____

• What information does CO tell us? What does it not tell us? _____

  _____

• What are average values for stroke volume and cardiac output in a man weighing 70 kg. while

  he is at rest?_____

• True or false? Defend your answer.
  The right side of the heart pumps blood only to the lungs, so its cardiac output is less than
  that of the left side of the heart, which must pump blood to many more tissues.

  _____

  _____

Heart Rate Is Controlled by Autonomic Neurons and Catecholamines

• Explain the antagonistic control of heart rate by sympathetic and parasympathetic neurons.
  (Fig. 14-28)

  _____

• If you were to block all autonomic input to the heart, what would happen to heart rate?

  _____

• Name the parasympathetic and sympathetic neurotransmitters and their receptors at the SA node.

  _____

Multiple Factors Influence Stroke Volume

• Stroke volume is directly related to the _____ generated by cardiac muscle during contraction.

• How is contractility different from the change in contractile force that occurs with changes in sarcomere length?

_____

Length-tension relationships and Starling's law of the heart
• As sarcomere length increases, what happens to force of contraction? _____

• In the intact heart we cannot measure sarcomere length directly. What parameter do we use as

   an indirect indicator of sarcomere length? _____

• State the Frank-Starling law of the heart. (Fig. 14-29)

_____

Stroke volume and venous return
• Define venous return. _____

• Venous return is best indicated by (end-systolic volume / end-diastolic volume ?).

• Explain pre-load and its relationship to venous return. _____

_____

• List and explain three factors that enhance venous return.

_____

_____

_____

Reflex control of contractility
• What is an inotropic agent? _____

• What effect does a positive inotropic agent have on the heart? Name two.

_____

• At the cellular level, an increase in contractility occurs when _____ increases.

• Explain how Fig. 14-30 (p. 417) shows that contractility is distinct from length-tension relationships in cardiac muscle.

_____

_____

TALK THE TALK

β_1 adrenergic receptor
acetylcholine (ACh)
action potential, autorhythmic cell
action potential, contractile cell
aorta
aortic valve
arrhythmia
artery
atrioventricular node (AV node)
atrioventricular valve
atrium
auscultation
autorhythmic cell
AV node delay
bicuspid valve
blood
bradycardia
bundle branch
bundle of His ("hiss")
Ca^{2+} permeability
Ca^{2+}-ATPase
Ca^{2+}-induced Ca^{2+} release
calcium
capillary
cardiac cycle
cardiac glycoside
cardiac output
cardiovascular system
catecholamine
chordae tendineae
complete heart block
contractility
coronary artery
cross-bridge formation
cyanosis
cyclic AMP
desmosome
diastole
digitoxin
driving pressure
ectopic pacemaker
Einthoven, Walter
electrocardiogram, or ECG
electroencephalogram, or EEG

electromyelogram, or EMG
end-diastolic volume (EDV)
end-systolic volume (ESV)
epinephrine
excitation-contraction coupling
fibrillation
first heart sound
flow
flow rate
foxglove
gap junction
graded contraction
Harvey, William
heart
heart attack
heart rate, control of
hepatic portal vein
hydrostatic pressure
hypoxia
I_f channel
inotropic agent
intercalated disk
internodal pathway
ischemia
isovolumic relaxation
isovolumic ventricular contraction
length-force relationship
mechanical events of cardiac cycle
millimeters of mercury (mm Hg)
mitral valve
myocardial infarction
myocardium
Na^+/K^+-ATPase
Na^+-Ca^{2+} exchanger
norepinephrine
ouabain
P wave
pacemaker potential
pacemakers, cardiac
papillary muscle
pericarditis
pericardium
phospholamban

Poiseuille's Law
portal system
pre-load
pressure
pressure gradient (P)
prolapse
pulmonary artery
pulmonary circulation
pulmonary trunk
pulmonary valve
pulmonary vein
Purkinje fiber
QRS complex
refractory period, cardiac muscle
resistance
respiratory pump
second heart sound
semilunar valve
septum
sinoatrial node (SA node)
Starling's law of the heart
stenosis
sternum
stroke volume
systemic circulation
systole
T wave
tachycardia
tetanus
tetrodotoxin
thoracic cavity
torr
tricuspid valve
troponin
vasoconstriction
vasodilation
vein
velocity of flow
vena cava
venous circulation
venous return
ventricle
viscosity
voltage-gated Na^+ channel
Wiggers diagram

ERRATA

p. 398, key to Fig. 14-11: The Na^+-Ca^{2+} cotransporter is also called a secondary active transporter.

p. 399, Fig. 14-12: The last row of boxes should read "contraction" instead of "concentration."

RUNNING PROBLEM - Heart Attack

Clot spotter. Does a test for fibrinogen predict risk of heart disease? (1991, June) *Scientific American.* 264(6):118-119.

Gene therapy: Cardiovascular gene therapy. (1997, July/August) *Science & Medicine.*

Gene therapy for clogged arteries passes test in pigs. (1994) *Science* 265(5): 738.

Lipoprotein(a) in heart disease. (1992, June) *Scientific American.* 266(6):54-60.

Physiologic effects and importance of exercise in patients with coronary artery disease. (1977, April) *Cardiovascular Medicine*, 365-386.

Serum cholesterol in young men and subsequent cardiovascular disease. (1993) *N. England J. of Med.* 328: 313-318.

Sudden impact. Why a little heart disease can be worse than a lot. (1989) *Scientific American* 261(4):35-36.

QUANTITATIVE THINKING

Flow $\propto \Delta P$ where $\Delta P = P_1 - P_2$

$F \propto 1/R$ where $R = 8L\eta/\pi r^4$ and in most cases, $R = 1/r^4$

Flow $\propto \Delta P/R$ or Flow $\propto \Delta P r^4$

stroke volume = EDV - ESV

cardiac output (CO) = heart rate x stroke volume

1. If the small intestines need more blood, what should happen to the resistance and radius of the blood vessels supplying their blood?

2. Compare the flow rates between tubes A and B below. L = length, r = radius, P1 = pressure at one end of the tube, and P2 = pressure at the other end

P1 = 125 mm Hg   P2 = 75 mm Hg
r = 2
L = 16

A

P1 = 180 mm Hg
r = 1   P2 = 108 mm Hg
L = 2

B

3. At rest, Juliet's heat rate is 72 beats/min and her cardiac output is 5.0 L/min. However, when she sees Romeo, her heart rate increases to 120 beats/min and her cardiac output reaches 15 L/min. What is Juliet's stroke volume before and after seeing Romeo?

PRACTICE MAKES PERFECT

1. Write out the full words for the following abbreviations:

ACh _____

AV (node or valves) _____

CO _____

ECG _____

EDV _____

ESV _____

SA (node) _____

SV _____

2. Describe the electrical and mechanical events associated with the electrocardiogram.

	Electrical Event	Mechanical Event
P wave		
QRS complex		
T wave		
PQ segment		
ST segment		
TP segment		

3. In a person at rest, only 20% of ventricular filling depends on atrial contraction. If heart rate speeds up, would this percentage become greater or less? Explain.

4. Compare the action potentials of cardiac and skeletal muscles.

	Skeletal muscle	Contractile myocardium	Autorhythmic myocardium
Membrane potential			
Events leading to threshold potential			
Rising phase of action potential			
Repolarization phase			
Hyperpolarization			
Duration of action potential			
Refractory period			

5. What happens to the kinetic energy of blood flow as it leaves the heart and travels through the blood vessels of the circulatory system?

6. You need to design two different drugs, one that increases and another that decreases the force of a cardiac muscle contraction. What molecular and cellular mechanisms of the cardiac muscle cell might you try to alter with these drugs? (There are several possible answers; be creative. You are not limited to what has already been developed.)

7. Explain how Starling's Law, venous return, and input from the autonomic nervous system affect cardiac output.

8. Below is an outline of the heart. Draw and label the atrioventricular node, atrial conducting system, bundle of His, Purkinje fibers, and sinoatrial node. Show the direction of current flow.

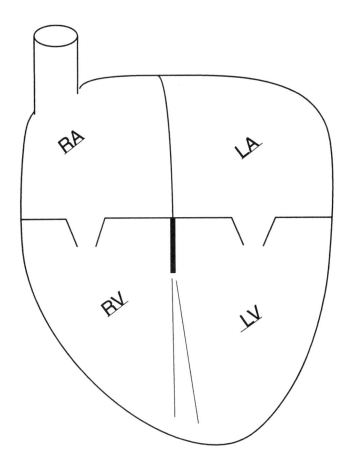

9. The total amount of blood in the circulatory system is about 5 liters. If the cardiac output of the left ventricle is 4.5 L/min at rest, what would be the cardiac output of the right ventricle? Explain your answer.

10. Answer the following questions using the pressure-volume loop shown in Fig. 14-26, p. 413.

• At point A, what can you say about atrial pressure relative to ventricular pressure? _____

• At what point on the curve does ventricular pressure match aortic pressure? _____

• Estimate the highest pressure in the aorta. _____

11. Draw a graph that has end-diastolic volume (EDV) on the x-axis and stroke volume on the y-axis. Draw a point that represents a stroke volume at rest of 70 mL with an EDV of 135 mL. Draw a curve that would represent an increase in stroke volume with the same EDV.

MAPS

1. Construct a flow chart using the following terms to trace a drop of blood through the cardio-vascular system:

aorta	left ventricle	right ventricle
arteriole	pulmonary artery	vein
artery	pulmonary vein	vena cava
capillary	right atrium	venule
left atrium		

2. The cardiovascular control center located in the medulla oblongata signals that an increase in cardiac output is required. Use the terms below to trace the sequence of events that would produce an increase in cardiac output.

acetylcholine	end-diastolic volume	parasympathetic activity
β_1 receptors	epinephrine	stroke volume
cardiac output	heart rate	sympathetic activity

3. Construct a diagram or map that provides a sequence of events that occurs during a cardiac cycle. Include in your answer the following: all electrical events, opening and closing of the valves, atrial and ventricular filling, atrial and ventricular systole and diastole, isovolumetric ventricular contraction, ventricular pressure increase, ventricular ejection, and isovolumetric relaxation.

4. Construct a flow chart or map that outlines the sequence of events leading to and occurring during excitation-contraction coupling in cardiac cells. Include membrane events, second messenger systems, and the role of ions and the contractile proteins in your answer. How is the contractile process modulated?

BEYOND THE PAGES

TRY IT:
William Harvey's Experiments
Duplicate William Harvey's experiment that led him to believe that blood circulated in a closed loop: (1) Take your resting pulse. (2) Assume that your heart at rest pumps 70 mL per beat and that 1 mL of blood weighs 1 gram. Calculate how long it would take your heart to pump your weight in blood. (1 kilogram = 2.2 pounds).

READING

Adenosine-a cardioprotective and therapeutic agent. (1993) *Cardiovascular Res.* 27: 2.

Alpha-adrenergic modulation of cardiac rhythm. (1991) *News in Physiological Sciences.* 6: 134-138 .

Amino acids as central neurotransmitters in the baroreceptor reflex pathway. (1994) *News in Physiological Sciences.* 9: 243-246.

An artificial heart inside the body. (1965, November) *Scientific American* 213(5):39-46.

Atrial tachycardias - A new look. (1990) *News in Physiological Sciences* 5: 187-191.

Beat-to-beat variations of heart rate reflect modulation of cardiac autonomic outflow. (1990) 5: 32-37.

Behavioral influences on coronary blood flow and susceptibility to arrhythmias. (1990) *News in Physiological Sciences* 5: 108-112.

Brunton's use of amyl nitrite in angina pectoris: An historic root of nitric oxide research. (1995, June) *News in Physiological Sciences* 10: 141-144.

Cardiac chloride currents. (1996, August) *News in Physiological Sciences* 11: 175-180.

The cardiac connection. (1991) *News in Physiological Sciences* 6: 34-38.

Cellular mechanisms for ventricular fibrillation. (1992) *News in Physiological Sciences* 7: 254-259.

Chasing myocardial calcium: A 35-year perspective. (1997, October) *News in Physiological Sciences* 12: 238-244.

The contributions of Willem Einthoven to physiology. (1989) *News in Physiological Sciences* 4: 162-165.

Does the heart's hormone, ANP, help in congestive heart failure? (1995) *News in Physiological Sciences* 10: 247-253.

The electrocardiographic inverse problem. (1996, November/December) *Science & Medicine.*

Elementary hemodynamic principles based on modified Bernoulli's equation. (1985) *The Physiologist* 28(1): 41-46.

Have a heart. Tissue-engineered valves may offer a transplant alternative. (1995, June) *Scientific American* 272(6):46.

The heart. (1957) *Scientific American* 196(5): 75-87.

The heart as a suction pump. (1969) *Scientific American* 254 (6):84-91.

The heart as an endocrine gland. (1986, February) *Scientific American* 254(2):76-81.

Intensive heart care. (1968, July) *Scientific American* 219(1):19-27.

Ionic currents in the human myocardium. (1990) *News in Physiological Sciences* 5: 28-31.

Is phospholipid degradation a critical event in ischemia- and reperfusion-induced damage? (1989) *News in Physiological Sciences* 4: 49-53.

Myocardial hibernation: Adaptation to ischemia. (1996, August) *News in Physiological Sciences* 11: 166-169.

Neural control of cardiac pacemaker potentials. (1991) *News in Physiological Sciences* 6: 185-190.

Neurosecretory hypothalamus-endocrine heart as a functional system. (1992) *News in Physiological Sciences* 7: 279-283.

The physiology of heart-lung transplantation in humans. (1990) *News in Physiological Sciences* 5: 71-74.

Radiofrequency ablation of cardiac arrhythmias. (1994, May/June) *Science & Medicine.*

Reflexes from the heart. (1991) *Physiological Reviews* 71(3): 617-653.

Regulation of myocardial adaptation. (1994, July/August) *Science & Medicine.*

The rise and fall of ischemic heart disease. (1980, November) *Scientific American* 243(5):53-59.

Short-term autoregulation of systemic blood flow and cardiac output. (1989) *News in Physiological Sciences* 4: 219-225.

Starling's "Law of the Heart": Rise and fall of the descending limb. (1992) *News in Physiological Sciences* 7: 134-137.

Sudden cardiac death. (1986) *Scientific American* 254 (5):37-43.

      An interesting and light reading article that would be enjoyable for pre-medical as well as non-science students. This article discusses heart disease and heart ailments that can sometimes cause a massive heart failure such as ventricular fibrillation. Equipment and techniques used to resuscitate victims are discussed.

Surgical treatment of cardiac arrhythmias. (1993, July) *Scientific American.*

The total artificial heart. (1981, January) *Scientific American* 244(1):74-80.

Using skeletal muscle for cardiac assistance. (1994, November/December) *Science & Medicine.*

Ventricular energetics. (1990) *Physiological Reviews* 70(2): 247-275.

CHAPTER 15

BLOOD FLOW AND THE CONTROL OF BLOOD PRESSURE

SUMMARY

Be sure to learn the pathways presented in this chapter. They'll be around for the rest of the book. Also become familiar with hydraulic and osmotic pressures and their roles in bulk fluid flow. Here are some concepts to watch for in this chapter.

• How do the different blood vessels vary?
• What factors determine blood pressure?
• Where is the primary site of variable resistance in the systemic circulation?
• What factors affect resistance?
• How is exchange at the capillaries accomplished?
• What are the functions of the lymphatic system?
• What are the components of the baroreceptor reflex?
• What are the risk factors for developing cardiovascular disease?

Blood vessels are composed of an inner layer of endothelium, surrounded by elastic and fibrous connective tissues and smooth muscle. Not all blood vessels have the same characteristics. For example, arterioles contain more smooth muscle than veins, and capillaries of the kidney are more porous than those of the brain. Movement across the endothelium of capillaries is accomplished by diffusion, transcytosis, and bulk flow. The lymphatic system plays an important role in restoring fluid lost through capillary filtration.

Various chemicals such as norepinephrine, epinephrine, and angiotensin II control vasodilation and vasoconstriction. The medullary cardiovascular control center (CVCC) in the medulla oblongata sends commands through the autonomic nervous system to adjust blood pressure in response to incoming sensory information from baroreceptors. However, paracrines can also have significant local control over blood pressure.

TEACH YOURSELF THE BASICS

• Trace a drop of blood from the vena cava to the aorta, naming all major structures the drop of blood passes.

_____

_____

_____

• If blood flow through the aorta is 5 L/min, what is blood flow through the pulmonary artery?

_____

THE BLOOD VESSELS

• Name the four different tissue layers found in blood vessel walls, starting with the layer closest to the lumen.

_____

Blood Vessels Contain Vascular Smooth Muscle

• What is vascular smooth muscle? _____

• What is the term for a decrease in the diameter of a blood vessel? _____

• What is the term for an increase in the diameter of a blood vessel? _____

• What is muscle tone? _____

• List five chemical factors that cause vasoconstriction and five that cause vasodilation. Tell whether each chemical is a paracrine, hormone, or neurocrine. (Table 15-1)

_____

_____

_____

• Name two paracrines secreted by the vascular endothelium. _____

• What is the source of Ca^{2+} for vascular smooth muscle contraction? [∫ p. 352]

_____

Arteries and Arterioles Carry Well-Oxygenated Blood to the Cells

• Describe the physical properties of the aorta and major arteries. (Fig. 15-2) _____

_____

• Blood flow from arteries to arterioles is best described as (divergent / convergent?).

• Describe the key characteristic of arterioles. _____

• How do metarterioles differ from arterioles? (Fig. 15-3) _____

• What is the function of metarterioles? _____

• What vessels make up the microcirculation? _____

Exchange between the Blood and Interstitial Fluid Takes Place in the Capillaries

• Describe the capillary wall and explain how this structure allows the capillaries to carry out their function.

_____

• How do the capillaries of the blood-brain barrier differ from those in the rest of the systemic circulation?

_____

Blood Flow Converges in the Venules and Veins

• Blood flow from capillaries to venules is best described as (divergent / convergent?).

• Compare the walls of veins with those of arteries. _____

_____

• How much of the blood in the circulatory system is found in the veins? _____

• Are the bluish blood vessels you see under the skin arteries or veins? _____

Angiogenesis

• Define angiogenesis. _____

_____

• What is the reason for angiogenesis in children? in adults? _____

_____

• If we can find a way to stop angiogenesis, why might this become useful in treating cancer?

_____

_____

• In coronary artery disease, what happens to the arteries? _____

_____

• Why would a drug that stimulates angiogenesis be useful for treating this condition?

_____

BLOOD PRESSURE

• Why are the arteries sometimes called a pressure reservoir for the circulatory system? (Fig. 15-4)

_____

• What property of artery walls plays a key role in the ability of arteries to sustain the driving pressure created by the heart?

_____

Blood Pressure in the Systemic Circulation Is Highest in the Arteries and Lowest in the Veins

• Why does blood pressure decrease as blood flows through the circulatory system? (Fig. 15-5)

_____

• Define systolic pressure and give an average value for systolic arterial pressure.

_____

• Define diastolic pressure and give an average value for diastolic arterial pressure.

_____

• True or false? Explain. The pulse is created by a wave of blood flowing through the arteries.

_____

• The pulse amplitude (increases / decreases?) over distance from the heart due to what factor(s)? (Fig. 15-5)

_____

• What is pulse pressure and how do you calculate it? _____

_____

• How can low-pressure venous blood in the feet flow "uphill" against gravity to get back to the heart? (Fig. 15-6)

_____

Arterial Blood Pressure Reflects the Driving Pressure for Blood Flow

• To what does the term "blood pressure" refer? _____

• Why is blood pressure such an important parameter to know? _____

• Define mean arterial pressure (MAP).

_____

• Write the calculation for MAP: _____

• What kinds of problems may result when blood pressure is too low? too high? _____

_____

Blood Pressure Is Estimated by Sphygmomanometry

• Explain how a sphygmomanometer is used to estimate arterial pressure of the radial artery. (Fig. 15-7)

_____

_____

• What makes Korotkoff sounds? _____

• Define systolic pressure. _____

• Define diastolic pressure. _____

• Explain a blood pressure of 100/70. _____

• What is an average value for blood pressure? _____

• Blood pressure is considered too high if systolic pressure is chronically over _____ mm Hg

   or diastolic pressure chronically exceeds _____ mm Hg. Another name for high blood

   pressure is_____.

Cardiac Output and Peripheral Resistance Are the Main Factors Influencing Mean Arterial Pressure

• What two main factors determine mean arterial pressure (MAP)? (Fig. 15-8)

_____

• If blood flow into the arteries increases but there is no change in blood flow out of the arteries, MAP will (increase / decrease ?).

• What happens to MAP if peripheral resistance increases? _____

Changes in Blood Volume Affect Blood Pressure

• If the volume of blood circulating through the system decreases, blood pressure (increases / decreases?).

☛ *As an analogy for blood volume exerting pressure, think of the tension on the wall of a water-filled balloon. One way to adjust pressure is to add or remove water from the balloon.*

• What organ is responsible for decreasing blood volume? _____

• The homeostatic regulation of blood pressure is accomplished by what two systems of the body? (Fig. 15-9)

_____

• True or false? If blood volume decreases, the kidneys can increase blood volume by reabsorbing water.

Explain. _____

• Name two ways the cardiovascular system tries to compensate if blood volume and therefore blood pressure decrease. [∫ p. 415]

_____

• Explain how the veins act as a volume reservoir that can be used to raise arterial blood pressure.

_____

_____

• If arterial pressure falls, venous constriction mediated through the (sympathetic / parasympathetic) division will have what effect on blood distribution and blood pressure?

_____

_____

RESISTANCE IN THE ARTERIOLES

• What vessels are the main site of variable resistance in the systemic circulation? _____

• What property of these vessels permits them to change resistance? _____

• Write the mathematical expression for the relationship between radius (r) and resistance (R).

_____

• What is the purpose of local control of arteriolar radius? _____

• What is the purpose of reflex control of arteriolar radius? _____

Myogenic Autoregulation Automatically Adjusts Blood Flow

• What is myogenic autoregulation?

_____

• What are the two theories of the mechanism of myogenic autoregulation? _____

_____

• When blood pressure in an arteriole increases, myogenic regulation causes the arteriole

to _____.

Paracrines Alter Vascular Smooth Muscle Contraction

• Tissue and endothelial paracrines locally control arteriole resistance. List some vasoactive
paracrines:

_____

• Write the pathway for active hyperemia. (Fig. 15-11a)

• Write the pathway for reactive hyperemia. (Fig. 15-11b)

The Sympathetic Division Is Responsible for Most Reflex Control of Vascular Smooth Muscle

• Most systemic arterioles are innervated by neurons of the _____

   division of the nervous system. These neurons release the neurotransmitter _____

   onto _____ receptors, causing vascular smooth muscle to _____ and

   peripheral resistance to _____.

• What branch of the nervous system controls arterioles involved in the erection reflex? _____

• Why isn't the diameter of brain arterioles regulated by the nervous system?

_____

• If only one division of the autonomic nervous system controls arteriolar radius, how can that one signal cause arterioles both to dilate and constrict? (Fig. 15-12)

_____

_____

• β_2 receptors are not innervated and respond primarily to circulating epinephrine. [∫ p. 314] In

   the cardiovascular system, where are β_2 receptors found? _____

• Briefly describe the fight-or-flight response. [∫ p. 308] _____

_____

DISTRIBUTION OF BLOOD TO THE TISSUES

• Why don't all tissues get equal blood flow at all times? _____

_____

• At rest, which four organ systems receive most blood flow? (Fig. 15-13)

_____

• Are arterioles arranged in series or parallel? (Fig. 15-1) _____

• At any given moment, the total blood flow through all arterioles = _____.

• Flow through individual arterioles depends on their _____ and _____.

• When resistance of an arteriole increases, its blood flow (increases / decreases?). (Fig. 15-14)

• When blood flow decreases through one set of arterioles, where does that blood go?

_____

• Capillary blood flow can be regulated by precapillary _____.
  (Fig. 15-15)

EXCHANGE AT THE CAPILLARIES

• Which materials can move through capillary pores? _____

• Which materials cannot pass through capillary pores? _____

• What determines capillary density in a tissue? _____

• Compare the structure and function of continuous capillaries and fenestrated capillaries. (Fig. 15-16)

_____

_____

The Velocity of Blood Flow Is Lowest in the Capillaries

• Explain the relationship between the total cross-sectional area and the velocity of blood flow in the circulatory system. (Fig. 15-17)

_____

Most Capillary Exchange Takes Place by Diffusion and Transcytosis

• For substances that diffuse freely across capillary walls, what factor is most important for

  determining the rate of diffusion? _____

• The pores of capillaries are too small to allow proteins to pass through them. How then do protein hormones and other essential proteins move out of the blood and into the interstitial fluid?

_____

Capillary Filtration and Reabsorption Take Place by Bulk Flow

• Define bulk flow. _____

• Distinguish between filtration and absorption in capillaries. _____

_____

• What forces create capillary bulk flow? _____

• Hydraulic pressure pushes fluid (in / out?) through capillary pores. This pressure decreases along the length of the capillary as energy is lost to _____.

• What creates the osmotic pressure gradient between the blood and the interstitial fluid?

_____

• What is colloid osmotic pressure ( )? _____

• What happens to colloid osmotic pressure along the length of the capillary? _____

• Is filtration in capillaries exactly equal to absorption? Explain. (Fig. 15-18a) _____

_____

THE LYMPHATIC SYSTEM

• Name the three systems with which the lymphatics interact and explain the role of the lymphatics in each system.

_____

_____

_____

• Compare the anatomy of the lymphatic system to that of the circulatory system. (Fig. 15-18b, 15-19)

_____

_____

_____

• Bulk flow of fluid, proteins, and bacteria is (into / out of ?) lymph capillaries.

• What is lymph? _____

• Where does lymph rejoin the blood? _____

• Name four factors that influence fluid flow through the lymphatics. (Does the lymph system have a pump like the heart?)

_____

• What is edema? _____

• Explain why disruption of the osmotic gradient between the plasma and the interstitial fluid causes edema.

_____

Edema: Disruption of Capillary Exchange

• List and explain the mechanism behind three different causes of edema. _____

_____

_____

_____

REGULATION OF BLOOD PRESSURE

• In what part of the brain is the neural control center for blood pressure homeostasis

found?_____

The Baroreceptor Reflex Is the Primary Homeostatic Control for Blood Pressure

• What type of sensory receptor responds to changes in blood pressure? _____

• Where are the two main receptors for blood pressure located? What is significant about these

locations?_____

• If you are monitoring the electrical activity of the sensory neurons linking these baroreceptors to the cardiovascular control center, would you observe any electrical activity when a person's blood pressure is in the normal range? Are these receptors tonic or phasic?

_____

• List the components of the baroreceptor reflex. (Fig. 15-20)

Stimulus: _____

Receptor(s): _____

Afferent path: _____

Integrating center: _____

All efferent pathways:_____

All effectors, matching the effectors to their efferent pathways: _____

_____

Responses of the effectors:_____

Systemic response: _____

• A decrease in blood pressure results in (increased / decreased?) sympathetic activity and

(increased / decreased?) parasympathetic activity.

• An increase in sympathetic activity will have what effect on heart rate, force of contraction, and

arteriolar diameter? _____

• An increase in parasympathetic activity will have what effect on heart rate, force of contraction,
and arteriolar diameter? _____

• Vasoconstriction will (increase / decrease?) peripheral resistance and (increase / decrease?)
blood pressure.

• Explain the integration of breathing and cardiac output. _____

_____

Orthostatic Hypotension Triggers the Baroreceptor Reflex

• Why does blood pressure initially fall when one stands up after lying flat? _____

_____

• What is the name given to this transient decrease in blood pressure? _____

• Map the reflex response to orthostatic hypotension. (Fig. 15-21) Be sure to include all the steps of the reflex pathway (stimulus, receptor(s), afferent path, integrating center, efferent pathways, all effectors matched to their efferent pathways, responses of the effectors, systemic response).

CARDIOVASCULAR DISEASE

• What is coronary artery disease (coronary heart disease) and why are millions of dollars yearly spent trying to find its cause and optimal treatment?

_____

Risk Factors for Cardiovascular Disease Include Gender, Age, and Inheritable Factors

• List the four uncontrollable risk factors for cardiovascular disease.

_____

_____

_____

• List five controllable risk factors for cardiovascular disease.

_____

_____

• What is atherosclerosis? _____

• What is low-density lipoprotein or LDL-cholesterol and what is its normal function? _____

_____

• What is high-density lipoprotein or HDL-cholesterol and what is its normal function? _____

_____

• Why is elevated LDL-cholesterol in the blood undesirable? _____

_____

• What role do macrophages play in the development of atherosclerosis?

_____

• What is a thrombus and why is thrombus formation in blood vessels dangerous?

_____

• Chemically, what happens to a molecule when it is oxidized? [∫ p. 86]

_____

Hypertension Represents a Failure of Homeostasis

• Hypertension is chronically elevated blood pressure, with systolic pressures greater

    than _____ mm Hg or diastolic pressures greater than _____ mm Hg.

• What is the cause of most hypertension? _____

• Explain why we say that hypertension represents failure of homeostasis. _____

_____

• How does the adaptation of sensory receptors [∫ p. 271] relate to hypertension?

_____

• Explain the association between hypertension and atherosclerosis. Use the concepts of cardiac output, resistance, radius, and work in your explanation.

_____

_____

_____

TALK THE TALK

absorption
active hyperemia
adenosine
alpha (α) receptor
angiogenesis
angiotensin II
arteriole
artery
atherosclerosis
atrial natriuretic peptide
barorecptor reflex
beta (β_2) receptor
bulk flow
capillary
cardiac output
cardiovascular disease
colloid osmotic pressure
continuous capillary
coronary artery disease
diastolic pressure
discontinuous capillary
edema

elastic connective tissue
endothelin
endothelium
epinephrine
fenestrated capillaries
fibrous connective tissue
filtration
high-density lipoprotein (HDL)
histamine
hydraulic pressure
hydrostatic pressure
hypertension
kinin
Korotkoff sound
low-density lipoprotein (LDL)
lymph
lymph node
lymphatic system
mean arterial pressure (MAP)
medullary cardiovascular control
      center (CVCC)
metarteriole

microcirculation
muscle tone
myogenic autoregulation
norepinephrine
orthostatic hypotension
peripheral resistance
pre-capillary sphincter
pulse pressure
reactive hyperemia
serotonin
smooth muscle
sphygmomanometer
stroke
systolic pressure
thrombus
transcytosis
vasoconstriction
vasodilation
vein
venule

ERRATA
p. 434, Table 15-2: For paracrine signal molecules, add kinins to the line for vasodilators and serotonin to the line for vasoconstrictors.

RUNNING PROBLEM - Essential Hypertension

Nitric oxide inhibition in hypertension. (1994) *News in Physiological Sciences*. 9: 268-271.

QUANTITATIVE THINKING

MAP = diastolic P + 1/3 (systolic P – diastolic P)

Pulse pressure = systolic pressure – diastolic pressure

MAP $\propto$ cardiac output $\times$ resistance$_{arterioles}$

Flow$_{arteriole}$ $\propto$ $1/R_{arteriole}$

1. At age 20 Missy had a blood pressure of 110/70. At 60 years old, she has a blood pressure of 125/82.

a) In what units are these blood pressures measured? _____

b) What are Missy's pulse pressures and mean arterial pressures at 20 and 60 years of age?

c) Missy does not smoke and does not have other controllable risk factors for cardiovascular disease. Why did her pulse and mean arterial pressures change with age?

2. If the radius of an arteriole increases from 2 to 3 millimeters, how does this affect resistance and blood flow? Explain your answer using qualitative and quantitative terminology.

3. If total peripheral resistance increases and cardiac output does not change, how is MAP affected?

4. Chris has been in training for a triathlon. The Kinesiology Department decided to study his endurance and put him through some tests. His end-systolic volume was 50 mL, his end-diastolic volume was 160 mL, his heart rate was 140 beats/min, and his arterial blood pressure was 135/78. What was his cardiac output?

PRACTICE MAKES PERFECT

1.   Match the blood vessel with its main characteristics.

_____ arteries	A. lots of smooth muscle
_____ arterioles	B. low compliance and high recoil
_____ capillaries	C. high compliance and high recoil
_____ veins	D. high compliance and low recoil
	E. contains endothelium only

2. Complete the table below. Your answer should include the relative amounts of the various types of tissues that each vessel contains.

Blood Vessel	Physical Characteristics	Function (s)
arteries		
arterioles		
capillaries		
veins		

3. A 45-year-old woman has a ventricular systolic pressure of 130. How high must you inflate the cuff of a sphygmomanometer on her left arm in order to stop blood flow through the brachial artery? Explain your reasoning?

4. Compare and contrast the response of a healthy artery with a diseased artery (arteriosclerosis) during systole and diastole. How would pulse pressure be affected by atherosclerosis?

5. Even though the radius of a single capillary is smaller than that of an arteriole, the peripheral resistance to blood flow through the capillaries is less than that of blood flow through the arterioles. Explain why this is true.

6. During exercise, blood flow to skeletal muscles is increased, but blood flow to the digestive system is decreased. How is this achieved?

7. Match the neurotransmitter/neurohormone and receptor with its target(s). Answers may be used more than once and more than one answer may apply to any target.

A - norepi on α receptors

B - ACh on nicotinic receptors   C - epi on β_2 receptors

D - ACh on muscarinic receptors

E - norepi on β_1

F - none of the above

SA node _____ ventricular myocardium _____ skeletal muscle capillary _____

cardiac vasculature _____ renal arterioles _____ brain arterioles _____

8. The arterioles of the kidneys constrict as a result of local control mechanisms. Assuming that no compensatory homeostatic mechanisms are triggered, what happens to each of the following state (*increases, decreases, no change?*). Be able to defend your answer.

blood flow through the kidneys? _____ mean arterial pressure? _____

blood flow through skeletal muscle arterioles? _____

cardiac output? _____ total peripheral resistance? _____

blood flow through the venae cavae? _____ through the lungs? _____

9. True/False? Explain. In a fight-or-flight reaction, epinephrine from the adrenal cortex will combine with β_1 receptors in the heart and cause vasodilation.

10. You are a doctor and have just prescribed a calcium channel-blocking drug for a patient with high blood pressure. The patient asks how the drug works at a molecular and cellular level. How would you answer?

11.The figure below shows the cardiovascular center (CVCC) in the medulla oblongata, the heart, aortic arch, carotid artery, carotid sinus, an arteriole with a capillary bed, and the adrenal medulla. Cardiac output can be influenced by reflexes that alter heart rate, force of contraction, and peripheral resistance. Draw the anatomical components of the reflex pathways, including sensory receptors, sensory neurons, integrating centers, and efferent neurons that control cardiac function and peripheral blood pressure. Use different colors to represent different parts of the system. Where neurons terminate on targets, write in the appropriate neurotransmitters and receptors.

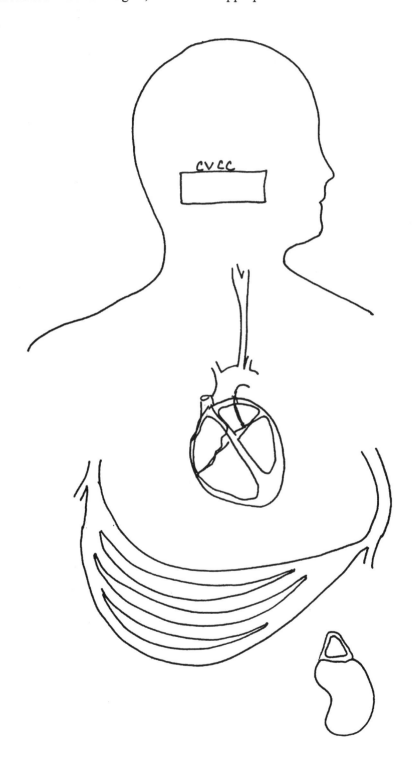

CVCC

12. Fill in the following reflex pathway.

Stimulus: Increased blood pressure

Receptor(s): _____

Afferent path: _____

Integrating center: _____

Efferent pathway 1: _____

Effector(s) 1: _____

Tissue/organ response(s): _____

Efferent pathway 2: _____

Effector(s) 2: _____

Tissue/organ response(s) 2: _____

Systemic response: ___decreased blood pressure_____

13. A man has developed thromboangiitis obliterans, a condition in which the arteries in his legs (only) have become calcified and partially obstructed. He comes to his physician complaining of pain when walking; the pain subsides when he stops walking and rests. The man's blood pressure, taken in his left arm, is normal.

a) What is the blood flow in his legs compared to normal? On the basis of this answer, explain why he has pain when walking but not at rest.

Explain what would happen to blood flow in this man's legs if you:

b) administer a peripheral vasodilator?

c) cut the sympathetic nerves innervating the blood vessels in his legs?

MAPS

1. Use the following terms and any you wish to add to create a map illustrating how atherosclerosis develops and how it affects cardiovascular function.

atherosclerosis	endothelium	peripheral resistance
baroreceptors activity (adaptation)	hypertension	plaque
blood flow	LDL-cholesterol	smooth muscle
blood vessel radius	macrophage	thrombus
cardiovascular control center	myocardial infarction	

2. A 22-year-old male was brought to the emergency room following a major automobile accident. His blood pressure was 80/25 and pulse was 135/min. His hands and feet were cold to the touch. Tests showed the presence of intra-abdominal bleeding. Draw a **COMPLETE reflex map** that explains all of this patient's physical findings, beginning with bleeding as the stimulus. Use the following terms and add any additional terms you wish to add.

α receptor	carotid artery	parasympathetic activity
β_1 receptor	decreased blood pressure	peripheral resistance
β_2 receptor	decreased blood volume	sensory neuron
acetylcholine	end-diastolic volume	stroke volume
aorta	heart rate	sympathetic activity
baroreceptor	hemorrhage	vasoconstriction
cardiac output	increased blood pressure	venous return
cardiovascular control center	norepinephrine	

BEYOND THE BASICS

TRY IT

Reactive Hyperemia

Reactive hyperemia, a temporary increase in blood flow into a tissue that has been deprived of flow, can be easily demonstrated without any equipment. Wrap the fingers of your right hand around the base of your left index finger. Squeeze tightly for at least one minute to shut off blood flow into the finger. It is important to shut off as much arterial flow as you can. After one minute, release the finger and watch for color changes. It should flush red for a few seconds, then gradually fade to normal color as the vasodilators are washed away by the restored blood flow.

The Baroreceptor Reflex

You can demonstrate the baroreceptor reflex easily with a friend. Find the subject's pulse at the radial artery of the wrist. While monitoring the pulse, have the subject find the pulse point in the carotid artery (just to the side of the Adam's apple) and press gently on it for a few seconds. You should notice a decrease in the subject's pulse as the increased pressure created by pressing on the carotid artery is sensed by the cardiovascular control center. The baroreceptor reflex is one reason that taking a carotid pulse in an exercise class is not the most accurate indicator of heart rate.

READING

About benefits and costs: Cholesterol screening. (1996, July/August) *Science & Medicine*.

About benefits and costs: Prevention of cardiovascular disease. (1994, July/August) *Science & Medicine*.

Angiotensin II and baroreflex control of heart rate. (1996, December) *News in Physiological Sciences* 11: 270-273.

Arterial plaque and thrombus formation. (1994, September/October) *Science & Medicine*.

Baroreflex abnormalities in congestive heart failure. (1993, April) *News in Physiological Sciences* 8: 87-90.

Benefits of lower cholesterol. (1994, May/June) *Science & Medicine*.

Cardiovascular effects of alcohol. (1995, March/April) *Science & Medicine*.

Cardiovascular effects of endothelin. (1997, June) *News in Physiological Sciences* 12: 113-116.

Endothelial cells as mechanoreceptors. (1993, April) *News in Physiological Sciences* 8: 55-56.

Endothelin: Receptors and transmembrane signals. (1992, October) *News in Physiological Sciences* 7: 207-211.

Excimer laser coronary angioplasty. (1996, January/February) *Science & Medicine.*

The heart as an endocrine gland. (1986, February) *Scientific American.*

How LDL receptors influence cholesterol and atherosclerosis. (1984, November) *Scientific American.*

Integrins and control of interstitial fluid pressure. (1997, February) *News in Physiological Sciences* 12: 42-48.

Intravascular ultrasound. (1995, September/October) *Science & Medicine.*

Lipoproteins in heart disease. (1992, June) *Scientific American.*

Magnesium in cardiovascular biology. (1995, May/June) *Science & Medicine.*

Mechanisms of obesity-induced hypertension. (1996, December) *News in Physiological Sciences* 11: 255-261.

Role of the lymph pump and its control. (1995, June) *News in Physiological Sciences* 10: 112-116.

Three-dimensional arterial imaging. (1996, March/April) *Science & Medicine.*

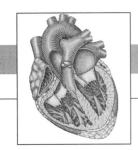

SUMMARY

What do you need to know about blood?

- Blood is composed of plasma, red blood cells (RBCs), white blood cells (WBCs), and platelets.
- All blood cells are descendants of multipotent progenitor cells, and the production of blood cells (hematopoiesis) is under chemical control.
- RBCs transport oxygen (O_2) from lungs to tissues and carbon dioxide (CO_2) from tissues to lungs.
- Hemoglobin (Hb) is the O_2-binding portion of the RBC.
- Hemostasis is achieved by activated platelets and the coagulation cascade.

This chapter gives some detail about our blood, its components and their functions. To help you organize these details, consider making a large map where the main headings are the components of blood (i.e., plasma, RBCs, WBCs, platelets).

Plasma resembles interstitial fluid except that it contains plasma proteins. Plasma proteins include albumins, globulins, fibrinogen, and immunoglobulins. Plasma proteins raise the osmotic pressure of plasma, thus pulling water from the interstitial fluid into the capillaries [∫ p. 439].

Red blood cells (RBC, erythrocytes) transport O_2 and CO_2 between the lungs and tissues with the help of hemoglobin (Hb). Hb production requires that iron be ingested in the diet. Mature RBCs are biconcave disks that lack a nucleus and membranous organelles. Thus, they are essentially membranous "bags" filled with enzymes and Hb. Because RBCs lack a nucleus, they cannot make new proteins or membrane molecules and therefore die after about 120 days. Their flexible membrane and flattened shape allow RBCs to change shape in response to osmotic changes.

White blood cells (WBC, leukocytes) consist of five mature cell types: lymphocytes, monocytes, neutrophils, eosinophils, and basophils. Neutrophils and monocytes (macrophage precursors) are collectively called phagocytes because of their ability to ingest foreign particles. Basophils, eosinophils, and neutrophils are called granulocytes because of cytoplasmic inclusions that give a granular appearance.

Platelets are anuclear cell fragments that have broken off from megakaryocytes in the bone marrow. Platelets are always present in the blood, but they are inactive unless there is damage to the circulatory system. Ruptured blood vessels expose collagen fibers that activate platelets, and the coagulation cascade is initiated to achieve hemostasis.

All blood cells are derived from multipotent progenitor cells that develop according to chemical signals into the mature cells already discussed. These chemical signals include cytokines [∫ p. 148], growth factors, and interleukins. Table 16-2 (p. 459) lists some of these chemicals and how they affect blood cell development. Other important examples include erythropoietin, which influences erythropoiesis; colony-stimulating factors, which influence leukopoiesis; and thrombopoietin, which influences platelet production.

TEACH YOURSELF THE BASICS

PLASMA AND THE CELLULAR ELEMENTS OF BLOOD

Plasma Is Composed of Water, Ions, Organic Molecules, and Dissolved Gases

• What is plasma? _____

• Describe the composition of plasma. _____

_____

• Compare plasma with interstitial fluid. _____

• Where are most plasma proteins made? _____

• Name three main groups of plasma proteins and give their functions. (Table 16-1)

_____

_____

• What role do plasma proteins play in capillary filtration? [∫ p. 439] _____

_____

The Cellular Elements Include Red Blood Cells, White Blood Cells, and Platelets

• List the three main cellular elements of the blood and describe their primary function(s).

_____

• Why do we call them ìcellular elementsî rather than "cells"? _____

_____

• What are the formal names for red blood cells and white blood cells?

_____

• What are the parent cells of platelets called? _____

• List the five mature WBCs found in blood and give the function(s) of each type.

_____

_____

_____

• Which of the WBCs are known as phagocytes, and why are they given that name? [∫ p. 126]

_____

• Why are lymphocytes also called immunocytes? _____

• Which WBCs are called granulocytes and why?

_____

BLOOD CELL PRODUCTION

Blood Cells Are Produced in the Bone Marrow

• What is the multipotent progenitor? (Fig. 16-3) _____

• What do multipotent progenitors develop into? _____

• Define hematopoiesis. Where does it take place in embryos, children, and adults?

_____

_____

• Describe red bone marrow. _____

• Describe yellow bone marrow. _____

• Of all blood cells produced, _____% will become RBCs and _____% will become WBCs.

• Compare the lifespans of RBCs and WBCs. _____

Hematopoiesis Is Controlled by Cytokines, Growth Factors, and Interleukins

• What are cytokines? [∫ p. 148] _____

• What role do cytokines play in hematopoiesis? _____

_____

• What are colony-stimulating factors (CSFs) and how did they get their name? _____

_____

• What is an interleukin? _____

☞ _Table 16-2 lists some cytokines involved in hematopoiesis_

Colony-Stimulating Factors Regulate Leukopoiesis

• Where are the CSFs that regulate reproduction and development of WBCs made?

_____

• What is leukemia? _____

• What is neutropenia? _____

Thrombopoietin Regulates Platelet Production

• What is thrombopoietin (TPO) and where is it made?

_____

Erythropoietin Regulates Red Blood Cell Production

• What is erythropoietin (EPO) and where is it made?

_____

• What is the stimulus for EPO synthesis and release? _____

RED BLOOD CELLS

• What is the red blood cell count of a microliter (μL) of whole blood? _____

• What is the hematocrit? (Fig. 16-2) _____

Mature Red Blood Cells Lack a Nucleus

• Name two immature forms of erythrocyte. How are they different from the mature RBC?

_____

_____

• Describe the structure and contents of the mature RBC. (Fig. 16-6a)

_____

• What is hemoglobin? _____

• Without mitochondria, RBCs cannot carry out (anaerobic / aerobic?) metabolism.

• What is the primary energy source for mature RBCs? _____

• How is the lack of a nucleus related to the limited life span of the RBC?

_____

• What holds the RBC in its unique shape? (Fig. 16-6) _____

• When placed in a hypotonic solution, a cell will do what? [∫ p. 134] _____

Hemoglobin Synthesis Requires Iron

• Describe the structure of hemoglobin (Hb). (Fig. 16-7) _____

_____

• Why must we have adequate iron in the diet in order to make hemoglobin? _____

_____

• How does fetal Hb differ structurally from adult hemoglobin?

_____

• What are the differences in function and location of transferrin and ferritin? _____

_____

• Where is heme made? _____

Red Blood Cells Live about Four Months

• What is the average life span of an RBC? _____

• How are old RBCs destroyed and what happens to the RBC components? _____

_____

• What is the relationship between heme, bilirubin, and bile? _____

_____

• How are bilirubin and its metabolites excreted? _____

• What is jaundice? _____

Red Blood Cell Disorders Decrease Oxygen Transport

• List four common causes of anemia. (Table 16-3) _____

_____

_____

• Why are people with anemia often weak and fatigued? _____

_____

• Anemias in which the RBCs are destroyed at a high rate are called _____ anemias.

• Lack of adequate iron in the diet results in _____ anemia.

• Smaller-than-normal RBCs are said to be _____, while paler-than-normal

  RBCs are said to be _____.

• What makes the hemoglobin of sickle cell disease abnormal? _____

_____

• In polycythemia vera, patients have higher than normal RBC production. Why is this harmful?
  [∫ p. 389]

_____

PLATELETS AND COAGULATION

• To repair broken blood vessels, blood flow through the vessel should (increase / decrease ?).

• What temporarily seals the hole in a broken blood vessel until it can be repaired? _____

Platelets Are Small Fragments of Cells

• Describe how platelets are formed from megakaryocytes. _____

_____

• What do platelets contain? _____

• What is the typical lifespan of a platelet? _____

• How do megakaryocytes get multiple nuclei? _____

• What prevents platelets from constantly forming clots throughout the circulatory system?

_____

Hemostasis Prevents Blood Loss from Damaged Blood Vessels

• Define hemostasis. _____

_____

• What is the difference between platelet adhesion and platelet aggregation?

_____

• Give the six main steps that take place when a blood vessel breaks and is repaired. (Fig. 16-10)

1. _____

2. _____

3. _____

4. _____

5. _____

6. _____

• Explain the relationship between thrombin, fibrin, fibrinogen, plasmin, coagulation, and fibrinolysis.

_____

_____

Platelet Aggregation Begins the Clotting Process

• Describe how platelets are activated. (Fig. 16-11)

_____

_____

• Activated platelets release chemicals that activate additional platelets. This is an example of what

kind of pathway? _____

• How is formation of the platelet plug restricted to the damaged region? _____

_____

Coagulation Converts the Platelet Plug into a More Stable Clot

• Define and describe coagulation. _____

_____

• How does initiation of the intrinsic pathway differ from that of the extrinsic pathway? (Fig. 16-12)

_____

• The intrinsic and extrinsic pathways unite at the common pathway to initiate _____ formation.

• Why is it necessary to cross-link fibrin? _____

Anticoagulants Prevent Coagulation

• What are anticoagulants? _____

• Describe two ways anticoagulants can work. _____

_____

TALK THE TALK

acetylsalicylic acid (aspirin)
adipocyte
albumin
anemia
anticoagulant
basophil
bile
bilirubin
blood
blood cell culture
blood sinus
bone marrow
clot
coagulation
coagulation cascade
collagen
colony-stimulating factor (CSF)
cytokine
cytoskeleton
differential white cell count
endothelium
eosinophil
erythroblast
erythrocyte
erythropoiesis
erythropoietin (EPO)
extracellular matrix
extrinsic path
factor
ferritin

fetal hemoglobin
fibrin
fibrinogen
fibrinolysis
fibroblast
globulin
granulocyte
Harvey, William
hematocrit
hematopoiesis
heme group
hemoglobin
hemolytic anemia
hemophilia
hemostasis
hereditary spherocytosis
hydroxyurea
hypertonic
hypochromic
hypotonic
immunocyte
immunoglobulin
infarct
interleukin
intrinsic path
iron-deficiency anemia
jaundice
leukemia
leukocyte
lymphocyte

macrophage
megakaryocyte
microcytic
monocyte
multipotent progenitor
neutropenia
neutrophil
nitric oxide
phagocyte
phagocytosis
plasma
plasma protein
plasmin
platelet
platelet adhesion
platelet aggregation
platelet plug
polycythemia vera
porphyrin ring
prostacyclin
recombinant DNA technology
reticular fiber
reticulocyte
sickle cell disease
stem cell
streptokinase
stroma
thrombin
thrombocyte
thrombopoietin

thrombus   transferrin   vasoconstriction
tissue factor   umbilical cord blood   venesection
tissue plasminogen activator   urokinase
   (TPA)

PRACTICE MAKES PERFECT

1. Describe two different major functions of blood.

2. Give at least two characteristics that are used to identify different types of leukocytes.

3. A blood sample from a patient shows normal white cell count, low red cell count, and more than normal reticulocytes present in the blood. Would you suspect that this personís problem is a result of a defect in the bone marrow or a problem with the circulating red blood cells? Defend your choice.

4. A woman has a total blood volume of 4.8 L and a hematocrit of 40%. What is her plasma volume?

5. Which ONE of the following is NOT TRUE?

   Neutrophils, eosinophils, and basophils:
   a) are leukocytes
   b) are lymphocytes
   c) are granulocytes
   d) are polymorphonuclear (have bi-lobed or tri-lobed nucleus)

6. Erythropoietin (EPO) is traditionally considered to be a hormone. Based on what you have learned about EPO, is it most accurately classified as a hormone or as a cytokine?

MAPS

1. Create a map showing hemostasis and coagulation. Include the terms below and any others you wish to add.

blood	extracellular matrix	intrinsic path	thrombin
clot	extrinsic path	plasmin	thrombocyte
coagulation	factor	platelet	thrombus
coagulation cascade	fibrin	platelet adhesion	tissue factor
collagen	fibrinogen	platelet aggregation	tissue plasminogen
cytokine	fibrinolysis	platelet plug	activator (TPA)
endothelium	hemostasis	prostacyclin	vasoconstriction

2. Create a map of the blood cells, their synthesis, and their basic functions, using the terms below and any others you wish to add.

basophil	erythrocyte	leukocyte	phagocytosis
blood	erythropoiesis	lymphocyte	platelet
bone marrow	erythropoietin (EPO)	macrophage	reticulocyte
colony-stimulating factor (CSF)	granulocyte	megakaryocyte	stem cell
	hematopoiesis	monocyte	stroma
cytokine	immunocyte	multipotent progenitor	thrombocyte
eosinophil	immunoglobulin	neutrophil	thrombopoietin
erythroblast	interleukin	phagocyte	umbilical cord blood

3. Create a map for red blood cell synthesis and destruction, using the terms below and any others you wish to add.

amino acids	erythrocyte	fetal hemoglobin	iron
bile	erythropoiesis	hematopoiesis	jaundice
bilirubin	erythropoietin (EPO)	heme group	protein
erythroblast	ferritin	hemoglobin	transferrin

BEYOND THE BASICS

READING

Antiplatelet therapy. (1996, July/August) *Science & Medicine.*

Blood substitutes. (1997, March/April) *Science & Medicine.*

Coagulation factor XI: Possible solution to a clinical enigma. (1997, April) *News in Physiological Sciences* 12: 95-96.

Discovery of heparin: Contributions of William Henry Howell and Jay McLean. (1992, October) *News in Physiological Sciences* 7: 237-242.

Hematopoietic stem cell transplantation. (1995, September/October) *Science & Medicine.*

Hormones that stimulate the growth of blood cells. (1988, July) *Scientific American.*

The molecular genetics of hemophilia. (1986, March) *Scientific American.*

Molecular mechanisms of mural thrombosis under dynamic flow conditions. (1995, June) *News in Physiological Sciences* 10: 117-121. The process by which platelets adhere and aggregate in damaged blood vessels.

Neutrophil activation: Control of shape change, exocytosis, and respiratory burst. (1992, October) *News in Physiological Sciences* 7: 215-219.

Sickle cell anemia. (1994, September/October) *Science & Medicine.*

The stem cell. (1991, December) *Scientific American.*

Vessel wall response to injury. (1996, March/April) *Science & Medicine.*

SUMMARY

Many basic principles governing blood flow in the circulatory system apply to air flow in the respiratory system. Look for themes to be repeated in this chapter.

- Air movement during breathing occurs because of pressure gradients created by volume changes.
- Gas exchange between lungs and blood or blood and cells takes place because partial pressure gradients are created as the body consumes oxygen (O_2) and releases carbon dioxide (CO_2).
- Total cross-sectional area is greater for smaller airways than for larger airways.
- Hemoglobin (Hb) binds O_2 and transports it to cells. Oxygen transport is affected by Po_2, temperature, pH, and metabolites.
- The reaction $CO_2 + H_2O \leftrightarrow H^+ + HCO_3^-$ is an essential part of CO_2 transport in the blood.
- A central pattern generator in the brain stem is responsible for breathing patterns, but we can consciously control our breathing to some extent.

Cellular respiration, which you studied in Chapter 4, refers to the metabolic processes that consume oxygen and nutrients and produce energy and CO_2. External respiration is the exchange of gases between the atmosphere and the cells. Ventilation (breathing) is the process by which air is moved into and out of the lungs. O_2 is transported via the blood to cells, and CO_2 is removed by the blood and taken to the lungs.

Air movement in the respiratory system highlights its anatomy: air goes from the nasopharynx to the trachea to the bronchi, the bronchioles, and finally the alveoli. The lungs are contained within the thoracic cage. Each lung is surrounded by a double-walled pleural sac, and the pleural fluid that exists between the pleural membranes holds the lungs against the thoracic wall. This pleural fluid also helps the membranes slip past each other as the lungs move during respiration. The diaphragm, a sheet of skeletal muscle, forms the bottom of the thoracic cage, and its movement creates volume changes that in turn create air movement. The other muscles involved in respiration include the intercostal muscles, the scalenes, the sternocleidomastoids, and the abdominals.

Air moves according to a set of physical laws collectively known as the gas laws. Boyle's Law describes how, in a closed system, volume increases as pressure decreases (and vice versa). Air moves from areas of higher pressure to areas of lower pressure. Therefore, when the thoracic volume is increased by respiratory muscle movement, the pressure in the thoracic cavity drops and air moves in down its pressure gradient (inspiration). Likewise, as thoracic volume decreases, the intrapulmonary pressure increases and air moves out of the body (expiration). Movement of individual gases depends on their partial pressure gradients. Just as water and solutes move down their concentration gradients, so do gases. Dalton's Law describes how the total pressure of a gaseous mixture is the sum of the pressures of the individual gases. The pressure of an individual gas is called a partial pressure, and gases move from higher partial pressures to lower partial pressures. Gas exchange is the result of simple diffusion down partial pressure gradients. The exchange surface in the lungs is the exchange epithelium of the alveoli.

Blood flow and respiration are closely matched by a variety of mechanisms to ensure efficient delivery of gases. Hemoglobin (Hb) is the main facilitator of gas transport in the blood. Hb, a protein component of RBCs, consists of an iron molecule surrounded by a porphyrin ring embedded within a 4-subunit protein. Hb binds O_2 and CO_2 and will carry either until a partial pressure gradient causes the gas to be released. Hemoglobin-binding ability is affected by pH, temperature, and 2,3-DPG. The relationship between P_{O_2} and Hb binding is shown by a hemoglobin saturation curve. Only 23% of CO_2 is transported bound to Hb. About 70% of CO_2 in venous blood is converted to H^+ and HCO_3 inside the RBCs. Carbonic anhydrase is the enzyme that catalyzes this reaction: $CO_2 + H_2O \rightleftarrows H^+ + HCO_3^-$.

A network of neurons in the pons and medulla oblongata controls respiration. This network includes the dorsal respiratory group and the ventral respiratory group, and is called a central pattern generator because it has intrinsic rhythmic activity. Chemical factors affect respiration. Central chemoreceptors respond to an increase in H^+ due to elevated P_{CO_2} by increasing ventilation. Peripheral chemoreceptors monitor blood pH and P_{O_2}. We can control our respiration consciously, but chemoreceptors override conscious control.

Other respiratory functions include pH regulation, vocalization, and protection from foreign substances. Be sure you understand the graphs of this chapter. They will help you create a visual explanation of respiratory function. Also, be sure you understand the ways in which the cardiovascular system and respiratory system are integrated.

TEACH YOURSELF THE BASICS

• List four key functions of the respiratory system.

_____

_____

• What is lost from the body through the respiratory system besides carbon dioxide?

_____

THE RESPIRATORY SYSTEM

• Distinguish between cellular respiration [∫ p. 91] and external respiration. _____

_____

• List the four processes of external respiration (three exchanges and one transport).

_____

• Distinguish between inspiration and expiration. _____

_____

•Name the three major components of the respiratory system. (Fig. 17-2)

_____

• Name the structures of the upper and lower respiratory system. (☛ *Note: The text on p. 475 is correct; the box in the Anatomy Summary on p. 478 is wrong: the trachea is part of the upper respiratory system.*)

_____

The Bones and Muscles of the Thorax Surround the Lungs

• What bones and muscles form the walls of the thoracic cage? the floor? (Fig. 17-2a)

_____

_____

• Name the two sets of muscles that lift the sternum and first two ribs when they contract.

_____

• Name the three sacs enclosed within the thorax. What is in each sac? (Fig. 17-2d)

_____

• What thoracic structures are not contained within these three sacs? _____

The Lungs Are Enclosed in the Pleural Sacs

• The lungs are light, spongy tissue mostly occupied by _____-filled spaces. (Fig. 17-2b)

• What is the relationship of the lungs, the pleura, and the pleural fluid? (Fig. 17-3)

_____

• What is the function of pleural fluid? _____

The Airways of the Conducting System Connect the Lungs with the Environment

• Following an oxygen molecule from the air to the exchange epithelium of the lung, name each structure the molecule passes. (Fig. 17-2b, e)

_____

_____

• As the molecule moves into the airways, the diameter of the airways gets progressively smaller, and the total cross-sectional surface area of the airways (increases / decreases?). (Fig. 17-4)

• The velocity of air flow is highest in the _____ and lowest in the _____.[∫ p. 390]

The Alveoli Are the Site of Gas Exchange

• Describe the structure of the alveoli. (Fig. 17-2f, g) _____

_____

• Describe and give the functions of the two types of epithelial cells in alveoli.

_____

_____

• Describe the composition of the alveolar walls and surrounding tissues. _____

_____

• Describe the association of the alveoli and the circulatory system. (Fig. 17-2f)

_____

The Pulmonary Circulation Is a High-Flow, Low-Pressure System

• Trace a drop of blood through the pulmonary circulation from the (left/ right ?) ventricle to the (left / right?) atrium.

_____

• Compare the following aspects of the pulmonary circulation to those of the systemic circulation:

Volume of blood in the pulmonary vessels _____

(Of this volume, how much is participating in gas exchange at any moment? _____)

Total blood flow through the lungs in liters per minute _____

Blood flow per gram of tissue weight _____

Pulmonary arterial pressure _____

• Give two reasons that the right ventricle does not need to create as much force as the left ventricle.

_____

• What effect does lower mean pulmonary blood pressure have on capillary fluid exchange? [p. 439]

_____

GAS LAWS   ☛ *Gas laws are given in Table 17-1*

• Although we can draw many comparisons between air flow and blood flow, air and blood differ
  significantly in what way?

_____

• When we use an atmospheric pressure of 0 mm Hg, what is that value equivalent to? _____

Air Is a Mixture of Gases

• State Dalton's Law. _____

_____

☛ *Table 17-2 summarizes partial pressures of atmospheric gases.*

• How do you calculate the partial pressure of a single gas in a mixture?

_____

• What happens to the partial pressures of individual gases if dry air is suddenly humidified?

_____

Gases Move from Areas of Higher Pressure to Areas of Lower Pressure

• What force creates bulk flow of air? _____

• Air flows from regions of _____ to regions of _____.

• What role does muscle contraction play in the creation of air flow in the respiratory system?

_____

• What must be present for individual gases to move from one region to another?

_____

• This process is similar to what process you have already studied? _____

Pressure-Volume Relationships of Gases Are Described by Boyle's Law

• What factors contribute to gas pressure in a closed container? _____

_____

• State Boyle's Law. (Fig. 17-5) _____

• For gases in a closed container, as volume (increases / decreases?), pressure (increases / decreases?).

• How does the respiratory system create volume changes? _____

The Solubility of a Gas in a Liquid Depends on the Pressure and Solubility of the Gas and on the Temperature

• When a gas is placed in contact with a liquid, what three factors determine how much gas will dissolve in the liquid?

_____

• True or false? If a liquid is exposed to a P_{CO_2} of 100 mm Hg and a P_{O_2} of 100 mm Hg, equal amounts of oxygen and carbon dioxide will dissolve in the liquid.

• The more soluble a gas is, the (greater / less?) the partial pressure needed to force the gas into solution.

• Gases move between liquid and gaseous phases until _____ is reached. (Fig. 17-6)

• At equilibrium, the (concentration / partial pressure / both ?) of a gas will be equal in the air and gas phases.

• Which is more soluble in body fluids: oxygen or carbon dioxide?

VENTILATION

• Define ventilation. _____

The Airways Warm, Humidify, and Filter Inspired Air

• Describe three ways the upper airways condition air before it reaches the alveoli. _____

_____

_____

• What is the source of heat and water in the airways? _____

• Explain how the airways filter air. _____

_____

• What kind of cell secretes mucus? [∫ p. 60]_____

• What is the mucus elevator? _____

• Why don't most pathogens trapped in the airway mucus make you sick?

_____

• Why do children with cystic fibrosis have so many lung infections?

_____

During Ventilation, Air Flows because of Pressure Gradients

• Name the primary muscles involved in quiet breathing. _____

_____

• What other muscles assist with forced breathing? _____

• As the pressure gradient increases, air flow _____.

• As resistance of the airways increases, air flow _____.

• What are the differences between intrapleural pressure and intrapulmonary pressure?

_____

_____

• What is a respiratory cycle? _____

• Compare the direction of air flow in the respiratory system to the direction of blood flow in the circulatory system.

_____

Inspiration Occurs When Intrapulmonary Pressure Decreases

• What division of the nervous system controls the muscles of ventilation? What neurotransmitter and receptor does this division use?

_____

• During inspiration, the diaphragm contracts and moves in which direction? _____

The _____ intercostal muscles contract and move the rib cage in which direction?

_____ Which contributes more to inspiration? _____

• Between breaths, is there air flow? _____ Therefore, intrapulmonary pressure must be equal

to_____. (Fig. 17-10, point A_1)

• When thoracic volume increases during inspiration, what happens to intrapulmonary pressure?

_____

• When thoracic volume decreases during expiration, what happens to intrapulmonary pressure?

_____

• At what point in the respiratory cycle is intrapulmonary pressure lowest? _____

• At what point in the respiratory cycle is intrapulmonary pressure highest? _____

Expiration Occurs When Intrapulmonary Pressure Exceeds Atmospheric Pressure

• Are the following muscles contracting in passive expiration?

    External intercostal muscles _____ Internal intercostal muscles _____

    Diaphragm _____ Abdominal muscles _____

• Are the following muscles contracting in active expiration?

    External intercostal muscles _____ Internal intercostal muscles _____

    Diaphragm _____ Abdominal muscles _____

• What property of the respiratory system is responsible for passive expiration?

_____

• What happens to intrapulmonary pressure during expiration? _____

• What happens to air flow? _____

• Internal and external intercostals are _____ muscle groups. [∫ p. 326]

Intrapleural Pressure Changes during Ventilation

• Explain why the intrapleural pressure is always subatmospheric. (Fig. 17-11a)

_____

• Explain why puncturing the pleural membrane causes the lung to collapse and the rib cage to move out. (Fig. 17-11b)

_____

• True or false? At the end of inspiration, intrapleural pressure is the same as the pressure at the beginning of inspiration. Explain

_____

Lung Compliance and Elastance May Change in Disease States

• Define compliance. _____

• A high-compliance lung (requires additional force to stretch it / is easily stretched?).

• Define elastance. _____

• True or false? A high compliance lung always has high elastance. Explain.

_____

• What happens to compliance and elastance in emphysema?_____

• Diseases in which compliance is reduced are called _____ lung diseases.

Surfactant Decreases the Work of Breathing

• What creates resistance to stretch in the lung? _____

• State the Law of LaPlace and relate it to surface tension in the alveoli. _____

_____

• According to the Law of LaPlace, if two alveoli have equal surface tension, the (larger / smaller ?) will have a higher internal pressure. (Fig. 17-12)

• What is the function of surfactants? _____

• Which will have a higher concentration of surfactant, a large alveolus or a small one? Explain.

_____

• What cells in the lung secrete surfactant? _____

• What type of biomolecule are lung surfactants? _____

• What happens in premature babies who have not produced surfactant? _____

_____

Resistance of the Airways to Air Flow Is Determined Primarily by Airway Diameter

•Explain the relationship between resistance to air flow, length of airways (L), viscosity of air (η), and radius of airways (r), using Poiseuille's Law [∫ p. 389]. (Table 17-3)

_____

• In the respiratory system, which of these factors is usually the most significant? _____

• Where in the airways does air flow normally encounter the highest resistance? _____

• What part of the respiratory system is the site of variable resistance? _____

• If the bronchioles constrict, what happens to resistance and air flow into the alveoli?

_____

• Tell what role each of the following plays in control of bronchiolar diameter (bronchodilation or bronchoconstriction?)

CO_2 _____   Histamine [∫ p. 147] _____

Parasympathetic neurons _____   What neurotransmitter and receptor? _____

Epinephrine _____   Binds to what type of receptor? _____

Pulmonary Function Tests Measure Lung Volume during Ventilation

• What does a spirometer do? (Fig. 17-13) _____

Lung volumes
• Name, give the abbreviation for, and describe the four lung volumes. (Fig. 17-14)

1. _____

2. _____

3. _____

4. _____

• Which one or ones can be measured with a spirometer? _____

Lung capacities
• What are lung capacities? _____

• Name, give the abbreviation for, and describe the four lung capacities. (Fig. 17-14)

1. _____

2. _____

3. _____

4. _____

Auscultation of breath sounds
• What is auscultation? _____

The Effectiveness of Ventilation Is Determined by the Rate and Depth of Breathing

• Define total pulmonary ventilation in words and give the mathematical expression for it.

_____

_____

Give normal average values for breathing rate _____ and tidal volume _____ in an adult male.

• Using these values, what is an average value for total pulmonary ventilation? (Give units!)

_____

• Define anatomic dead space. (Fig. 17-15) _____

• How is alveolar ventilation different from total pulmonary ventilation?

_____

• Why is it a more accurate indicator of breathing efficiency?

_____

☛ *Table 17-4 compares alveolar ventilation differences with different breathing patterns. Table 17-5 describes various patterns of ventilation.*

Gas Composition in the Alveoli Varies Little during Normal Breathing

• What are the P_{O_2} and P_{CO_2} of the atmosphere and the alveoli during normal breathing?

P_{O_2} atmosphere _____ P_{O_2} alveoli _____

P_{CO_2} atmosphere _____ P_{CO_2} alveoli _____

• What happens to P_{O_2} and P_{CO_2} with increased alveolar ventilation? (Fig. 17-16)

_____

• What happens to P_{O_2} and P_{CO_2} with decreased alveolar ventilation? (Fig. 17-16)

_____

• During normal breathing, partial pressures in the alveoli remain constant. Why?

_____

Ventilation Is Matched to Alveolar Blood Flow

• An alveolus with a normal P_{O_2} is no guarantee that adequate oxygen will get to the cells. Name two other factors that must be normal in order for oxygen to be picked up by blood at the alveoli.

_____

• Explain what is meant by the expression "Ventilation is matched to perfusion in the lungs."

_____

• How are pulmonary capillaries different from other capillaries?

_____

• How does this property help the body meet a demand for additional oxygen, such as during exercise?

_____

_____

• Local homeostatic mechanisms attempt to keep ventilation and perfusion matched in each section of the lung. (Table 17-6)

When P_{CO_2} of expired air increases, bronchioles (dilate / constrict ?). When P_{CO_2} of expired air decreases, bronchioles (dilate / constrict ?).

When the tissue P_{O_2} around pulmonary arterioles decreases, the arterioles (dilate / constrict ?). When the tissue P_{O_2} around pulmonary arterioles increases, the arterioles (dilate / constrict ?).

• Compare this response of pulmonary arterioles to that of systemic arterioles. _____

_____

• Will these homeostatic responses always be able to correct the initial disturbance? Explain.

_____

GAS EXCHANGE IN THE LUNGS

• By what mechanism do gases move between the alveoli and the plasma? _____

• List the four rules for diffusion of gases. [∫ p. 117]

1. _____

2. _____

3. _____

4. _____

The Partial Pressure Gradient Is the Primary Factor Influencing Gas Exchange

• Give the following partial pressures in a normal person at sea level: (Fig. 17-18)

☞ Remember: Unless otherwise specified, the terms "arterial blood" and "venous blood" refer to blood in the systemic circulation.

P_{O_2}: Alveoli = _____ Arterial blood = _____ Resting cells = _____ Venous blood = _____

P_{CO_2}: Alveoli = _____ Arterial blood = _____ Resting cells = _____ Venous blood = _____

• What two cell layers must gases cross to go from the alveoli to the plasma?

_____

• If the alveolar P_{O_2} is 98 mm Hg, what will the plasma P_{O_2} be? _____

• If alveolar P_{O_2} is low, what two factors might have caused the decrease?

_____

• Explain the relationship between altitude and P_{O_2}. _____

_____

• List three common pathological changes that might cause a decrease in alveolar ventilation. (Fig. 17-19)

_____

Gas Exchange Can Be Affected by Changes in the Alveolar Membrane

• Describe two changes in the alveoli that might cause decreased oxygen exchange between the alveoli and the blood.

_____

_____

• Explain how emphysema can result in a loss of alveolar surface area.

_____

• Explain how fibrotic lung diseases can cause decreased oxygen exchange between the alveoli and the blood.

_____

• How much of the exchange epithelium must be incapacitated before arterial P_{O_2} drops? _____

• What is pulmonary edema and how does it alter gas exchange? _____

_____

• Explain why some patients with pulmonary edema have low arterial P_{O_2} but normal arterial P_{CO_2}.

_____

• Define hypoxia and hypercapnia. _____

_____

GAS EXCHANGE IN THE TISSUES

• What causes oxygen to move from plasma into cells and carbon dioxide to move from cells into plasma?

_____

GAS TRANSPORT IN THE BLOOD

• List two ways that gases are transported in the blood. _____

Hemoglobin Transports Most Oxygen to the Tissues

• Total blood oxygen content = _____ + _____.

• Only _____ mL of O_2 will dissolve in plasma of 1 L of blood, but _____ mL will be carried bound to hemoglobin. (Fig. 17-20)

• Why are oxygen-binding pigments such as hemoglobin essential for larger animals?

_____

• Compare the body's oxygen consumption at rest with the delivery of dissolved oxygen to the cells. Assume a cardiac output of 5 L/min.

_____

• How much additional oxygen per minute can be delivered by hemoglobin if each liter of blood carries 197 mL O_2 bound to hemoglobin?

_____

• The amount of O_2 bound to Hb depends on what two factors?

_____

• List three factors that establish the arterial P_{O_2}.

_____

• What determines the number of binding sites for oxygen? _____

• What is mean corpuscular Hb? _____

Each Hemoglobin Molecule Binds Up to Four Oxygen Molecules

• Hemoglobin molecules are composed of (how many?) _____ protein subunits, each with an O_2-binding _____ group. [∫ p. 460] This group is based around the element _____ that binds weakly to oxygen. Hb bound to O_2 is called _____ $(HbO_2)_{1-4}$.

• Trace the steps followed by an oxygen molecule as it goes from the alveoli to its binding site on hemoglobin.

_____

• As dissolved O_2 diffuses into RBCs, what happens to the P_{O_2} of the surrounding plasma?

_____

• Therefore, when O_2 binds with Hb, (more / less?) O_2 can diffuse from the alveoli into plasma.

• The amount of O_2 bound to Hb at any given PO_2 is shown as a percentage, called the

_____

• At 100% saturation, all possible binding sites are (bound / free ?).

• At the cells, dissolved O_2 in the plasma enters the cells because _____.

This disturbs the _____, so O_2 dissociates from hemoglobin, obeying

the Law of _____. [∫ p. 83] When equilibrium is restored, the P_{O_2} of the plasma

reflects the P_{O_2} of the _____.

The Oxygen-Hemoglobin Dissociation Curve Shows the Relationship between P_{O_2} and Hemoglobin Binding of Oxygen

• In the oxygen-hemoglobin dissociation curve (Fig. 17-21), the (P_{O_2} / percent saturation of Hb)

   determines the (P_{O_2} / percent saturation of Hb).

• Adaptively, why is it important that the slope of the curve flattens out at P_{O_2} values above 60 mm Hg?

_____

• Below P_{O_2} of 60 mm Hg, where the curve is steeper, small changes in P_{O_2} cause relatively (small / large?) releases of O_2 from hemoglobin.

• The arterial blood leaving the lungs is _____ saturated. The venous blood leaving a cell is 60% saturated. How much of the oxygen being transported by hemoglobin has been released? (Express as a percent of the maximum carrying capacity.)

_____

Temperature, pH, and Metabolites Affect Oxygen-Hemoglobin Binding

• Any factor that alters the hemoglobin protein may alter O_2-binding ability.

   An increase in pH (increases / decreases ?) hemoglobin's affinity for oxygen.

   An increase in temperature (increases / decreases ?) hemoglobin's affinity for oxygen.

   An increase in P_{CO_2} (increases / decreases ?) hemoglobin's affinity for oxygen.

   The metabolite 2,3-DPG (increases / decreases ?) hemoglobin's affinity for oxygen.

   Fetal Hb has a/an (increased / decreased ?) affinity for oxygen. (Fig. 17-22)

• A change in the O_2-binding affinity of hemoglobin is reflected by a shift in the O_2-Hb dissociation curve.

    A left shift in the curve indicates (increased / decreased ?) binding affinity.
    A right shift in the curve indicates (increased / decreased ?) binding affinity.

• Using your answers from the question above, tell which way the curve shifts with an increase in pH, P_{CO_2}, temperature, or 2,3-DPG (Fig. 17-23), or with fetal hemoglobin.

_____

• Why is it significant that a shift in the O_2-Hb dissociation curve is more pronounced at low P_{O_2} and lass pronounced at higher P_{O_2}?

_____

• A right shift in the hemoglobin dissociation curve is also known as the _____ effect.

• Under what conditions is the cellular production of 2,3-DPG increased? _____

Carbon Dioxide Is Transported Dissolved in Plasma, Bound to Hemoglobin, and as Bicarbonate Ions

• Give two reasons why abnormally elevated P_{CO_2} can be toxic. _____

_____

• Explain the three ways that CO_2 is transported in the blood. (Fig. 17-25)

_____

_____

• The name for hemoglobin bound to CO_2 is _____.

• Write the equation in which CO_2 is converted in bicarbonate ion ($HCO3^-$) and H^+.

_____

• What enzyme catalyzes this reaction and where in the blood is it found? _____

• Explain why CO_2 forms HCO_3^- and H^+ in the systemic capillaries, but HCO_3^- and H^+ form CO_2

in pulmonary capillaries. (Fig. 17-25) _____

_____

• What is a buffer? [∫ p. 26] _____

• Name two buffers found in the blood. _____

• What is the chloride shift and why does it occur? _____

_____

• Can bicarbonate leave the RBC by simple diffusion? Explain. _____

_____

• Constant removal of CO_2 from plasma (increases / decreases ?) P_{CO_2} and allows (more / less ?) CO_2 to leave cells.

• How does CO_2 move from cells to plasma and from plasma into the alveoli? _____

Summary of Gas Transport

☞ *O_2 and CO_2 transport are summarized in Fig. 17-26.*

REGULATION OF VENTILATION

• Compare the rhythmicity and control of breathing to that of the heart beat. _____

_____

_____

• Compare the types of efferent neurons leaving the respiratory control center and the cardiovascular control center.

_____

_____

• What is a central pattern generator? [∫ p. 374] _____

• What role does each of the following play in the control of breathing?

medulla oblongata _____

pons _____

cerebrum _____

Respiratory Neurons in the Medulla Control Inspiration and Expiration

• Compare the functions of the dorsal and ventral respiratory groups of neurons in the medulla. (Fig. 17-27)

_____

_____

• Which group is active during forced expiration? _____

Carbon Dioxide, Oxygen, and pH Influence Ventilation

• List three chemical factors that affect ventilation and tell where the sensory receptors are located for each.

_____

_____

• What is the primary chemical stimulus for changes in ventilation? _____

• Using the oxygen-hemoglobin dissociation curve in Fig. 17-21 (p. 502), explain the adaptive significance of the fact that the peripheral chemoreceptors do not respond to decreases in P_{O_2} until the P_{O_2} drops below 60 mm Hg.

_____

_____

• Explain the strategic significance of the location of the peripheral chemoreceptors.

_____

• Explain how the central chemoreceptors respond to elevated blood P_{CO_2}.

_____

_____

The carotid and aortic bodies (peripheral receptors)

• To what chemical signals do the carotid and aortic bodies respond?

_____

• Where do the sensory neurons leading from these receptors send their signals? _____

• A decrease in pH will trigger a/an (decrease / increase ?) in ventilation.

• A decrease in arterial P_{O_2} below _____ mm Hg will trigger a/an (decrease / increase ?) in ventilation.

• An increase in P_{CO_2} will cause a/an (decrease / increase ?) in pH, which in turn will trigger a/an

(decrease / increase ?) in ventilation.

Central chemoreceptors
• Central H+ receptors in the medulla mediate ventilation changes in response to _____. (Fig. 17-30)

• A decrease in P_{CO_2} will trigger a/an (decrease / increase ?) in ventilation.

• If P_{CO_2} is chronically elevated, the sensory receptors will _____ and the ventilation rates will (increase / decrease ?) compared to normal.

• If a person has chronic hypercapnia and hypoxia, is CO_2 the primary chemical drive for

  ventilation? _____

• What will happen to ventilation if this person is given pure O_2 to breathe? Explain.

_____

Mechanoreceptor Reflexes Protect the Lungs from Inhaled Irritants

• Write the reflex response to an inhaled irritant.

  Receptor _____

  Afferent pathway _____

  Integrating center _____

  Efferent pathway (include chemicals and receptors) _____

  Effector _____

  Tissue response _____

  Systemic response _____

• Describe the Hering-Breuer inflation reflex. _____

_____

Higher Brain Centers Affect Patterns of Ventilation

• Give two examples of how higher brain centers can alter ventilation.

_____

_____

TALK THE TALK

2,3-diphosphoglycerate
(2,3-DPG)
acidosis
active expiration
aerobic metabolism
airway
alveolar macrophage
alveolar ventilation
alveoli
anatomic dead space
angiotensin
antagonistic muscle groups
asbestos
atmospheric pressure
auscultation of breath sounds
beta2 receptors, bronchioles
blood substitutes
blood-brain barrier
Bohr effect
Boyle's Law
bronchiole
bronchoconstriction
bronchodilation
buffer
carbaminohemoglobin
carbon dioxide
carbonic anhydrase
cardiac output
carotid and aortic bodies
cellular respiration
central chemoreceptor
central pattern generator
cerebrospinal fluid
chemical control of ventilation
chest wall
chloride shift
chronic hypoxia
chronic obstructive pulmonary
disease
ciliated epithelium
collagen
compliance
conditioning of inspired air
cross-sectional area, airway
cystic fibrosis

Dalton's Law
dehydration
diaphragm
dipalmitoylphosphatidylcholine
dopamine
dorsal respiratory group
elastance
elastin
emphysema
endothelium
exchange epithelium
expiration
expiratory reserve volume
external respiration
fetal hemoglobin
fibroblast
fibrotic lung disease
forced breathing
friction rub
functional residual capacity
gas exchange
glomus cell
goblet cell
growth factor
hemoglobin
Hering-Breuer inflation reflex
high-altitude acclimatization
histamine
humidity
hypercapnia
hyperventilation
hypoventilation
hypoxia
inspiration
inspiratory capacity
inspiratory muscles
inspiratory neuron
inspiratory reserve volume
intercostal muscle
intrapleural pressure
intrapulmonary pressure
irritant receptor
larynx
Law of LaPlace
lower respiratory tract

lung
lung capacities
lung volumes
maximum voluntary ventilation
medulla oblongata
mucus escalator
mucus layer
myasthenia gravis
obstructive lung disease
oxygen
oxygen consumption
oxygen-hemoglobin dissociation
curve
oxyhemoglobin
parasympathetic innervation of
bronchioles
partial pressure
passive expiration
percent saturation of hemoglobin
perfusion, lung
peripheral chemoreceptor
pH regulation
pharynx
pleural fluid
pleural membrane
pleural sac
pneumothorax
Poiseuille's Law
polio
pons
primary bronchi
protective reflex
pulmonary arterial blood pressure
pulmonary arteriole
pulmonary circulation
pulmonary edema
pulmonary function test
pulse oximeter
residual volume
resistance, pulmonary circulation
resistance, to air flow
respiratory cycle
respiratory distress syndrome
respiratory system
restrictive lung disease

scalene muscle	surfactant	upper respiratory tract
secondary bronchi	thoracic cage	velocity of air flow
simple diffusion	thorax	ventilation
solubility	tidal volume (VT)	ventral respiratory group
solubility, carbon dioxide	total lung capacity	viscosity
solubility, oxygen	total pulmonary ventilation	vital capacity
spirometer	trachea	vocal cords
sternocleidomastoid muscle	type I alveolar cell	
surface tension	type II alveolar cell	

ERRATA

p. 479, part (g): alveoli misspelled in caption.

p. 504, High Altitude box: The correct abbreviation for meters is a lower case "m."

p. 505, Fig. 17-25: In the systemic capillary on the left, the arrowhead showing Cl coming into the RBC is missing. In the pulmonary capillary, the dot for the transporter should be on the membrane, not in the cytoplasm.

p. 752, answer to concept check question on p. 496: When bronchioles constrict, they shunt AIR, not blood.

QUANTITATIVE THINKING

1. The diagram below represents a spirometer tracing.

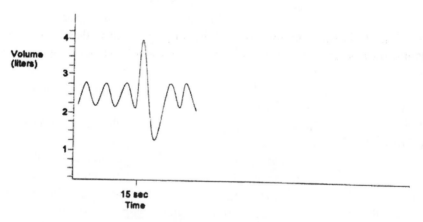

Using the tracing above, calculate the following:

Total lung capacity _____   Vital capacity _____

Expiratory reserve volume _____   Residual volume _____

Total pulmonary ventilation, normal breathing (L/min) _____

2. If atmospheric pressure is 720 mm Hg and nitrogen is 78% of the atmosphere, what is the partial pressure of nitrogen?

3. A student breathes according to the following schedule (assume an anatomical dead space of 150 mL):
     tidal volume = 300 mL / breath   breath rate = 20 breaths / min
Calculate her pulmonary ventilation rate and her alveolar ventilation rate.

4. Patient A is breathing 12 times a minute with a tidal volume of 500 mL. Patient B is breathing 20 times a minute with a tidal volume of 300 mL. Which patient has better alveolar ventilation? Explain.

5. You are given the following data on a person:
     A. Arterial plasma P_{O_2} = 95 mm Hg

     B. Blood volume = 4.2 liters

     C. Hematocrit = 38%

     D. Hemoglobin concentration = 13 g/dL whole blood

     E. Maximum oxygen-carrying capacity of hemoglobin = 1.34 ml oxygen/g hemoglobin

     F. At a P_{O_2} of 95 mm Hg, plasma contains 0.3 mL oxygen per deciliter (dL) and hemoglobin
        is 97% saturated

Using the data above, calculate the total amount of oxygen that could be carried in the person's blood.

HINTS:

Total blood oxygen = amount dissolved in plasma + amount bound to hemoglobin
To determine amount of oxygen dissolved in plasma:
     What is this person's plasma volume? (use hematocrit to determine)
     What is the solubility of oxygen in plasma?
To determine the amount of oxygen bound to hemoglobin:
     How much hemoglobin is in this person's blood?
You are given total blood volume, hematocrit, and hemoglobin content/dL whole blood. Which of
     these parameters will you use?
Maximum oxygen-carrying capacity represents what percent saturation?
What is the percent saturation in this person's blood?

6. During exercise, a man consumes 1.8 L of oxygen per minute. His arterial oxygen content is 190 mL/L and the oxygen content of his venous blood is 134 mL/L. What is his cardiac output?

TEACH YOURSELF THE BASICS

1. Gas concentrations in air and body fluids, in mm Hg. Fill in this table.

	Atmospheric air	Alveolar air	Arterial blood	Venous blood	Cells, resting	Expired air
Oxygen						
Carbon dioxide						

2. Spell out the words for the following abbreviations:

2,3-DPG _____ V_T_____

P_{O_2}_____   RV_____

IRV_____   Hb _____

In the following questions, mark each answer as either true or false.

3. Oxygen
      a. is mainly transported in the blood while bound to the hemoglobin in red blood cells. _____
      b. is as soluble as carbon dioxide in plasma. _____
      c. is the primary chemical drive for ventilation. _____

4. Carbon dioxide
      a. is primarily transported as a gas dissolved in the plasma. _____
      b. binds to hemoglobin in erythrocytes. _____
      c. is converted to carbonic acid through the action of carbonic anhydrase. _____

5. The P_{O_2} of the blood

        a. is a measure of the amount of oxygen dissolved in the plasma. _____

        b. is the most important factor determining the percent saturation of hemoglobin. _____

        c. is normal in anemia. _____

        d. is an accurate indicator of the total oxygen content of blood. _____

        e. determines P_{O_2} of the alveoli. _____

<u>Pick the single best answer:</u>

6. Which of the following would decrease the ability of oxygen to diffuse across the alveolar/capillary membrane? (circle all that are correct)

        a. An increase in thickness of the alveolar membrane

        b. Increased hemoglobin concentration in erythrocytes

        c. An increase in the partial pressure of oxygen in the alveoli

        d. CNS depression by drugs or alcohol

        e. A decrease in the surface area of the alveoli

7. Compare the following pairs of items. Choose the correct symbol to put below in the space.

    greater than >   less than <   same as or equal =

A - intrapleural pressure at the end of expiration
B - intra-alveolar pressure at the end of inspiration   A _____ B

A- oxygen released from hemoglobin at a cell whose P_{O_2} is 40 mm Hg when the plasma is at pH 7.4

B - oxygen released from hemoglobin at a cell whose P_{O_2} is 40 mm Hg when the plasma is at pH 7.2   A _____ B

A - arterial oxygen transport in a person with a hemoglobin of 10 g Hb/dL blood and P_{O_2} of 140 mm Hg
B - arterial oxygen transport in a person with a hemoglobin of 11 g Hb/dL blood and P_{O_2} of 100 mm Hg   A _____ B

A -blood flow in peripheral arterioles when surrounding tissue P_{O_2} is 70 mm Hg
B -blood flow in pulmonary arterioles when interstitial P_{O_2} is 70 mm Hg   A _____ B

A - arterial P_{O_2} in a person with anemia
B - arterial P_{O_2} in a normal person   A _____ B

A - resistance to air flow in the bronchioles
B - resistance to air flow in the trachea   A _____ B

A - compliance in alveoli with surfactant
B - compliance in alveoli without surfactant   A _____ B

8. Match the neurotransmitter/neurohormone and receptor with its target(s). Answers may be used more than once and more than one answer may apply to any target.

A - norepi on α receptors   B - ACh on nicotinic receptors   C - epi on β_2 receptors

D - ACh on muscarinic receptors   E - norepi on β_1

diaphragm _____   bronchioles _____   external intercostals _____

9. Blood flow to a small region of lung is blocked due to a blood clot in a small pulmonary artery.

a) What happens to the P_{O_2} and P_{CO_2} of the alveoli that are associated with that artery?

b) What happens to the tissue P_{O_2} and P_{CO_2} around the arterioles distal to the blockage?

c) What is the response of the bronchioles and the arterioles in this region? Will either or both of these responses be effective in compensating for the blocked artery? Explain.

10. Peter Premed inhales some oxygen molecules that are transported directly to a liver cell where they undergo aerobic metabolism and are converted to carbon dioxide molecules. In detail, anatomically trace the oxygen molecules from the air to the liver. Then trace the carbon dioxide molecules from the liver to the atmosphere. Use words connected by arrows, not sentences.

11. You are an astronomer who has been invited with colleagues to work for a week at the observatory on the summit of Mauna Kea on the big island of Hawaii. The summit of this extinct volcano is 13,796 feet above sea level. Describe how each of the following parameters will change by the end of your journey to the summit, and explain the stimulus and pathway for each change.

a) Partial pressure of oxygen in the air _____

b) Barometric pressure _____

c) Arterial P_{O_2} _____

d) Arterial P_{CO_2} _____

e) Arterial pH _____

f) When you arrive at the observatory, you meet some resident astronomers who have been living there for years. How does the hemoglobin content of their blood compare with that of you and your colleagues?

g) Within a day of arrival, one of your colleagues complains of having difficulty breathing and having a severe headache. The emergency oxygen tank has run out of oxygen. What should you do?

h) When you first arrive at the observatory, you notice that you begin breathing more rapidly. But within a day, your breathing gets even deeper although you would think that you are beginning to acclimate. How could you explain this phenomenon?

12. A woman hyperventilates. What effect will this have on the total oxygen content of her blood? Explain.

13. Alveolar air has an average P_{O_2} of 100 mm Hg but expired air has an average P_{O_2} of 120 mm Hg. If the lungs are taking oxygen into the body, why is there more oxygen in the expired air?

MAPS

1. Compile the following terms into a map of ventilation.

air flow

contract

diaphragm

expiratory muscles

external intercostals

forced breathing

in, out, from, to

inspiratory muscles

internal intercostals

P_{alv}

P_{atm}

$P_{intrapleural}$

quiet breathing

relax

scalenes

↑, ↓

greater than (>), less than (<)

2. Compile the following terms into a concept map showing the relationships between them. The major concept of your map is **total arterial O$_2$ content**. This should be your starting point.

2,3-DPG
adequate perfusion of alveoli
airway resistance
alveolar surface area
alveolar ventilation
amount of interstitial fluid
composition of inspired air

diffusion distance
dissolved in plasma
hemoglobin (Hb) content
lung compliance
membrane thickness
number (#) of Hb binding sites
number of RBCs

O$_2$ diffusion between
  alveoli and blood
pH
P$_{O_2}$
rate and/or depth of breathing
temperature

BEYOND THE PAGES

Normal values in pulmonary medicine

Lung volumes and capacities (liters)	Men	Women
Tidal volume	0.5	0.5
Inspiratory reserve volume	3.3	1.9
Expiratory reserve volume	1.0	0.7
Residual volume	1.2	1.1
Total lung capacity	6.0	4.2

Total pulmonary ventilation   6 L/min   Total alveolar ventilation   4.2 L/min
Max. voluntary ventilation   125-170 L/min   Respiration rate   12 -20 breaths/min

Blood gases
Arterial P_{O_2}- 95 mm Hg (85-100)*   Arterial P_{CO_2}- 40 mm Hg (37-43)   Arterial pH - 7.4
Venous P_{O_2}- 40 mm Hg   Venous P_{CO_2}- 46 mm Hg   Venous pH - 7.38

* Although we are considering arterial P_{O_2} to be equal to alveolar P_{O_2}, in reality the P_{O_2} drops slightly after leaving the pulmonary capillaries. This is because a small amount of deoxygenated venous blood from the non-exchange portions of the respiratory tract and from the coronary circulation combines with oxygenated blood as it returns to the left side of the heart. The actual arterial P_{O_2} value is closer to 95 mm Hg.

TRY IT

Lung volumes and capacities
Calculate your lung volumes and capacities using the table below. What will happen to your predicted vital capacity when you are 70 years old?

H = height in cm, A = age in years (Source: Medical Physiology Syllabus, Univ. Texas Medical Branch, Galveston)

Lung volume (L)	Subject	Formula
Vital capacity	Men	$(0.06 \times H) - (0.0214 \times A) - 4.65$
	Women	$(0.0491 \times H) - (0.0216 \times A) - 3.95$
Total lung capacity	Men	$(0.0795 \times H) + (0.0032 \times A) - 7.333$
	Women	$(0.059 \times H) - 4.537$
Functional residual capacity	Men	$(0.0472 \times H) + (0.009 \times A) - 5.29$
	Women	$(0.036 \times H) + (0.0031 \times A) - 3.182$
Residual volume	Men	$(0.0216 \times H) + (0.0207 \times A) - 2.84$
	Women	$(0.0197 \times H) + (0.0201 \times A) - 2.421$

Demonstrations for chemical control of ventilation

Using what you have learned about the chemical control of ventilation, first make these predictions.

In which case can you hold your breath the longest?
       a) after normal breathing
       b) after hyperventilating
       c) after breathing into a paper bag

Explain your reasoning: _____

_____

Will your breathing rate increase or decrease after doing the following? (circle one answer)
       a) hyperventilating   increase   decrease
       b) breathing into paper bag   increase   decrease

Explain your reasoning: _____

_____

You can hyperventilate by increasing breathing rate or by taking deeper breaths at your usual breathing rate.

After which type of hyperventilation, do you think that you will be able to hold your breath longer?
       a) increasing breathing rate
       b) taking deeper breaths at your usual breathing rate

Explain your reasoning: _____

_____

With a partner, do the following exercises. If you have time, repeat the sequence three times and take the average. Compare your results with your predictions above.

1. Normal breathing or **eupnea**
    Breathe normally. Have your partner count and record the number of breaths you take per minute for three one-minute intervals.

2. No breathing or **apnea**
    Breathe normally for several minutes. After a normal inspiration, time how long you can hold your breath.

3. **Hyperventilation**: increased ventilation with no change in metabolic rate
  a) Increased rate
      Breathe normal volumes rapidly for two minutes. Record the rate. Immediately after hyperventilation, hold your breath from the end of a normal inspiration for as long as possible. Time.

b) Increased volume

Breathe maximum volumes for two minutes, trying to keep the rate as close to your normal rate as possible. *If you get dizzy, STOP! Note the time and extrapolate to two minutes.* Record the rate.

Immediately after hyperventilation, hold your breath from the end of a normal inspiration for as long as possible. Time.

4. **Depressed blood carbon dioxide levels** (P_{CO_2})

Breathe deeply and rapidly for one minute. Stop sooner if you start to get dizzy. At the end of the one minute test, breathe naturally (normal volumes) for three to four minutes. Record the rate in the first 90 seconds of normal breathing. Try not to regulate the rate in any way.

5. **Elevated blood carbon dioxide levels**

Breathe normally into and out of a paper bag held over your nose and mouth. After one minute, record the breathing rate for the second minute. Then hold your breath for as long as possible. Time your breath-holding.

	Test #1	Test #2	Test #3	Average rate	Time for breath holding
Normal breathing					
Increased rate, normal volume					
Increased volume, normal rate					
Increased rate and volume					
Breathe into paper bag					

READING

Airway inflammation in asthma. (1995, March/April) *Science & Medicine.*

Breath tests in medicine. (1992, July) *Scientific American.*

Breathing rhythm generation: Focus on the rostral ventrolateral medulla. (1995, June) *News in Physiological Sciences* 10: 133-140.

Cystic fibrosis. (1995, December) *Scientific American.*

Endocytosis of polypeptides in rabbit nasal respiratory mucosa. (1997, October) *News in Physiological Sciences* 12: 219-225. The rabbit epithelium uses transcytosis to absorb polypeptides, probably those that act as antigens in allergic responses.

Gas exchange in vertebrates through lungs, gills, and skin. (1992, October) *News in Physiological Sciences* 7: 199-202.

The human voice. (1992, December) *Scientific American.*

Mountain sickness. (1992, October) *Scientific American.*

Multi-drug-resistant tuberculosis. (1994, May/June) *Science & Medicine.*

The physiology of decompression illness. (1995, August) *Scientific American.*

Pulmonary blood-gas barrier: A physiological dilemma. (1993, December) *News in Physiological Sciences* 8: 249-252.

Pulmonary edema fluid clearance pathways. (1995, June) *News in Physiological Sciences* 10: 107-111.

The pulmonary surfactant system. (1994, February) *News in Physiological Sciences* 9: 13-19.

Respiratory chemoreception: The acid test. (1995, February) *News in Physiological Sciences* 10: 49-50.

Sodium transport in cystic fibrosis. (1996, December) *News in Physiological Sciences* 11: 299.

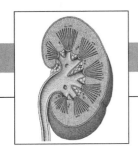

SUMMARY

Key points to learn in this chapter:

- The nephron is the functional unit of the kidney.
- The four basic urinary processes are: filtration (F), reabsorption (R), secretion (S), and excretion (E). $E = F - R + S$
- The filtration rate can be controlled by altering the blood flow through the arterioles. There are three levels of control: myogenic response, tubuloglomerular feedback, and reflex control.
- Reabsorption and secretion involve transport and therefore exhibit saturation, specificity, and competition.
- Clearance is an abstract way of determining renal handling of a substance based on blood and urine analysis. Spend some time becoming familiar with the concept and with the math involved.

The urinary system consists of the kidneys, bladder, and accessory structures. The system produces urine and eliminates it to help the body maintain fluid and electrolyte balance. The kidneys have six functions: regulation of extracellular fluid volume, regulation of osmolarity, maintenance of ion balance, homeostatic regulation of pH, excretion of wastes and foreign substances, and production of hormones. There are four basic processes in the urinary system: filtration, reabsorption, secretion, and excretion.

The kidney is composed of nephrons, and each nephron is composed of vascular and tubular elements. Following the path of blood through the nephron, the anatomy is as follows: renal arteries, afferent arteriole, glomerulus, efferent arteriole, peritubular capillaries. Fluid filters out of the glomerulus into Bowman's capsule. It then moves into the proximal tubule, loop of Henle, distal tubule, collecting duct, and renal pelvis. After this point, the fluid will not be altered again, and it can be called urine. Urine drains into the urinary bladder via the ureters and is then eliminated in a process called urination (micturition).

One of the most important concepts to take from this chapter is that the amount of fluid excreted (E) is equal to the amount filtered (F), minus the amount reabsorbed (R), plus the amount secreted (S). Expressed mathematically, this becomes: $E = F - R + S$. Remembering this equation can often help you if you're in a bind trying to solve a problem.

Filtration is the movement of fluid from the blood of the glomerulus into the nephron lumen at Bowman's capsule. Filtered fluid composition is equal to that of plasma, minus blood cells and most proteins. Filtration occurs because the hydraulic pressure exceeds the osmotic pressure, and the net driving force is 10 mm Hg in the favor of filtration. Specialized epithelial cells in the capsule, called podocytes, form slits that can be manipulated to change the glomerular filtration rate (GFR). GFR is the amount of fluid that filters into Bowman's capsule per unit time; the average GFR is 125 mL/min (180 L/day). Control of GFR includes the myogenic response, tubuloglomerular feedback, and reflex control. In the myogenic response, smooth muscle of the arteriole stretches as blood pressure increases, and this ultimately causes vasoconstriction. Tubuloglomerular feedback involves chemical communication between the macula densa cells and JG cells (together called the juxtaglomerular apparatus) that controls arteriolar diameter. Reflex control involves sympathetic neurons and alpha receptors on afferent and efferent arterioles.

Reabsorption is the movement of filtered material from the nephron lumen back into the blood supply. Bulk reabsorption takes place in the proximal tubule, but regulated reabsorption takes place in later tubule segments. Most reabsorption involves transepithelial transport (movement across the apical and basolateral membranes). Reabsorption of water and solutes depends on both active and passive transporting mechanisms. The transport of Na^+ into the extracellular fluid creates concentration gradients that allow movement of other substances. Transport involves protein-substrate interaction, so saturation, competition, and specificity are also involved. The renal threshold for a substance is the plasma concentration of that substance at which saturation occurs. At concentrations above renal threshold, substances that are normally reabsorbed are excreted in the urine.

Secretion is the transfer of molecules from the extracellular fluid into the nephron lumen. Secretion depends mostly on active membrane transport. As with reabsorption, secretion shows saturation, competition, and specificity.

Excretion is the result of the other three processes. It depends on the filtration rate and on whether reabsorption or secretion is involved. Clearance is an abstract concept describing how many milliliters of plasma passing through the kidneys have been totally cleared of a substance in a given period of time. Clearance is used clinically to assess renal handling of a substance based only on the analysis of the blood and urine. Spend some valuable time working through the examples in the chapter to get a grasp of this concept mathematically. It can be a little tricky, so be sure you give yourself plenty of time to work through it.

Micturition is the elimination of urine from the bladder. Two sphincters close off the opening between the ureters and the bladder. The external sphincter is skeletal muscle and can be consciously controlled. Micturition is a simple spinal reflex initiated by stretch in the bladder wall.

TEACH YOURSELF THE BASICS

FUNCTIONS OF THE KIDNEYS

• State the law of mass balance. [∫ p. 6] _____

• List the six functions of the kidneys and put a star next to the most important.

1. _____   2. _____

3. _____   4. _____

5. _____   6. _____

ANATOMY OF THE URINARY SYSTEM

The Urinary System Consists of Kidneys, Ureters, Bladder, and Urethra (Anatomy summary, Fig. 18-1)

• Starting at the kidneys, follow a drop of urine to the external environment. (Fig. 18-1a)

_____

• Explain the term "retroperitoneal." _____

• Describe the vascular supply to the kidneys. _____

_____

• What is the function of the urinary bladder? _____

The Nephron Is the Functional Unit of the Kidney

• The medulla is the (outer/inner?) layer and the cortex is the (outer/inner ?) layer of the kidney. (Fig. 18-1c)

• What is a nephron? _____

Vascular elements of the nephron
• Trace a drop of blood through the nephron from a renal artery to a renal vein. (Fig. 18-1e)

_____

• Describe the portal system of the nephron. _____

• What takes place at the glomerulus? _____

Tubular elements of the nephron
• Trace a drop of fluid through the tubule of the nephron, ending in the renal pelvis. (Fig. 18-1d, 18-2)

_____

• What occurs in the renal corpuscle? _____

• Describe the relationship of the different segments of the nephron to the cortex and medulla of the kidney.

_____

• Describe the juxtaglomerular apparatus. (Fig. 18-8)

• Fluid is considered urine when it enters the _____.

PROCESSES OF THE KIDNEYS

• How much plasma on average enters the nephrons per day? _____

• How much urine on average leaves the body per day as urine? _____

• What happens to the fluid that doesn't leave in the urine? _____

The Four Processes of the Kidney Are Filtration, Reabsorption, Secretion, and Excretion

• Name the four processes of the kidney and describe them. (Fig. 18-2)

1. _____

2. _____

3. _____

4. _____

• In which of these processes is fluid entering the external environment? _____

• Which of these processes could be considered bulk flow? _____

• Which of these processes uses transporting epithelia? [∫ p. 57] _____

• What is the technical term for urination? _____

• The amount of a substance excreted in urine reflects renal handling (Fig. 18-3):

Amt. _____ (E) = Amt. _____ (F) Amt. _____ (R) + Amt. _____ (S)

☞ *Remember this expression:*   *[E = F – R + S]*

Volume and Osmolarity Change as Fluid Flows through the Nephron

• Fluid entering Bowman's capsule is nearly _____ with plasma = 300 mOsM [∫ p. 454].

• At the end of the proximal tubule, _____ % of the filtered fluid remains and the osmolarity

of the fluid is _____ mOsM. Based on these numbers, we say that the proximal tubule is

the primary site of (filtration / reabsorption /excretion?).

• Fluid leaving the loop of Henle is usually (isosmotic / hyperosmotic /hyposmotic ?) to plasma, so

we say that the loop is the primary site for production of (concentrated / dilute / salty?) urine.

• By the time fluid passes through the distal tubule and collecting duct, its volume is down

to _____ L and its osmolarity can range from _____ to _____ mOsM.

Therefore we say that the distal nephron is the site of the fine regulation of salt and water balance.

• Salt and water balance in the nephron is regulated by _____.

• The final volume and concentration of the urine reflect what need(s) of the body?

_____

FILTRATION

• Describe the composition of the filtrate that enters the lumen of the nephron.

_____

The Renal Corpuscle Consists of the Glomerulus and Bowman's Capsule

• List the three layers a water molecule will pass through as it travels from the blood into the lumen of Bowman's capsule.

_____

• Describe the structure of glomerular capillaries and their pores. [∫ p. 438] (Fig. 18-4d)

_____

• Describe the structure and function of the basal lamina. _____

_____

• Describe the portion of Bowman's capsule that surrounds the capillaries of the glomerulus.

_____

• How do podocytes alter the glomerular filtration rate? _____

_____

• Define filtration fraction. _____

Filtration Occurs because of Hydraulic Pressure in the Capillaries

• Glomerular filtration occurs because (Fig. 18-6):

1. _____ pressure of blood forces blood out of the leaky capillary epithelium.

2. _____ pressure in capillaries is greater than the same pressure within Bowman's capsule.

3. Fluid pressure created by _____

   (opposes / enhances ?) fluid movement into Bowman's capsule.

4. The net driving force is 10 mm Hg in the direction from _____ to _____.

• Compare the pressures involved in glomerular filtration to filtration out of systemic capillaries. [∫ p. 439]

_____

Glomerular Filtration Rate Averages 180 Liters per Day

• Define GFR. _____

• What two factors influence GFR the most? _____

• An average value for GFR is _____ L / day or _____ mL/min.

• The total body plasma volume is _____ L, which means that the kidneys filter the plasma

_____ times each day.

• What is the relationship between blood pressure and GFR? (Fig. 18-7) _____

_____

GFR Is Regulated by Paracrines, Hormones, and the Nervous System

• What is autoregulation? _____

Myogenic response
• What is the myogenic response? _____

• Describe the myogenic response to the nephron to increased blood pressure.

_____

• Why is a decrease in GFR when blood pressures fall below normal an adaptive response?

_____

Tubuloglomerular feedback
• How does the distal tubule communicate with arterioles? (Fig. 18-8) _____

• What does the abbreviation "JGî stand for? _____

• Describe tubuloglomerular feedback as a result of <u>decreased</u> blood pressure. (Fig. 18-9)

_____

_____

Reflex control of GFR

• In neural control of GFR, (sympathetic / parasympathetic ?) neurons release (ACh / norepi) onto

(α, β_1, β_2, nicotinic, muscarinic ?) receptors, causing (vasodilation / vasoconstriction) of renal

arterioles.

• Vasoconstriction of the afferent arteriole will (increase / decrease) its resistance, will (increase /

decrease) hydraulic pressure in the glomerulus, and will (increase / decrease) GFR.

• Vasoconstriction of the efferent arteriole will (increase / decrease) its resistance, will (increase /

decrease) hydraulic pressure in the glomerulus, and will (increase / decrease) GFR.

• Hormones that influence arteriolar resistance and GFR include _____,

a potent vasoconstrictor, and _____, which are vasodilators.

REABSORPTION

• Bulk reabsorption in the nephron takes place in the _____.

• Why is the kidney designed to filter such large volumes if 99% of what is filtered is reabsorbed?

_____

• Compare fluid movement across peritubular capillaries to fluid movement across the glomerulus.

_____

Reabsorption May Be Active or Passive

• To reabsorb molecules against their concentration gradient, the transporting epithelia of the nephron

must use what process? _____

• What one ion plays a key role in bulk reabsorption in the proximal tubule? _____

Active transport of sodium

• Filtrate entering proximal tubule has [Na^+] similar to that of _____ and

that is (higher / lower?) than the [Na^+] inside cells.

Secondary active transport: symport with sodium

• List some molecules that are transported using Na+-linked secondary active transport. [∫ p. 129]

_____

Passive reabsorption: urea reabsorption

• Urea in the proximal tubule can only move by diffusion. If the urea concentration of filtrate is equal to the urea concentration of plasma, what creates the urea concentration gradient needed for diffusion?

_____

Transcytosis: plasma proteins

• Some of the smaller peptide hormones and enzymes can cross the filtration barrier at the glomerulus.

What happens to those that are not reabsorbed intact? _____

• For each of the following substances, tell how it crosses the apical and basolateral membranes of the proximal tubule cell and what form of transport it uses at each membrane (primary active, secondary active, facilitated diffusion, simple diffusion).

Molecule	Apical membrane	Type of transport?	Basolateral membrane	Type of transport?
Na$^+$				
glucose				
water				
urea				
proteins				

Saturation of Renal Transport Plays an Important Role in Kidney Function

• List the three properties of mediated transport seen in all protein-substrate interactions. [∫ p. 120]

_____

• Explain what happens when transport is saturated. _____

_____

• Below saturation, the rate of transport is proportional to _____.

• The rate of transport at saturation is also known as the _____.
   (Fig. 18-13)

• The plasma concentration of a substance at which the tubule reaches saturation is known as the

_____. (Fig. 18-14)

• Filtration (does / doesn't ?) exhibit saturation. (Fig. 18-14a)

• Normal plasma glucose concentrations are (lower than / equal to / higher than ?) the renal threshold

   for glucose. Therefore, normally all glucose filtered is (excreted / secreted / reabsorbed ?).

• What happens when the plasma glucose concentration exceeds the renal threshold? (Fig. 18-14b)

_____

• What is the term for "glucose in the urine"? _____

• If filtration of a substance is greater than reabsorption, then the excess substance is

_____.

☞ Remember! $E = F - R + S$

SECRETION

• In renal secretion, molecules move from the _____ to the _____.

• By what means of transport is most secretion accomplished? _____

• What is the adaptive significance of secreting a substance in addition to filtering it?

_____

Competition and Penicillin Secretion

• Describe competition in a mediated transport system, using penicillin and probenecid as your
   example.

_____

• When probenecid is given at the same time as penicillin, what happens to the excretion rate of
   penicillin?

_____

EXCRETION

• How is excretion different from secretion? _____

_____

• Does looking at the composition of urine tell us if a substance has been filtered? _____

   reabsorbed? _____ secreted? _____ excreted? _____

• To analyze how the renal tubule is handling a substance, we must know what information

   about a substance? _____

• Define clearance and give its units. _____

_____

• Write the mathematical equation for clearance: _____

• To calculate the urinary excretion rate of a substance, you must know what two pieces of

   information? _____

• Explain why the clearance rate of inulin is equal to the GFR. _____

_____

• What is creatinine? [∫ p. 338] _____

• Why is creatinine used clinically for estimating GFR rather than inulin? _____

_____

• If the filtration rate of a substance is less than its excretion rate, you know that the substance is being

   _____ by the renal tubule. (Table 18-2)

Clearance Can Be Used to Determine Renal Handling of a Substance

• Clearance rate for a substance can be compared with inulin/creatinine clearance. If clearance of χ

   is greater than inulin/creatinine clearance, the tubule is _____ χ.

☛ *Clearance tells you only about the net handling of a substance. It does not tell you if a molecule is both reabsorbed and secreted.*

MICTURITION

• The urinary bladder can hold about _____ mL.

• What prevents urine from leaving the bladder? _____

• Describe the tonic control exerted by the CNS over the bladder. (Fig. 18-16a)

_____

• Fill in the steps of the involuntary micturition reflex.

Stimulus _____

Receptor _____

Afferent pathway _____

Integrating center _____

Efferent pathway(s) _____

Effectors _____

Tissue responses _____

Systemic response _____

• What neurotransmitters and receptors are involved in the tissue responses?

_____

• What role do higher brain centers play in micturition? _____

TALK THE TALK

active transport, kidney
afferent arteriole
ascending limb, loop of Henle
basal lamina
benzoate
Bowman's capsule
clearance
collecting duct
cortex, renal
creatinine
descending limb, loop of Henle
diabetes mellitus
distal tubule
efferent arteriole
erythropoietin
excretion
extracellular fluid volume
fenestrated capillary
filtration
filtration fraction
filtration slit
Fleming, Alexander
fluid pressure
foot process
functional unit
GFR, autoregulation of
glomerular filtration rate (GFR)
glomerulus

glucosuria (glycosuria)
hemoglobin, in urine
hydrostatic pressure
inulin
juxtaglomerular apparatus
juxtaglomerular cell
kidney
law of mass balance
loop of Henle
macula densa
medulla, renal
mesangial cell
micturition
micturition reflex
myogenic response
Na+/K+-ATPase
Na+-glucose co-transporter
nephron
osmotic pressure
parasympathetic neurons, bladder
passive reabsorption
penicillin
peritoneum
peritubular capillary
plasma volume
podocyte
probenecid
proximal tubule

renal artery
renal corpuscle
renal pelvis
renal threshold
renal vein
renin
retroperitoneal
saccharin
saturation
secondary active transport
secretion
sphincters, bladder
transcytosis, kidney
transepithelial transport
transport maximum (Tm)
transporting epithelium
tubuloglomerular feedback
urea
ureter
urethra
uric acid
urinary bladder
urinary system
urinary tract infection
urine
urine reabsorption
uroscopy
vasoconstriction

ERRATA

p. 527, Fig. 18-6: The PH arrow should read 55, not 50mm Hg.

p. 537, Fig. 18-15d: "Clearance of penicillin = 150 mL/miL" should read "Clearance of penicillin = 150   mL/min."

RUNNING PROBLEM - Gout

Gout: Fresh insights into an ancient disease. (1996, July/August) *Science & Medicine*.

Gout is a disease that arose due to the mutation of a gene during primate evolution, when primates lost the ability to convert uric acid to allantoin, a more soluble metabolite. Birds and terrestrial reptiles excrete uric acid crystals to conserve water. In humans, gout is particularly common in Pacific Islanders such as Filipinos and Samoans, and relatively rare in other Asians. About 90% of the cases are in men over 30; women are not usually affected until menopause.

What is the rationale for limiting alcohol intake in the management of gout? Ethanol metabolism leads to the degradation of adenosine triphosphate (ATP) to adenosine monophosphate (AMP). When AMP is formed, it liberates adenlytic acid, which in turn prompts the breakdown of purines. Limiting alcohol ingestion limits the degradation of ATP and AMP and the formation of adenlytic acid, thus slowing purine metabolism. Decreased purine metabolism leads to decreased levels of uric acid.

ETHICS IN SCIENCE

Kidney transplants are now a very common procedure. How does organ availability affect our lives? Think about the "urban myth" that has spread through the Internet: a man visiting New Orleans was drugged following a wild party. The next morning he was found unconscious and bleeding, and doctors discovered that he was missing a kidney, removed for sale on the black market. Although this is just a folk tale, it brings up the question of whether people should be allowed to sell their organs. There has been news recently about prison officials in China selling organs "to order" from prisoners who are to be executed. In this country, the newest controversy is whether young women should be allowed to sell their ova (eggs) to infertile couples.

QUANTITATIVE THINKING

Excretion = Filtration – Reabsorption + Secretion

Filtration rate of χ = GFR $\times$ plasma concentration of χ

Clearance of χ = excretion rate of χ / plasma concentration of χ

Excretion rate of χ = urine flow rate $\times$ urine concentration of χ

1. Tamika goes in for a routine physical examination. Her urinalysis shows proteinuria but she has no other abnormalities, according to her physical exam. She weighs 60 kg and is 159 cm tall. Her lab data show:

serum creatinine: 1.8 mg/dL   urine creatinine: 276 mg/dL
urine volume: 1100 mL in 24 hours

Calculate Tamikaís creatinine clearance and GFR.

2.   Plasma concentration of inulin   1 mg/mL
Plasma concentration of X   1 mg/mL
GFR 125 mL/min

What is the filtration rate of inulin? of X?
What is the excretion rate of inulin? What is the excretion rate of X?

3. An alien willingly allows you to test its renal function. Answer the following questions about the alien's kidney function (Keep in mind: alien kidneys donít necessarily follow the same rules as human kidneys).

a) Tests show that the alien kidney freely filters glucose. Once in the lumen of the nephron, glucose is reabsorbed but not secreted. The renal threshold is determined to be 500 mg glucose/100 mL plasma, and the alienís glucose transport maximum is 90 mg/min.

Can you calculate the alienís GFR from this information? If so, what is it? If not, what other information do you need?

b) Additional tests show that creatinine gives an accurate GFR in this alien species. The alien transport maximum for the reabsorption of phenol red is 40 mg/min. Look at the following data:

     GFR: 25 mL/min
     Urine: creatinine = 5 mg/mL; phenol red = 5 mg/mL
     Plasma: CREATININE = 6 mg/mL; phenol red = 2 mg/mL
     Urine flow: 2 mL/min

What is this alien's creatinine clearance? phenol red clearance?

4. Graphing question:

You are given a chemical Z and told to determine how it is handled by the kidneys of a mouse. After a series of experiments, you determine that (1) Z is freely filtered, (2) Z is not reabsorbed, (3) Z is actively secreted, and (4) the renal threshold for Z secretion is a plasma concentration of 80 mg/mL plasma, and the transport maximum is 40 mg/min. The mouse GFR is 1 mL/min. On the graph below, show how filtration, secretion, and excretion are related. One axis will be plasma concentration of Z (mg/mL) with a range of 0-140, and the other axis will show rates of kidney processes (mg/min) with a range of 0-140.

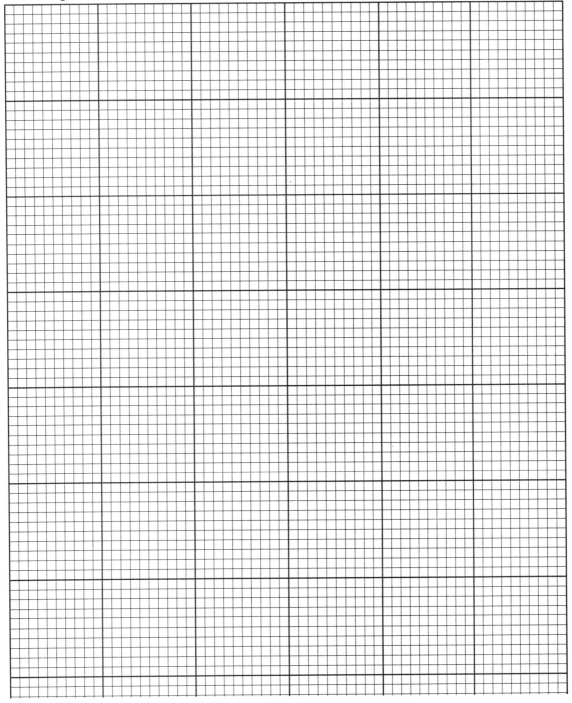

PRACTICE MAKES PERFECT

1. What has gone wrong with the nephron if a person has proteinuria (protein in the urine)?

_____

2. Glucose is easily filtered by the kidneys. Why is glucose normally absent in the urine?

_____

3. Diagram a nephron and label it. Next to each part name, write the processes or functions of that part.

4. Compare the hydraulic pressure, fluid pressure, osmotic pressure, and net direction of fluid flow in the glomerular, peritubular, systemic, and pulmonary capillaries.

	Hydraulic pressure	Fluid pressure	Osmotic pressure	Net direction of fluid flow
Glomerular capillaries				
Peritubular capillaries				
Systemic capillaries				
Pulmonary capillaries				

5. Define and give units for:

a) renal threshold _____

_____

b) clearance _____

_____

c) GFR _____

spell out "GFR"_____

6. Compare and contrast the following concepts:

a) clearance and glomerular filtration rate

b) renal threshold and transport maximum

7. Using what you know about K^+ concentrations in the filtrate and proximal tubule cells, figure out how the cell could reabsorb K^+ from the tubule lumen. See Table 5-3 (p. 124) for a list of common transporters.

	Apical membrane	Type of transport?	Basolateral membrane	Type of transport?
K^+				

8. Trace a water molecule from a capillary in the hand directly to the kidney and out through the urine. Name all anatomical structures through which the water molecule will pass. Do not write sentences. Use only names connected by arrows.

9. Inulin clearance when mean arterial pressure = 100 mm Hg is (greater than / less than / the same as ?) inulin clearance when mean arterial pressure = 200 mm Hg.

10. Below is a graph of phenol red excretion in the bullfrog. You can assume that the structure of the frog kidney is like that of humans and that all four basic processes of human kidney function can occur. Based on this graph, answer the following questions:

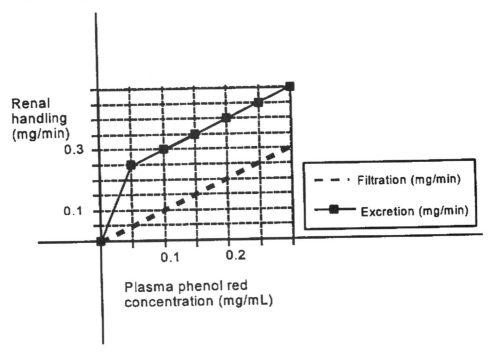

a) Circle the process or processes that allow excretion of phenol red by the bullfrog kidney:
      filtration   reabsorption   secretion

On the graph above, draw in lines for any missing processes. (This line(s) should not be estimated but rather calculated based upon the given data.)

b) What causes the slope of the excretion line at a plasma phenol red concentration of 0.05 mg/mL to change?

_____

c) If a phenol red transport inhibitor was administered concurrently with phenol red to this bullfrog, the clearance rate of phenol red compared to normal would:

increase   decrease   stay the same

BEYOND THE PAGES

☛ You can see a retroperitoneal kidney the next time you buy a whole chicken at the grocery. The abdominal organs are removed before the chicken is sold, but if you separate the chicken into breast and back halves by cutting through the ribs, you can see the paired kidneys lying alongside the back-bone underneath the membranous lining of the abdomen.

READING

About benefits and costs: Prevention of end-stage renal disease. (1994, November/December) *Science & Medicine*.

Cellular and molecular mechanisms of renal peptide transport (review). (1997) *American Journal of Physiology* 273: F1-F8.

Epithelial cell polarity and disease (review). (1997) *American Journal of Physiology* 272: F434-F442.

Microalbuminuria. (1996, May/June) *Science & Medicine*.

Renal proximal tubular albumin reabsorption: Daily prevention of albuminuria. (1998, February) *News in Physiological Sciences* 13.

A role for tubuloglomerular feedback in chronic partial ureteral obstruction. (1993, April) *News in Physiological Sciences* 8: 74-78.

Structure and function of renal organic cation transporters. (1998, February) *News in Physiological Sciences* 13.

CHAPTER 19

INTEGRATIVE PHYSIOLOGY II: FLUID AND ELECTROLYTE BALANCE

SUMMARY

Here are some key points in this chapter:

- The kidneys regulate water and solute balance in the body by altering the concentration of the urine excreted.
- Vasopressin (ADH), from the posterior pituitary, controls water reabsorption in the collecting duct of the nephron.
- Aldosterone, from the adrenal cortex, controls Na^+ reabsorption in the distal nephron and collecting duct.
- Direct control of aldosterone release is due to increased K^+. Indirect control is due to renin release.
- When dealing with a fluid-solute challenge, the body deals with osmolarity first.
- An essential equation in acid-base balance is $CO_2 + H_2O \leftrightarrow H^+ + HCO_3^-$. Know how each case of acid-base balance affects the above equation.

Fluid and electrolyte balance are under homeostatic control involving the renal, respiratory, and cardiovascular systems. Respiratory and cardiovascular mechanisms are primarily under nervous control and are therefore more rapid than renal mechanisms, which are primarily under hormonal control.

Body osmolarity is homeostatically maintained at around 300 mOsM. Hypothalamic osmoreceptors sense changes in osmolarity and signal the kidneys to adjust urine concentration to correct for the changes. Kidneys must reabsorb water to make concentrated urine, but they must reabsorb Na+ and not reabsorb water to make dilute urine. Water reabsorption takes place in the collecting duct of the neuron and is under the control of ADH released from the posterior pituitary. The presence of ADH increases the permeability of the collecting duct to water by inserting water pores into the apical membrane of the duct cell. Sodium reabsorption takes place in the distal tubule and collecting duct, and it is under the control of aldosterone from the adrenal cortex. Aldosterone initiates synthesis of Na^+/K^+-ATPase pumps on the basolateral membranes of P cells and thus allows Na+ to be reabsorbed into the ECF.

Control of aldosterone release is complex. Secretion is controlled directly at the adrenal cortex by increased K^+ levels; secretion is inhibited by high osmolarity. Indirect control of aldosterone secretion involves renin produced in the juxtaglomerular cells of the kidney nephron. Renin secretion is directly or indirectly triggered by low blood pressure. When secreted, renin converts angiotensinogen into angiotensin I (AGI). AGI is converted by angiotensin-converting enzyme (ACE) into angiotensin II (AGII). AGII causes release of aldosterone and increased blood pressure.

Changes in salt and water balance must be corrected, but it isn't always possible to correct the imbalances perfectly. In some instances, such as dehydration, integrating centers receive conflicting signals from different sensory pathways. When the body is faced with conflicting signals, the rule of thumb is to correct osmolarity imbalances first. Work through pp. 559-563 to get a full understanding of integrated salt and water balance. It would be especially helpful to your understanding for you to learn the integration of control well enough to be able to draw a map similar to the one on p. 562 without looking at any references.

Potassium balance is important to cell function, and is closely associated with Na^+ balance as well as with pH balance. Hyperkalemia and hypokalemia can cause problems with excitable tissues, especially the heart.

Acid-base balance is another homeostatic parameter that is crucial to body function. Remember that enzymes and other proteins are sensitive to pH changes. Acidosis is more common than alkalosis, but each requires compensation to restore homeostasis. The most important contributor to acidosis is CO_2 from respiration. Remember the equation: $CO_2 + H_2O \leftrightarrow H^+ + HCO_3^-$. There are three compensations that the body uses to combat acid-base imbalances: buffers, ventilation, and renal excretion of H^+ and HCO_3^-. Ventilation removes CO_2 and therefore decreases H^+ production. The kidneys remove the excess acid or base directly by excreting H^+ or HCO_3^- into the urine. Intercalated cells of the collecting duct are responsible for the excretion of ions: type A cells excrete H^+ and reabsorb HCO_3^- during acidosis; type B cells excrete HCO_3^- and reabsorb H^+ during alkalosis.

Acid-base disturbances are classified according to the pH change they induce and by how the imbalance originated. There are four possibilities: respiratory acidosis, respiratory alkalosis, metabolic acidosis, and metabolic alkalosis. When learning about acid-base disturbances, pay careful attention to how each of the four possibilities change the CO_2, H^+ equation. Also, learn the hallmarks associated with each condition and its compensations. If you can remember how the equation is manipulated, then you will have an easier time working problems concerned with acid-base balance. Make charts, graphs, or whatever helps you the most.

TEACH YOURSELF THE BASICS

• How are pH and CO_2 associated in the body? _____

_____

HOMEOSTASIS OF VOLUME AND OSMOLARITY

• List two ways the kidneys alter fluid volume and osmolarity. _____

_____

• When the body needs to excrete excess water, what happens to the volume and osmolarity of the urine?

_____

• Urine osmolarity can range from _____ to _____   mOsM.

• Water excretion is controlled primarily by the hormone _____,

a (steroid / peptide / amine ?) neurohormone made in the _____ and

secreted from the _____. [$\int$ p. 185]

• What is the target tissue of this hormone? _____

• When ADH is present in high concentrations, the kidney will have (maximal / minimal?) water reabsorption.

• Explain why Na$^+$ balance in the body is closely linked to ECF volume and osmolarity. _____

_____

• What kinds of receptors does the body use to monitor Na$^+$? _____

_____

WATER BALANCE AND THE REGULATION OF URINE CONCENTRATION

• In a 70-kg man, how much water is in his entire body? _____ intracellular fluid (ICF) ? _____

   extracellular fluid (ECF) ? _____ plasma? _____ interstitial fluid? _____ [∫ p. 132]

• What is an approximate normal value for osmolarity? _____ [∫ p. 134]

• Explain how fluid balance is associated with blood pressure. [∫ p. 431] _____

_____

Daily Water Intake Is Balanced by Water Excretion

• List the normal routes of water input and water loss for the body. (Fig. 19-1) _____

_____

• Which are the most significant? Put a star next to them.

• What is insensible water loss? _____

• Give some examples of pathological water loss. _____

Kidneys Conserve Water

• True or false? When body osmolarity goes up, the kidneys reabsorb water and bring osmolarity back to normal. Explain. (Fig. 19-2)

_____

• When water is lost from the body, how can it be restored? _____

Changes in Blood Pressure and Osmolarity Are the Stimuli for Water Balance Reflexes

• The two primary stimuli that trigger renal reflexes are _____

_____

• Osmolarity changes are sensed by what type of sensory receptor? _____

• Where are these receptors found? _____

• To what integrating center(s) are these receptors linked? _____

• What efferent pathway(s) are linked to the integrating center(s)? _____

• When osmolarity goes above _____ mOsM, ADH release is (increased / decreased ?), leading

 to (increased / decreased ?) renal water reabsorption.

• What receptors respond to changes in blood volume or blood pressure?

_____

• When blood pressure or volume increases, what kind of signal do these receptors send to the ADH-

 secreting cells? _____

☞ *Fig. 19-4 shows integrated responses to blood volume/pressure changes.*

Extracellular Fluid Osmolarity Affects Cell Volume

• Why is the regulation of body osmolarity important? _____

_____

• When exposed to a hypertonic ECF, cells (lose / gain ?) water. When exposed to a hypotonic ECF,
 cells (lose / gain ?) water. [p. 134]

How Does the Nephron Establish the Concentration of the Urine?

• To produce a dilute urine, the kidney tubules must reabsorb (water / solute ?) and not reabsorb
 (water / solute ?).

• The region of the kidney where this process occurs is the _____.

• For the tubule to produce a dilute fluid, the tubule must have what properties?

_____

• To produce a concentrated urine, the kidney tubules must reabsorb (water / solute ?) and not reabsorb
 (water / solute ?).

• The region of the kidney where this process occurs is the _____.

• For the tubule to produce a concentrated fluid, the tubule must have what properties and the interstitial fluid around the tubule must have what properties?

_____

• The water permeability of the collecting duct is regulated by _____.

The Loop of Henle Is a Countercurrent Multiplier

• Describe a countercurrent system. _____

_____

• What are the two functions of the countercurrent system in the loop of Henle? (Fig. 19-7)

_____

• The descending limb of the loop is permeable to (water / solutes ?) and impermeable to (water / solutes ?).

The ascending limb is permeable to (water / solutes ?) and impermeable to (water / solutes ?). As a result of the countercurrent system, fluid leaving the loop is (hypo- / iso- /hyper-) osmotic to the blood.

Role of the vasa recta
• What is the vasa recta? _____

• What causes water and solutes leaving the loop of Henle to pass into the vasa recta? (Fig. 19-7)

_____

• What is the function of the vasa recta? _____

Urine Concentration Is Regulated by Antidiuretic Hormone

• Fluid leaving the loop of Henle is _____osmotic to the blood.

• To produce a dilute urine, the epithelium of the distal tubule and collecting duct must be _____ to water.

• Water permeability of the distal regions of the nephron is regulated by _____. (Fig. 19-8)

• How does ADH increase water permeability of collecting duct cells? _____

_____

_____

• In the presence of ADH, the kidney will produce (dilute / concentrated ?) urine.

• Urine leaving the collecting duct has a maximum osmolarity of _____ mOsM.

• What two stimuli control ADH secretion? (Table 19-1)

_____

• Where are the osmoreceptors for ADH secretion located and what is their threshold?

_____

• True or false? Water reabsorption by the kidney will always decrease osmolarity and increase blood
  volume.Explain.

_____

• Where are the ADH receptors that monitor blood pressure located?

_____

• When blood pressure goes up and these receptors increase signaling, what is the response of the

  hypothalamus? _____

• Explain how this response is homeostatic. _____

_____

SODIUM BALANCE AND THE REGULATION OF EXTRACELLULAR FLUID VOLUME

• What are the primary stimuli for reflexes that regulate salt and water balance?

_____

• Na^+ regulation is mediated through the RAAS. What do the initials RAAS stand for?

_____

Sodium Balance Is Controlled by the Hormone Aldosterone

• Fill in the following information for aldosterone. [∫ p. 180]

Aldosterone is a (peptide / amine / steroid ?) hormone synthesized in the (what organelle?) _____

_____ of the (what gland?) _____.

It is secreted into the blood by (mechanism?) _____ and transported in

the plasma (dissolved / bound to carrier proteins ?). Aldosterone receptors in the (what portion of

the renal tubule?) _____ are located primarily (on the cell mem-

brane/ inside the cell ?) of the _____ cell. The combination of aldosterone

with its receptors initiates what events?

_____

• The net result of aldosterone activity is Na$^+$ (secretion / reabsorption ?) and K$^+$ (secretion / reabsorption ?).

• Describe the location of the Na$^+$/K$^+$-ATPase and Na$^+$ and K$^+$ leak channels on P cells . (Fig. 19-11)

_____

• To increase Na$^+$ reabsorption, what happens to the Na$^+$/K$^+$-ATPase and to the leak channels?

_____

Aldosterone Secretion Is Influenced by Blood Pressure, Osmolarity, and K+

• Increased K$^+$ (stimulates / inhibits ?) aldosterone secretion.

• Increased osmolarity (stimulates / inhibits ?) aldosterone secretion.

The renin-angiotensin-aldosterone pathway (RAAS)
• Arrange the following terms into a map of the RAAS pathway. (Fig. 19-12)

active   angiotensin I (AGI)   liver
adrenal cortex   angiotensin II (AGII)   plasma protein
afferent arteriole   angiotensinogen   renin
aldosterone   endothelium
angiotensin converting enzyme   inactive
        (ACE)   JG cells

• List three stimuli for renin secretion. (Fig. 19-13) _____

_____

• How does the macula densa communicate with the JG cells? _____

• True or false? Na+ retention immediately raises low blood pressure. Explain.

_____

Angiotensin II Helps Regulate Blood Pressure

• In addition to aldosterone secretion, AGII has significant blood pressure-raising effects (Fig. 19-12):

1. Increased _____ secretion leads to increased water reabsorption.

2. AGII stimulates _____ and water intake.

3. AGII is potent _____ , which increases blood pressure, but not blood volume.

4. AGII stimulates the _____ , resulting in increased sympathetic output to heart and blood vessels. [∫ p. 443]

• ACE inhibitors were developed to block AGII production and treat _____.

Atrial Natriuretic Peptide

• What is the overall effect of atrial natriuretic peptide (ANP)? _____

• What cells synthesize and release ANPs? _____

• ANP is secreted when atria _____ due to increased blood volume.

• ANP enhances Na$^+$ _____ by the kidney. (Table 19-3)

• ANP increases glomerular filtration rate (GFR) by _____.
   [∫ p. 525]

• ANP decreases NaCl and water reabsorption in the _____.

• ANP (inhibits / stimulates ?) renin, aldosterone, and ADH release.

• ANP in the brain affects the _____ and (raises / lowers ?) blood pressure.

POTASSIUM BALANCE

• K^+ balance is important despite its (high / low ?) ECF concentration. [∫ p. 223, Fig. 8-23]

• Hyperkalemia (increases / decreases ?) the K^+ concentration gradient across cell membranes and

(depolarizes / hyperpolarizes ?) cells. This leads to (increased / decreased ?) excitability in excitable

tissues and can lead to cardiac arrhythmias.

• How does the body compensate for increased K^+ ingestion? _____

BEHAVIORAL MECHANISMS IN SALT AND WATER BALANCE

• What are the two key behavioral responses for the maintenance of normal fluid and electrolyte
balance?

_____

Thirst

• Hypothalamic _____ receptors trigger thirst at plasma osmolarities above 280 mOsM.

• How can the act of drinking alleviate thirst? Describe a possible pathway. _____

_____

• What is the adaptive value of the oral reflex? _____

_____

Salt Appetite

• What is salt appetite? _____

• In humans, salt appetite centers are in the _____ near the _____ centers.

Avoidance Behaviors

• Give two examples of osmoregulatory avoidance behaviors in humans or other animals.

_____

_____

INTEGRATED CONTROL OF VOLUME AND OSMOLARITY

Disturbances of Salt and Water Balance

• List four common routes of fluid loss in the body.

_____

• Volume and osmolarity each have three possible states (Fig. 19-14):

_____

• In each of the following situations, describe in general terms the appropriate homeostatic response(s).

    1. Increased volume, increased osmolarity: Eat salty foods, drink liquids
        a. Compensation:

_____

    2. Increased volume, unchanged osmolarity: Ingested fluid equals isotonic saline
        a. Compensation:

_____

    3. Increased volume, decreased osmolarity: Drink pure water, no solutes
        a. Compensation:

_____

        b. This will be imperfect compensation because kidneys cannot _____ all solutes

    4. Normal volume, increased osmolarity: Eat salty foods, drink no liquids
        a. Compensation:

_____

        b. After water addition, situation becomes like (1) or (2) above?

    5. Normal volume, decreased osmolarity: A dehydrated person ingests pure water
        a. Compensation:

_____

        b. Will this compensation be perfect or will additional compensatory reflexes be initiated?

_____

6. Decreased volume, increased osmolarity: Dehydration from sweating or diarrhea
   a. Compensation:

_____

7. Decreased volume, no osmolarity changes: Hemorrhage
   a. Compensation:

_____

8. Decreased volume, decreased osmolarity: Uncommon, could arise from incomplete compensation of another situation

The Homeostatic Response to Dehydration

☛ *Table 19-4 summarizes pathways involved in volume/osmolarity homeostasis.*

• When integrating centers receive conflicting information about regulation of volume and osmolarity,

   correction of _____ has priority.

• In compensation for severe dehydration (Fig. 19-15):

Carotid, aortic baroreceptors (increase / decrease ?) firing as blood pressure decreases.

The cardiovascular control center (increases / decreases ?) parasympathetic output and

(increases / decreases ?) sympathetic output in an effort to (increase / decrease ?) blood pressure.

• What effects do these autonomic neurons have on different effectors? Name the neurotransmitter and receptor as well as the specific target.

_____

_____

_____

• What effect does severe dehydration have on GFR? _____

• List all the stimuli that will cause renin release in this pathway. _____

_____

• What other pathway(s) will be initiated? _____

_____

☛ *Redundancy in control pathways ensures that four main compensatory mechanisms are activated.*

• List the four main compensatory mechanisms:

1. _____

2. _____

3. _____

4. _____

ACID-BASE BALANCE

• Define pH. [∫ p. 25, Appendix A]

_____

• What are the normal average and range of normal pH in the human body? _____

• A change of 1 pH unit = a _____-fold change in [H^+].

• An alkaline solution has a (higher / lower ?) H^+ concentration and a (higher / lower ?) pH than an acid solution.

Enzymes and the Nervous System Are Particularly Sensitive to Changes in pH

• Which body fluid will be most indicative of the body pH? _____

• If the pH range compatible with life is 7.0-7.7, how can we survive stomach juices with a pH as low as 1 or urine with a pH ranging from 4.5 - 8.5?

_____

• If pH falls outside the normal range, what kinds of things go wrong? _____

_____

•Acidosis (increases / decreases ?) neuron excitability; alkalosis (increases / decreases ?) neuron excitability.

• Disturbances in pH balance are often associated with changes in what other important ion (besides HCO_3^-)?

_____

• Name the membrane transporter that links movement of H^+ and this ion in the kidney. _____

• This transporter moves the ions in (the same / opposite ?) directions, so that in acidosis, when H^+ is

excreted, _____ is reabsorbed, leading to the condition known as _____emia.

• In alkalosis, when H^+ is excreted, _____ is reabsorbed, leading to the condition known as

_____emia.

Acids and Bases in the Body Come from Many Sources

The principle of _____ states that acid intake and production in
the body must be balanced with acid excretion (Fig. 19-16).

Acid input
• List some metabolic intermediates and foods that contribute H^+ to the body.

_____

• What are the common metabolic acids in the body? _____

• The biggest daily cause of acid production is _____.

• Write the equation for $CO_2 + H_2O \rightleftarrows$ _____.

• What enzyme catalyzes the reaction above? [∫ p. 505] _____

Base input
• What are the primary sources for bases in the body?

_____

• Which is more common: disturbances in the direction of alkalosis or acidosis? _____

• The body compensates for pH changes using three different mechanisms. List and briefly describe
  these mechanisms.

1. _____

2. _____

3. _____

Buffer Systems in the Body Include Proteins, Phosphate Ions, and Bicarbonate Ions

• Define a buffer. [∫ p. 26] _____

• Where in the body are buffers found? _____

• Intracellular buffers include [∫ p. 505, Fig. 17-25] _____

• _____ produced from metabolic CO_2 is the most important extracellular buffer.

• What buffers the matching H^+ made from CO_2 and water? _____

• Write the equation showing the relationship between CO_2, HCO_3^-, and H^+ in plasma.

_____

• According to the Law of Mass Action [∫ p. 86; workbook p. 4-13], an increase in CO_2 will cause an/a (increase / decrease ?) in H^+ and HCO_3^-.

• What effect does an increase in $[H^+]$ due to production of metabolic acids have on pH, HCO_3^-, and CO_2?

_____

• According to the Law of Mass Action, while HCO_3^- concentrations change, these changes are sometimes not noticed in clinical settings. Explain why this is true.

_____

• In metabolic acidosis, plasma P_{CO_2} is often normal although you would predict from the Law of Mass Action that it would be increased. Explain why this is true.

_____

Respiratory Compensation in Acid-Base Disturbances Is Triggered by H^+ or CO_2

• The P_{CO_2} of plasma is a measure of its _____.

• If ventilation increases, plasma P_{CO_2} (⇑ or ⇓ ?), H^+ (⇑ or ⇓ ?), and HCO_3^- (⇑ or ⇓ ?).

• If ventilation decreases, plasma P_{CO_2} (⇑ or ⇓ ?), H^+ (⇑ or ⇓ ?), and HCO_3^- (⇑ or ⇓ ?).

• What two chemical changes associated with a disturbance in acid-base balance will initiate a reflex change in ventilation? _____

• What receptors are sensitive to these changes? _____

• When arterial P_{CO_2} increases, ventilation (increases / decreases ?).

Renal Compensation in Acid-Base Disturbances Takes Place through Excretion or Reabsorption of H^+ and HCO_3^-

• List two ways the kidneys compensate for pH changes. _____

• During acidosis, the kidney secretes _____ and reabsorbs _____ in the proximal tubule and distal tubule (Fig. 19-18a).

• Ammonia and phosphate ions act as _____ in the urine.

• What is the role of bicarbonate reabsorbed by the kidney? _____

• In alkalosis, the kidney secretes _____ into lumen and reabsorbs _____. (Fig. 19-18b)

• Renal compensation for acid-base disturbances is (faster/slower?) than respiratory compensation.

• Name all the transport proteins that:

secrete H^+ across the apical membrane into the lumen. _____

reabsorb HCO_3^- across the basolateral membrane. _____

reabsorb H^+ across the basolateral membrane into the lumen. _____

secrete HCO_3^- across the apical membrane. _____

• Name the energy source used by each of the transport proteins in the question above.

_____

The proximal tubule: hydrogen ion secretion and bicarbonate reabsorption
• Filtered bicarbonate does not cross the apical membrane of the proximal tubule cell. How then is filtered HCO_3^- reabsorbed ? (Fig. 19-19)

_____

_____

• The net result of this process is that Na^+ is also (reabsorbed / secreted ?) and H^+ is (reabsorbed / secreted?).

The distal nephron: H^+ and HCO_3^- handling depend on the acid-base state of the body

• What is the role of intercalated cells (I cells) in the distal nephron?

_____

• What is the difference between type A and type B intercalated cells? (Fig. 19-20)

_____

• What enzyme is found in high concentrations in I cells? _____

• Why are disturbances in potassium balance often associated with disturbances in acid-base balance?

_____

Disturbances of Acid-Base Balance

• Describe the classification of acid-base disturbances. (Table 19-5)

_____

• To distinguish a respiratory acidosis from a respiratory alkalosis, you should look primarily at the relative concentrations of (H^+ / CO_2 / HCO_3^- ?).

• To distinguish a respiratory acidosis from a metabolic acidosis, you should look primarily at the relative concentrations of (H^+ / CO_2 / HCO_3^- ?).

• To distinguish a respiratory alkalosis from a metabolic alkalosis, you should look primarily at the relative concentrations of (H^+ / CO_2 / HCO_3^- ?).

• Disturbances with respiratory causes can use what type(s) of compensation? _____

• Disturbances with metabolic causes can use what type(s) of compensation? _____

Respiratory acidosis

• What alteration in ventilation will cause respiratory acidosis? _____

• Name some conditions/situations in which this might occur. _____

• What happens? (fill in ⇑ or ⇓)   _____ $CO_2 + H_2O \Rightarrow\Rightarrow$ ____ H^+ + _____ HCO_3^-

• The hallmark of respiratory acidosis is decreased plasma pH and elevated _____ .

• What compensation mechanisms are available in this state? _____

• As compensation occurs, the pH will (increase / decrease) and the HCO_3^- levels will be

(greater than / less than / the same as ?) predicted by Pco_2 levels.

Metabolic acidosis
• Name some causes of metabolic acidosis. _____

• What happens? (fill in ⇑ or ⇓)   _____ CO_2 + H_2O ⬅⬅ ____ H^+ + _____ HCO_3^-

• The hallmark of metabolic acidosis is decreased _____ .

• What compensation mechanisms are available in this state? _____

• The change in P_{CO_2} in this condition does what to ventilation? _____

• What do the kidneys do to compensate for this condition? _____

• Because compensation is almost instantaneous in this condition, the P_{CO_2} levels will often be

   (greater than / less than / the same as ?) normal.

Respiratory alkalosis
• What are the most common causes of respiratory alkalosis? _____

• What happens? (fill in ⇑ or ⇓)   _____ CO_2 + H_2O ⬅⬅ ____ H^+ + _____ HCO_3^-

• What compensation mechanisms are available in this state? _____

• The hallmark of respiratory alkalosis is decreased _____ and increased _____ .

Metabolic alkalosis
• What could cause metabolic alkalosis? _____

• What happens? (fill in ⇑ or ⇓)   _____ CO_2 + H_2O ⇒⇒⇒ ____ H^+ + _____ HCO_3^-

• The hallmark of metabolic alkalosis is increased _____ accompanied by increased _____.

• What compensation mechanisms are available in this state? _____

• The change in P_{CO_2} in this condition does what to ventilation? _____

• What do the kidneys do to compensate for this condition? _____

TALK THE TALK

acidosis
adrenal cortex
aldosterone
alkalosis
ammonia
angiotensin converting enzyme
(ACE)
angiotensin II (AGII)
antidiuretic hormone (ADH)
asthma
atrial natriuretic peptide
atrial stretch receptor
atriopeptin
bedwetting
bicarbonate
buffer
carbonic anhydrase
cardiovascular control center
carotid and aortic baroreceptor
carotid and aortic chemoreceptor
central chemoreceptor
chloride shift
chronic obstructive pulmonary
disease
CO2 and acid-base balance
countercurrent heat exchanger
countercurrent multiplier
daily water intake
desmopressin
diarrhea
diuresis
ECF volume, role of sodium in
determining

emphysema
enuresis
feedforward reflex, drinking
fibrosis
fluid and electrolyte balance
guanylin and uroguanylin
H+-ATPase
H+-K+-ATPase
HCO3 -Cl antiport
hemoglobin
hemorrhage
hyperkalemia
hyperventilation
hypokalemia
hypoventilation
insensible water loss
intercalated cell (I cell)
intravenous (IV) injection
juxtaglomerular granular cell
(JG cell)
ketoacid
ketoacidosis
lactic acidosis
law of mass action
law of mass balance
leak channels for Na+ and K+
loop of Henle
macula densa
medullary chemoreceptor
membrane recycling
metabolic acidosis
metabolic alkalosis

Na+/K+-ATPase
Na+-H+ antiport
Na+-HCO3 symport
natriuresis
oropharynx receptor
osmolarity
osmoreceptor
pH
phosphate ion
pneumonia
potassium homeostasis
principal cell (P cell)
regulation of cell volume
renin
renin-angiotensin-aldosterone
system (RAAS)
respiratory acidosis
respiratory alkalosis
respiratory compensations to pH
changes
salt appetite
severe dehydration
sweating
sympathetic neurons and nephron
thirst
tonicity
urine concentration
vasa recta
vasopressin
vomiting
water excretion

ERRATA

p. 557, right column, second paragraph, last line: Should read ANP instead of ANF.

RUNNING PROBLEM - Diabetes insipidus

Signal peptide and neurophysin II gene mutations in hereditary central diabetes insipidus. (1997, April) *News in Physiological Sciences* 12: 67-71.

Molecular analysis of hereditary diabetes insipidus: The understanding of water drinkers! (1997, December) *News in Physiological Sciences* 12: 296-297.

QUANTITATIVE THINKING

Osmotic Diuresis

When a diabetic's plasma glucose concentration exceeds the renal threshold for glucose reabsorption, the unreabsorbed glucose is excreted in the urine. The presence of this excess solute in the tubule lumen will cause additional water loss and create an **osmotic diuresis**. In the clinics, it is sometimes said that "the glucose holds the water in the urine." Physiologically, osmotic diuresis is based on simple amount/concentration relationships. The nephron can only concentrate urine to a maximum of 1200 mosmoles/liter. If there is more solute in the lumen (i.e., greater amount of solute) but the osmolarity cannot increase above 1200 mOsM, there will be a greater urine volume. To understand why this is so, work through the problem below.

The diagram shows a nephron. The numbers inside the nephron at points A, B, and C represent the concentration of the filtrate at those points. Assume for this example that vasopressin is present in amounts that allow maximal concentration of the urine and that there is no solute reabsorption in the distal parts of the nephron.

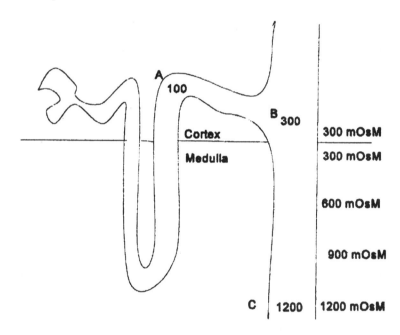

Assume that in a normal individual, 150 mosmoles of NaCl per day pass from the loop of Henle into the urine. Given the concentrations shown at points A-C, what volume of fluid passes those points in one day?

Point A _____   Point B _____ Point C _____

Now suppose the person suddenly becomes diabetic and is unable to reabsorb all filtered glucose. The 150 mosmoles of NaCl in the distal nephron are joined by 150 mosmoles of unreabsorbed glucose. Now the <u>amount</u> of solute leaving the loop of Henle has increased from 150 to 300 mosmoles.

Do you think the volume of fluid leaving the loop of Henle will be the same, greater or smaller?

Now do the calculations:

	Osmolarity	Volume
Point A	_____	_____
Point B	_____	_____
Point C	_____	_____

What happened to the volume of urine when the <u>amount</u> of solute entering the distal nephron doubled?

_____ . This shows you the process behind fluid loss in osmotic diuresis.

PRACTICE MAKES PERFECT

1. Compare the following pairs of items. Put the symbols below in the space.
   greater than >   less than <   same as or equal =

A - urine osmolarity in a normal person with maximal ADH
B - urine osmolarity in a diabetic who is excreting glucose with maximal ADH   A _____ B

A - aldosterone secretion when osmolarity is high and blood pressure low
B - aldosterone secretion when osmolarity and blood pressure are both low   A _____ B

A - plasma P_{CO_2} in respiratory acidosis
B - plasma P_{CO_2} in respiratory alkalosis   A _____ B

A - renal reabsorption of HCO_3^- in acidosis
B - renal reabsorption of HCO_3^- in alkalosis   A _____ B

A - ventilation in metabolic alkalosis
B - ventilation in metabolic acidosis   A _____ B

A - renin secretion when blood pressure is high
B - renin secretion when blood pressure is low   A _____ B

A - renal filtration of HCO_3^- in acidosis
B - renal filtration of HCO_3^- in alkalosis   A _____ B

2. What effect does a decrease in mean arterial blood pressure have on each item below?

(a) Na^+ reabsorption in the proximal tubule _____

(b) blood pressure in the afferent arteriole _____

(c) atrial stretch _____

(d) aldosterone secretion _____

(e) angiotensinogen levels in the plasma _____

(f) blood volume _____

3. The labels were left off the axes on the following graph. One axis is vasopressin concentration and one axis is plasma osmolarity. Which is which? Defend your answer.

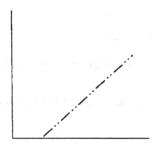

4. It's the night before your physiology test and you awake from a crazy dream to find yourself still at your desk in a pile of books and papers. You remember that in your dream you were walking through a peach orchard. As you were walking down the path, you noticed that all the trees on your left were turning into CO_2 and H_2O. Looking to your right, you noticed that all the trees on that side were turning into H^+ and HCO_3^-. Having thoroughly studied acid-base balance, you realized that you were standing in the middle of the important acid-base equation. As you realized this, a funny new tree popped up. On one side of the tree there were green fruit, and on the other side there were red fruit. You were hungry from all your studying, so your subconscious picked a green fruit from the tree.

As you ate the green fruit, something strange happened (as if this wasn't strange enough already!). The left side of the path (the $CO_2 + H_2O$ side) grew and grew until it looked as if it would topple on top of you. You suddenly realized that you have a touch pad in your hand that would enable you to set the proper compensatory mechanisms in action. What did you do to prevent your impending doom?

Write the equation of interest, show the imbalance that occurred, and describe what physiological mechanisms you employed to compensate for the imbalance. Why did you choose the way you did?

5. In a patient with an acid-base disturbance characterized by a plasma P_{CO_2} of 80 mm Hg and a plasma bicarbonate concentration of 33 mEq/L (normal = 24), you would expect to find
(Circle all correct answers):

         a) virtually complete renal bicarbonate reabsorption despite the elevated plasma bicarbonate

         b) urine pH less than 6.0

         c) stimulation of distal tubule H^+ secretion due to elevated plasma P_{CO_2}

         d) hypokalemia

6. Mr. Osgoode arrives at the hospital hyperventilating and disoriented. You order tests and receive these results: plasma pH 7.31; arterial P_{CO_2} = 30 mm Hg; plasma HCO_3^- = 20 mEq/L (normal = 24). What are the normal values for pH and P_{CO_2}? What is Mr. Osgoode's acid-base state, and how did you come to that conclusion?

7. Ari wants to play a trick on his friends, so he gets a long tube and hides in the bottom of a pond, breathing through the tube. After a few minutes, he feels that he is having trouble getting air and considers surfacing. If you tested his plasma pH, plasma HCO_3^-, and arterial P_{CO_2} at this time, would you expect each of them to be normal, above normal, or below normal? Explain your answers.

MAPS

1. Create a summary map showing the compensations for acid-base disturbances, using the following terms and any others you wish to add.

acidosis
alkalosis
ammonia
bicarbonate
buffer
carbonic anhydrase
carotid and aortic chemoreceptor
central chemoreceptor
chloride shift
CO_2
excretion
filtration

H^+-ATPase
H^+-K^+-ATPase
HCO_3^- -Cl antiport
hyperkalemia
hyperventilation
hypokalemia
hypoventilation
intercalated cell (I cell)
law of mass action
medullary chemoreceptor
metabolic acidosis

metabolic alkalosis
Na^+/K^+-ATPase
Na^+-H^+ antiport
Na^+-HCO_3^- symport
pH
phosphate ion
reabsorption
respiratory acidosis
respiratory alkalosis
respiratory compensation
secretion

2. Use the map on the renin-angiotensin – aldosterone pathway (Fig. 19-12, p. 556) as the basis for a large map. Fill in all the mechanisms and anything else you can think of.

3. Take each of the following situations. Figure out:
        a. what changes will occur in the volume, osmolarity, and sodium concentration of the body.
        b. what homeostatic responses these changes will trigger and how.
Then
        c. describe *in as much detail as possible* the homeostatic response.
        d. explain how the homeostatic response will bring the altered condition(s) back to normal.
On large pieces of paper, map the detailed homeostatic pathways for each situation below.

Example 1: Person ingests a large amount of salt (NaCl) but no water.
Example 2: Person ingests a large amount of salt and drinks enough water so that the result is as if he/she drank a large volume of isotonic saline
Example 3: Person drinks a large volume of pure water
Example 4: Person loses water and salt, but more water than salt through sweating (loses hyposmotic fluid)
Example 5: Person loses a large volume of blood (hemorrhage)
Example 6: Person drinks a large volume of water and ingests enough salt to equal a hyperosmotic solution

HINTS FOR MAPPING VOLUME AND OSMOLARITY PROBLEMS

1. What are the possible stimuli for reflex responses? The primary stimuli are blood pressure and osmolarity (increased, decreased, or no change . . . see the matrix on pg. 560). You should also consider $[Na^+]$, $[K^+]$, and blood volume.

2. Make a list of all receptors/tissues that respond to the stimuli in #1. Don't forget tissues that respond directly to the stimuli, such as the atrial cells that secrete ANP and the aldosterone-secreting cells of the adrenal cortex. Don't forget GFR autoregulation and blood pressure homeostasis via the cardiovascular control center.

3. Once you have established which pathways will be stimulated/inhibited, follow your pathways through to the desired response, which should be the opposite of the stimulus. If you get to the end and have a response that is the same as the stimulus (i.e., the stimulus was increased BP but the response increases BP), go back . . . you've done something wrong or forgotten a link.

4. Sometimes two pathways will have opposite effects on a single integrating center (e.g., one stimulates aldosterone, the other inhibits). Assume that the response that opposes the stimulus is dominant. Alternately, remember that the body defends osmolarity before it defends volume.

BEYOND THE PAGES

TRY IT

Kitchen Buffers

Your kitchen contains the necessary ingredients for a simple demonstration of the bicarbonate buffer system. Pour a couple of tablespoons of white or cider vinegar into a small bowl. Taste the vinegar and notice the sour or tart taste. The sensation of sour taste is directly proportional to the acidity of a solution. Now put a tablespoon of baking soda into another small bowl. Baking soda is sodium bicarbonate, the sodium salt of the bicarbonate buffer.

Remember the equation in which bicarbonate buffers acid:

$$H^+ + HCO_3^- \rightleftarrows CO_2 + H_2O$$

Based on this equation, what do you predict will happen when you pour the vinegar into the baking soda? If you taste the resulting solution, do you think the taste will be the same as it was before? Explain.

Now conduct your experiment. What happens when the vinegar and baking soda mix? This reaction has been used in baking for centuries. Allow the excess baking soda to settle to the bottom of the bowl and taste the remaining vinegar solution. Is it as sour as before? (see answers)

READING

ACE inhibitors: Almost too good to be true. (1994, July/August) *Science & Medicine.*

Activation and supply of channels and pumps by aldosterone. (1996, June) *News in Physiological Sciences* 11: 126-132.

Atrial natriuretic peptide as a regulator of transvascular fluid balance. (1996, June) *News in Physiological Sciences* 11: 138-142.

Control of vasopressin release: An old but continuing story. (1996, February) *News in Physiological Sciences* 11: 7-12.

Endocrinology: People and Ideas (1988) American Physiological Society, pub. Chapters on vasopressin and atrial natriuretic factor.

Guanylin and uroguanylin: Intestinal peptide hormones that regulate epithelial transport. (1996, February) *News in Physiological Sciences* 11: 17-23.

Hepatorenal and hepatointestinal reflexes in sodium homeostasis. (1996, June) *News in Physiological Sciences* 11: 103-107.

Membrane receptors for aldosterone: A new concept of nongenomic mineralocorticoid action. (1993, December) *News in Physiological Sciences* 8: 241-244.

Membrane receptors for aldosterone: A novel pathway for mineralocorticoid action. (1992) *American Journal of Physiology* 263 (5 Pt 1): E974-979.

Molecular physiology of urinary concentrating mechanism: regulation of aquaporin water channels by vasopressin (review). (1997) *American Journal of Physiology* 272: F3-F12.

Natriuretic peptides and receptors. (1994, March/April) *Science & Medicine.*

Progress in oral rehydration therapy. (1991, May) *Scientific American.*

Regulation of atrial natriuretic peptide secretion. (1993, December) *News in Physiological Sciences* 8: 261-265.

Renal vasa recta: Passive filters or active participants? (1996, August) *News in Physiological Sciences* 11: 191-192.

Role of nitric oxide in the control of sodium excretion. (1996, April) *News in Physiological Sciences* 11: 62-66.

Understanding how cells sense their volume. (1996, June) *News in Physiological Sciences* 11: 146-147.

Water channels in renal and nonrenal tissues. (1995, February) *News in Physiological Sciences* 10: 12-17.

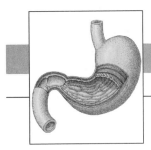

SUMMARY

Here are some points to focus on in this chapter:

- There are four digestive processes: digestion, absorption, motility, and secretion.
- Organize the details of digestion and absorption according to biomolecule class. This information would work well in a chart.
- Learn the muscle contraction patterns and the factors that affect motility.
- The digestive system secretes many different substances. You should recognize the ion secretion processes as being similar to those in the kidneys. As for enzyme secretion, remember that some enzymes are secreted as zymogens. Other than that, it all boils down to memorization (Table 20-1).
- Learn the reflex patterns for GI function and how the enteric nervous system functions.
- What are the three families of digestive hormones? (Table 20-2) How do the hormones participate in regulation of GI function?
- Learn the distinction between different secretory cells in the stomach.
- What are the gastric events following the ingestion of a meal?

If you trace food through the digestive system, it passes through the following structures: mouth, pharynx, esophagus, stomach (fundus, body, antrum), small intestine (duodenum, jejunum, ileum), large intestine, rectum, and anus. Exocrine secretions are added by the salivary glands, pancreas, and liver. The GI tract contains the largest collection of lymphoid tissue, called the gut-associated lymphoid tissue (GALT). The enteric nervous system integrates and initiates GI reflexes.

There are four digestive processes: digestion, absorption, motility, and secretion. Digestion of biomolecules takes place at various locations throughout the GI tract. Carbohydrates are digested in the mouth and small intestine by amylases and disaccharidases. Proteins are digested in the stomach and small intestine by proteases and peptidases. Fat digestion begins in the mouth and small intestine with the help of lipases, but most fat digestion takes place in the small intestine via pancreatic lipase and colipase. Bile from the liver emulsifies fats, increasing the surface area exposed to enzyme action.

Absorption takes place mostly in the small intestine. Because the intestinal wall is composed of transporting epithelium, transepithelial transport mechanisms similar to those in the kidneys are observed. Glucose is transported across the apical membrane by a Na^+-glucose symporter and across the basolateral membrane by a facilitated diffusion carrier. Amino acids are absorbed by a Na^+-dependent cotransporter, while small peptides are absorbed by H^+-dependent cotransporters or by transcytosis. Fat absorption is primarily by simple diffusion. In epithelial cell cytoplasm, monoglycerides and fatty acids are moved to the smooth ER where they combine with cholesterol and proteins to form chylomicrons. Chylomicrons are then transported out of the cells by exocytosis and moved into the lymphatic system ithelial cell cytoplasm, monoglycerides and fatty acids are moved to the smooth ER where they combine with cholesterol and proteins to form chylomicrons. they are delivered into the liver.

Motility is the movement of material through the digestive system. The muscles of the GI tract are single-unit smooth muscle whose cells are connected by gap junctions. Muscle contraction can be generated by nervous or chemical control, and some muscle cells generate spontaneous slow wave potentials. The wall of the intestinal tract is composed of an outer layer of longitudinal muscle and an inner layer of circular muscle. Peristaltic contractions are progressive waves of contraction that propel material from the esophagus to the rectum. Segmental contractions are mixing contractions that churn the material without propelling it forward. Material in the large intestine is moved forward by mass movement and is removed from the body by means of the defecation reflex. Each segment of the GI tract is separated by muscular sphincters that are tonically contracted except when food must move into the next segment or out of the body.

Various components of the digestive system contribute mucus, enzymes, hormones, and paracrines that make up the 7 L secreted by the body into the GI lumen. Digestive enzymes are secreted by either exocrine glands or epithelial cells in the mucosa of the stomach or small intestine (Table 20-1). Because enzymes are proteins, they are synthesized ahead of time, stored in secretory vesicles, and released on demand. To prevent autodigestion and provide an additional level of control, some enzymes are secreted as inactive zymogens that must be activated in the GI lumen. Mucus forms a protective coating over the GI mucosa and provides lubrication. Large amounts of water and ions (Na^+, K^+, Cl, H^+, and HCO_3^-) are also secreted. For example, the stomach secretes H^+-rich fluid while the pancreas secretes HCO_3^-—rich fluid.

GI tract processes are governed by two types of neural reflexes: short reflexes, which take place entirely in the enteric nervous system, and long reflexes, which are integrated in the CNS. Stimuli for long reflexes may originate either inside or outside the GI tract. Hormones controlling GI processes can be grouped into three categories: gastrin family, secretin family, and those not belonging to either. Digestive hormones can have trophic effects or can directly affect brain regions to influence behavior.

TEACH YOURSELF THE BASICS

FUNCTION AND PROCESSES OF THE DIGESTIVE SYSTEM

• What is the primary function of the GI tract? _____

• List two important challenges the GI tract must overcome to carry out this function. _____

_____

• List and define the four basic processes of the digestive system. (Fig. 20-1)

1. _____

2. _____

3. _____

4. _____

• What happens to nutrients brought into the body through the digestive system?

_____

• Why does the GI tract have the most immune tissue of any organ?

_____

• What is the name given to the GI system's immune tissue? _____

ANATOMY OF THE DIGESTIVE SYSTEM

• What are the first steps in digestion?

_____

• The GI tract is a long tube lined with _____ epithelium and surrounded by _____

  muscle. It is closed off by _____ muscle sphincters at the ends.

• The combination of ingested food and digestive secretions forms a mixture known as _____.

The Digestive System Consists of the Gastrointestinal Tract and Accessory Glandular Organs

• Trace a piece of food that enters the mouth through the digestive system, and follow its undigested portion until it is excreted. Name all sphincters the food passes. (Fig. 20-2)

_____

_____

_____

_____

• List the three sections of the stomach. _____

• List the three sections of the small intestine. _____

• The digestive waste that leaves the body is called _____.

The Gastrointestinal Tract Wall Has Four Layers

• List the four layers of the GI wall in stomach and intestines, from inside to outside. (Fig. 20-2c,f).

_____

The mucosa

• List and describe the three layers of the mucosa. _____

_____

• What anatomical modifications increase surface area facing the lumen of the stomach and intestine? Use the proper names given to these modifications.

_____

• The tubular invaginations of the lumen that extend into supporting connective tissue are called

_____ in the stomach and _____

in the small intestine.

• The surface area on the apical membrane of epithelial cells is increased by the presence of

_____. (Fig. 20-2h) [∫ p. 47]

• In the stomach, four types of cells are modified for secretion (Fig. 20-2d):

a. _____ cells secrete mucus   b. _____ cells secrete pepsin

c. _____ cells secrete HCl   d. _____ cells secrete gastrin

• What is the role of absorptive cells in the intestine? _____

• List the three other types of intestinal cells and give their functions.

_____

How do the cell-to-cell junctions of the stomach and intestine differ?

_____

• What is found in the layer known as the lamina propria? _____

_____

• What are Peyer's patches? _____

• Describe the muscularis mucosa. _____

_____

The submucosa
• What structures are found in the submucosa? _____

_____

• Describe the enteric nervous system and explain why it is unique. [∫ p. 202]

_____

Musculature and serosa
• The outer intestinal wall consists of two smooth muscle layers. (Fig. 20-2c,e,f) Contraction of the

_____ layer decreases lumen diameter, while contraction of the _____ layer

shortens the length of the tube. The stomach has an incomplete third muscle layer, arranged

_____.

• What is the myenteric plexus and where is it found?

_____

• What is the serosa and where is it found? _____

_____

• What are the peritoneum (peritoneal membrane) and mesentery? _____

_____

MOTILITY
• What are the two functions of motility? _____

_____

Gastrointestinal Smooth Muscle Contracts Spontaneously
• Most of the intestinal tract is composed of _____-unit smooth muscle whose

cells are electrically connected by _____ junctions. [∫ p. 350]

• Pacemaker muscle cells generate _____ wave potentials (Fig. 20-3a) that do what when

they depolarize to threshold? _____

• Explain pharmacological coupling in smooth muscle. [∫ p. 354] _____

_____

_____

GI Smooth Muscle Exhibits Different Patterns of Contraction

• Name and describe the three main patterns of muscle contraction in the gut.

1. _____

2. _____

3. _____

• Name the seven sphincters in order beginning in the mouth. _____

_____

• What is a bolus? _____

• What do muscles in the receiving segment do? _____

• Describe the peristaltic reflex. _____

_____

• This reflex is mediated primarily through which division of the nervous system?

_____

Movement of Food through the Gastrointestinal Tract

• What is mastication? _____

• What is the function of saliva? _____

• What is deglutition? _____

• Describe the steps of swallowing a bolus of food. (Fig. 20-5) _____

_____

• The upper esophageal sphincter is normally (open / closed ?). The lower esophageal sphincter is normally (open / closed ?).

• What is reflux esophagitis? _____

• Describe how food moves through the stomach and into the small intestine. _____

_____

• What are mass movement contractions? _____

_____

• Describe the defecation reflex. Where is this reflex integrated? _____

_____

SECRETION

• List the sources and volumes of fluid input into the GI tract. _____

_____

• List the routes and volumes of fluid removal from the GI tract. _____

_____

• Explain why continuing untreated diarrhea will become a medical emergency. _____

_____

Digestive Enzymes Are Secreted in the Mouth, Stomach, and Intestine

• List the cells, glands, and organs that secrete digestive enzymes. (Table 20-1)

_____

• Describe the synthesis, storage, and release of digestive enzymes. [∫ p. 127] _____

_____

• What is the brush border? _____

• What distinguishes brush border enzymes? _____

• Enzymes secreted in inactive form are known collectively as _____.

• Explain how these inactive enzymes are activated. (Fig. 20-7) _____

_____

• How do you recognize the name of an inactive enzyme? _____

• The secretion of saliva is under (neural / hormonal ?) control. Secretion of gastric and intestinal enzymes is under neural (sympathetic / parasympathetic ?) and hormonal control.

Mucus Is Secreted by Specialized Cells

• Mucus is composed of glycoproteins called _____.

• Name two functions of mucus. _____

• Mucus is made by specialized cells: _____ cells in the stomach, _____

cells in the intestine, and _____ cells in the mouth.

• List three different signals for mucus release. _____

The Digestive System Secretes Substantial Amounts of Fluid and Ions

• List the five major ions found in digestive secretions. _____

• Movement of water and ions in the GI tract is very similar to water and ion movement in what other organ?

_____

Sodium, potassium, and chloride movement

• List the carriers and channels that participate in transepithelial transport of Na^+.

_____

• Describe one form of Na^+ transport that does not occur in the kidney.

_____

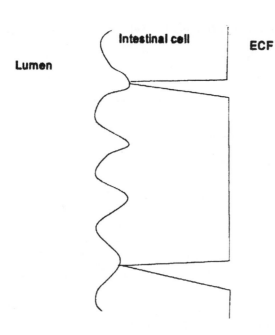

• Arrange the following transporters on the epithelial cell at left so that it secretes Cl^- with Na^+ and water following passively. (Fig. 20-8a)

Na^+-K^+-$2Cl^-$ symporter

CFTR Cl^- channel

K^+ leak channel

Na^+/K^+-ATPase

• Now arrange the same transporters so that they secrete fluid rich in KCl.

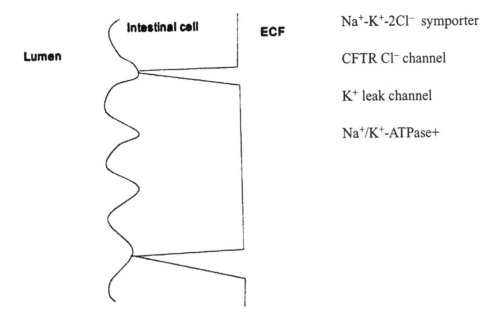

Na⁺-K⁺-2Cl⁻ symporter

CFTR Cl⁻ channel

K⁺ leak channel

Na⁺/K⁺-ATPase+

Hydrogen ion and bicarbonate ion transport

• Secretion of H⁺ and HCO_3^- in the GI tract is similar to the acid-base handling processes of what

organ? [∫ p. 567] _____

• Arrange the following items on the cell below so that it secretes acid and reabsorbs bicarbonate.

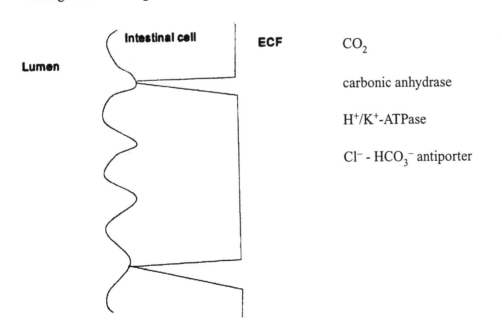

CO_2

carbonic anhydrase

H⁺/K⁺-ATPase

Cl⁻ - HCO_3^- antiporter

DIGESTION AND ABSORPTION
Overview of Enzymatic Digestion

• Define digestion. _____

• What is the role of enzymes in digestion? _____

• How does the chewing and churning of chyme assist enzymatic digestion? _____

• Enzymes in the stomach work best at a (high / low ?) pH, while those in the intestine work best at a (high / low ?) pH.

Overview of Nutrient Absorption

• Where does most nutrient absorption take place? _____

• Name two substances absorbed in the stomach. _____

• Describe the anatomical route followed by most nutrients after they reach the blood.

_____

• What is a portal system? [∫ p. 185] _____

• Why does the liver have two different sets of capillaries, one from the hepatic artery and one from the hepatic portal vein?

_____

• Why can't most digested fats enter the hepatic portal system? _____

Carbohydrates Are Absorbed as Monosaccharides

• In what two locations of the GI tract does most carbohydrate digestion take place? (Fig. 20-11)

_____

• Name the enzyme that is responsible for the following reactions. What part(s) of the GI tract secretes it?

maltose ⇒ monosaccharides _____

starch ⇒ maltose _____

sucrose ⇒ monosaccharides _____

lactose ⇒ monosaccharides _____

• In what monosaccharides are the following disaccharides digested?

maltose ⟹ _____   sucrose ⟹_____

lactose ⟹ _____

• Why are chloride ion's an important component of saliva? _____

• In glucose absorption, movement across the apical membrane uses a _____

(transporter), and movement across the basolateral membrane uses a _____

carrier. [∫ Fig. 5-26, p. 129 and 18-11, p 531]

• Describe fructose absorption. _____

• The monosaccharide facilitated transporters belong to the _____ transporter family.

• Intestinal epithelial cells use _____ as their main energy source.

Proteins Are Digested to Small Peptides and Amino Acids

• Name the type of enzyme that is responsible for each of the following reactions. What part(s) of the GI tract secretes it? (Fig. 20-12, 20-13)

Peptides ⟹ smaller peptides, amino acids _____

Proteins ⟹ smaller peptides _____

Breaks interior peptide bonds to make smaller peptides _____

Breaks exterior peptide bonds to make single amino acids _____

• What prevents these enzymes from digesting the cells that make them? _____

_____

• What protease works best in the acidic environment of the stomach? _____

• Describe how single amino acids are absorbed. (Fig. 20-14)

_____

• Why is it necessary for the cells to have multiple amino acid transport systems?

_____

• Describe how small peptides are absorbed.

_____

• Describe how some larger peptides can be absorbed intact. [∫ p. 130] _____

_____

Fat Digestion Is Assisted by Bile

• Name five common forms of fat or fat-related molecules in the Western diet.

_____

• Why is it necessary to emulsify fats and make them into micelles? _____

_____

• What is the role of the following in fat digestion? From what organ/gland/cell is each secreted?

bile _____

lipase (Fig. 20-15) _____

colipase _____

phospholipase _____

• How is cholesterol digested? _____

• Digestion of a triglyceride yields _____

• How are fats absorbed into intestinal epithelial cells? _____

The roles of bile and colipase
• What is in bile ? _____

• What is bilirubin? _____

• Follow the route taken by bile as it moves from the hepatocyte into the lumen of the intestine.

_____

• How do bile acids interact with fats to make micelles? _____

• What is a micelle? [∫ p. 109] _____

• How is cholesterol absorbed into the epithelial cell? _____

• What is a chylomicron and what is the fate of chylomicrons? _____

_____

• Trace the anatomical route followed by a chylomicron as it leaves the intestinal epithelium cell. [∫ p. 440]

_____

• What happens to bile salts that are secreted into the intestine? _____

_____

Nucleic Acids Are Digested into Bases and Monosaccharides

• What are nucleic acids digested into? [∫ Fig. 2-17, p. 32] _____

• Describe the absorption of nucleic acids. _____

_____

Vitamins and Minerals Are Absorbed Along with Nutrients

• Compare the absorption of fat-soluble and most water-soluble vitamins.

_____

• Name the fat-soluble vitamins. _____

• Describe the absorption of vitamin B_{12}. _____

_____

• Mineral absorption usually occurs by _____ transport.

• Name two minerals whose absorption increases when body concentration decreases. _____

Water and Electrolyte Absorption Is Similar in the Kidney and Intestine

• Describe the basic pattern for solute and water absorption in the intestine.

_____

Very Little Digestion Occurs in the Large Intestine

• Name three actions of bacteria in the large intestine.

_____

REGULATION OF GI FUNCTION

• What is unique about the enteric nervous system (ENS)? _____

_____

• Describe the relationship between the ENS and the CNS. _____

_____

The Enteric Nervous System Is Known as the Little Brain

• Compare short and long reflexes. (Fig. 20-19) _____

_____

_____

• In the _____ phase of digestion, GI reflexes originate completely outside the ENS.

• Name some examples of stimuli that initiate a reflex in this phase.

_____

• What is the adaptive significance of this phase of digestion? _____

_____

• In general, cholinergic parasympathetic innervation of the GI tract is (excitatory / inhibitory ?) while

  adrenergic sympathetic innervation is (excitatory / inhibitory ?).

• Describe four ways that the enteric nervous system is anatomically and functionally like the cephalic brain.

_____

_____

_____

_____

• What does it mean if a neuron is described as being "non-adrenergic, non-cholinergic?"

_____

Digestive Hormones Control GI Function, Metabolism, and Eating Behavior

• Name the two physiologists who discovered the first digestive hormone.

_____

• Give two reasons that research on GI hormones lagged behind research on other hormones.

_____

Classification of GI hormones
• Name the three groups of GI hormones.

_____

• What do the following abbreviations stand for?

CCK _____ VIP _____

GIP _____

Synthesis sites of GI hormones
• Where are the following GI hormones synthesized/secreted?

gastrin _____ secretin _____

CCK _____ GIP _____

motilin _____ VIP _____

Physiological effects of GI hormones
• Hormones made in cells of the digestive tract are secreted into the _____

and reach their target cells by what process? _____

• What is satiety? _____

• What is a trophic hormone? [∫ p. 182]

_____

• For each of the following hormones, tell what stimulates its release and list its effects on secretion, motility, and other GI functions. (Table 20-2)

CCK _____

_____

_____

Gastrin _____

_____

_____

VIP _____

_____

_____

Secretin _____

_____

_____

GIP _____

_____

_____

Motilin _____

_____

_____

Signals for hormone release
• In general, what is the primary stimulus for GI hormone secretion? _____

☛ *Learn the stimuli for hormone release and use those to remember the effects of the hormone, as the two are usually associated.*

• Predict the stimuli for hormone release by looking at hormone function:

CCK causes bile release: Release is stimulated by _____

GIP causes insulin release: Release is stimulated by _____

Secretin triggers pancreatic HCO_3^- secretion: Release is stimulated by _____ in small intestine

• An increase in (parasympathetic / sympathetic ?) activity following a meal stimulates gastrin release.

Paracrines in the GI Tract Have Diverse Effects on Digestion

• Describe two routes by which GI paracrines can reach their targets.

_____

• Describe two examples of GI paracrines. _____

_____

INTEGRATION OF GI FUNCTION: THE STOMACH

☞ *Figure 20-21 is a summary figure for GI processes.*

Secretion Is a Major Function of the Stomach

• Describe the secretions of the following cells in the stomach. (Table 20-3)

Mucus cells _____

     What is the function of these secretions? _____

Chief cells _____

     What is the function of these secretions? _____

Parietal cells _____

     What is the function of these secretions? _____

G cells _____

     What is the function of these secretions? _____

Enterochromaffin cells _____

     What is the function of these secretions? _____

D cells _____

      What is the function of these secretions? _____

Gastric Events Following Ingestion of a Meal

• Digestion is divided into what three phases? _____

The cephalic phase of gastric function
• Fill in the following steps of a cephalic phase reflex:

Stimuli _____

Receptor(s) _____

Integrating center _____

Efferent path: parasympathetic neurons that travel to the GI tract in the _____
nerve.

Effector 1: _____ Response 1: secrete HCl

Effector 2: _____ Response 2: release histamine

Effector 3: _____ Response 3: secrete gastrin

The gastric phase of gastric function
• Fill in the following steps of a gastric phase reflex:

Stimuli _____

Receptor(s) _____

Integrating center _____

Efferent path: neurons of the enteric nervous system

Effector 1: _____ Response 1: increased motility

Effector 2: _____ Response 2: pepsinogen secretion

Effector 3: _____ Response 3: acid secretion

• What is the net effect of the gastric phase reflexes? _____

_____

The intestinal phase of gastric function and feedback signals

• What initiates the intestinal phase? _____

• The overall goal of coordinating gastric and intestinal function is to _____

_____.

• Acidic chyme in the duodenum causes release of the hormone _____,

which in turn (excites / inhibits ?) gastric motility, (excites / inhibits ?) acid secretion, and (excites

/ inhibits ?) pancreatic HCO_3^- secretion.

• If a meal contains fats, the hormone _____ is secreted. This hormone

(slows / increases ?) gastric motility and (slows / increases ?) acid secretion.

• If a meal contains carbohydrates, the hormone _____ is secreted.

•The mixture of acid, and enzymes, digested food known as _____ is usually

_____osmotic to body osmolarity.

• Intestinal osmoreceptors respond to this osmolarity by (exciting / inhibiting ?) gastric emptying.

TALK THE TALK

5-hydroxytryptamine (5-HT)
absorption
absorptive cell
acid lipase
alcohol
amylase
anal sphincter
antrum
anus
Beaumont, William
bicarbonate secretion, stomach
bile
bile acid
bile canaliculi
bile salt
bilirubin
bolus
brush border
cell-to-cell junctions
cellulose

cephalic phase
chief cell
chloride secretion
cholecystokinin (CCK)
cholera toxin
cholesterol transport
chylomicron
chyme
Cl-HCO_3^- antiporter
cobalamin
colipase
common bile duct
crypt
cystic fibrosis
cystic fibrosis transmembrane
                regulator (CFTR channel)
D cell
DDAVP (1-deamino-8-D-arginine
                vasopressin)
defecation reflex

deglutition
diarrhea
digestion
disaccharidase
disaccharide
duodenum
emulsion
endopeptidase
enteric nervous system
enterochromaffin cell
enteroglucagon
enteropeptidase
enterostatin
enterotoxin
erythropoiesis
Escherichia coli
esophageal sphincters
esophagus
exopeptidase
feces

fistula
flatus
food allergies
fructose
fructose absorption
fundus
G cell
gallbladder
gap junction
gastric gland
gastric inhibitory peptide
gastric lipase
gastric phase
gastrin
gastrin-releasing peptide
gastritis
gastrointestinal tract
GIP
glucagon
glucose
glucose transport
glucose-dependent insulinotropic
       peptide (GIP)
GLUT transporter
gluten allergy
glycogen
goblet cell
gut-associated lymphoid tissue
       (GALT)
H^+-K^+-ATPase
H^+-peptide cotransport
heartburn
Helicobacter pylori
hepatic artery
hepatic portal system
hepatocyte
histamine
histamine receptor (H2 receptor)
ileocecal sphincter
ileum
intestinal phase
intrinsic factor
jejunum
lactase
lactose
lactose intolerance
lamina propria

large intestine
leaky epithelium
lingual lipase
lipid digestion
liver
lobule
long reflex
maltase
maltose
mass movement
mastication
mesentery
micelle
microvilli
monosaccharide
motilin
motility
mucin
mucosa
mucus cell
muscularis mucosa
myenteric plexus
Na^+-K^+-$2Cl^-$ symporter
Na^+-K^+-ATPase
Na^+-amino acid cotransport
Na^+-glucose symporter
nitric oxide (NO)
non-adrenergic, non-cholinergic
       neurotransmitter
non-steroidal anti-inflammatory
       drugs (NSAIDs)
nucleic acid digestion
Olestra
oral cavity
osmoreceptor
pancreas
pancreatic polypeptide
paracrines in the GI tract
parietal cell
pepsin
pepsinogen
peptic ulcer disease
peptidase
peptide YY
peristaltic contraction
peristaltic reflex
peritoneal membrane

pernicious anemia
Peyer's patch
pharmacomechanical coupling
plicae
protease
pyloric sphincter
pylorus
receiving segment
rectum
reflux esophagitis
rugae
saliva
satiety
secretin
secretion
segmental contraction
serosa
serotonin
short reflex
single-unit smooth muscle
sinusoid
slow wave potential
small intestine
somatostatin
sphincter
sphincter of Oddi
starch digestion
stomach
submucosal gland
submucosal plexus
sucrase
sucrose
tonic contraction
transcytosis
triglyceride
trypsin
trypsinogen
vagal reflex
vasoactive intestinal peptide
       (VIP)
villi
vitamin
vitamin B_{12}
Zollinger-Ellison syndrome
zymogen

ERRATA

p. 584, right column, first sentence under header "Digestive enzymes...": Digestive enzymes are secreted (not selected).

p. 598, right column, #2 in list: GIP was not found to block acid secretion (not secretin)...

p. 600, Table 20-2: First topic under ìEffectsî should be endocrine secretion. Last column head is misspelled, should be somatostatin.

RUNNING PROBLEM - Peptic Ulcer Disease

An interesting correlation has been observed between blood type, duodenal ulcers, and the ability of *Helicobacter pylori* to bind to host cell receptors in the presence of blood group antigens. *H. pylori* does not bind to the receptor in the presence of blood type A or B antigens. This observation may be related to the fact that people with blood group O (no A or B antigens) have a 30% increased risk of developing duodenal ulcers. For a discussion and references, see Letter. (1994, June 3) *Science* 264: 1387-1388.

PRACTICE MAKES PERFECT

1. Number the following structures of the gastrointestinal tract in the order in which food passes:

_____ Stomach   _____ Ileum   _____ Esophagus

_____ Ascending colon   _____ Pyloric sphincter _____ Duodenum

2. Why doesn't salivary amylase work in the stomach?

3. Pepsin is produced by the _____, and trypsin is produced by the _____.

Pepsin is active only at _____ H^+ concentrations, while trypsin is active at _____ H^+ concentrations.

4. How does the digestive system prevent autodigestion?

5. True or false? Be able to defend your answer.

a) The pathway from the intestinal lumen to the circulating blood for a short-chain fatty acid (less than 10 carbon atoms) is intestinal mucosa cell to chylomicrons to capillary to systemic venous blood in the hepatic portal vein.

b) Most of the bile salts secreted by the liver into the lumen of the intestine are excreted in the feces.

c) The amount of water that is reabsorbed from the intestinal tract in a day is equal to the amount of water that is ingested in a day

d) If the blood supply to the small intestine decreases dramatically so that the cells become hypoxic, the absorption of glucose will decrease.

6. The parietal cells of the stomach secrete hydrochloric acid.

a) Based on your knowledge of acid secretion in the kidney, draw the mechanism of HCl secretion in the parietal cell.

b) Based on the model you just drew, what kind of acid-base disturbance would the excessive vomiting of stomach contents cause? In one sentence, defend your answer.

7. Why is the enteric nervous system considered by some to be a third division of the nervous system that is equivalent to the CNS and peripheral nervous system?

8. Why do physicians recommend that parents not feed infants gluten-based cereals until they are several months of age?

9. Fill in the reflex pathway below. The stimulus and response are given.

Stimulus: test anxiety

Receptor(s): _____

Afferent pathway:_____

Integrating center: _____

Efferent pathway: _____

Effector(s): _____

Cellular response: _____

Tissue response: _____

Systemic response: acid stomach syndrome

10. In no more than two phrases per pair, compare/contrast the four processes of the urinary system to the four processes of the digestive system.

11. You have been doing some experiments with lingual and gastric lipase, but someone marked the test tubes with a marker that rubbed off. You know that one tube has the gastric enzyme and one tube has the lingual enzyme. What test could you do with the enzymes to tell which is which?

MAPS

1. Start with some sucrose and starch molecules in a candy bar. Follow these molecules through ingestion, digestion, and absorption as they move through the digestive tract.

2. Map digestive smooth muscle contraction, using the following terms and any others you wish to add.

circular layer	paracrines in the GI tract	short reflex
gap junction	peristaltic contraction	single-unit smooth muscle
longitudinal layer	peristaltic reflex	slow wave potential
motility	pharmacomechanical coupling	tonic contraction
muscularis mucosa	receiving segment	
myenteric plexus	segmental contraction	

3. Map the long and short reflexes of the gastrointestinal system.

4. Create a map of stomach function as described on pp. 601-605.

5. Compile information from the entire chapter to create a map of small intestine function.

BEYOND THE PAGES

Try It
Digestion of starch by salivary amylase

You can sense the changes that take place during starch digestion by chewing on a soda cracker and holding it in your mouth rather than swallowing it. For best results use a starchy, unflavored soda cracker like a Saltine®. Put it in your mouth and start chewing. What happens to the taste as you chew? Rather than swallowing the chewed cracker right away, keep it in your mouth for a minute or two and see if the taste continues to change. Your salivary amylase acts on the starch in the cracker, breaking it up into smaller glucose polymers and the disaccharide maltose. Maltose has about 40% of the sweetness of sugar, so you should notice the taste of the cracker changing as you chew.

Terminology
GI physiology is full of technical words for functions that go by much more common names. Students enjoy learning about them:

mastication	= chewing	deglutition	= swallowing
emesis	= vomiting	eructation	= burping
flatulence	= having intestinal gas (flatus)		
borborygmi	= rumbling noises in the GI tract from intestinal gas		

Organizing study
This is another chapter that is top heavy with terminology and information that needs to be memorized. There are several ways that you can organize the information in the chapter for study. One is to go anatomically, section by section, and outline everything that happens in each section. Within a section, you can subdivide into the four processes of motility, secretion, digestion, and absorption. For the latter two sections, you can further subdivide according to nutrients. Another way to organize is to take the three major classes of biomolecules and follow them one at a time from ingestion through absorption. Whichever way you organize it, maps are a great method for putting lots of integrated information into a relatively compact space. You may want to buy some poster board for the maps in this chapter, though!

Here is an example of a chart to make:

	Carbohydrate	Fat	Protein
List ALL places in the digestive tract where significant digestion of this takes place. What enzymes?			
In what form is this group of foods absorbed? (i.e., What is the product of digestion?)			
Describe the process of absorption.			

READING

Alcohol and the liver. (1995, March/April) *Science & Medicine.*

The artificial liver. (1995, May/June) *Science & Medicine.*

The bacteria behind ulcers. (1996, February) *Scientific American.*

Barrett's esophagus. (1994, November/December) *Science & Medicine.*

Bugs and barriers: Enteric pathogens exploit yet another epithelial function. (1995, August) *News in Physiological Sciences* 10: 160-165.

Cholecystokinin type A and type B receptors and their modulation of opiate analgesia. (1997, December) *News in Physiological Sciences* 12: 263-267.

The enteric neuroimmune system. (1997, October) *News in Physiological Sciences* 12: 245-246.

Gastric enterochromaffin-like cells and the regulation of acid secretion. (1996, April) *News in Physiological Sciences* 11: 57-61.

The gastrointestinal tract in growth and reproduction. (1989, July) *Scientific American.*

Genetics of colon cancer. (1997, July/August) *Science & Medicine.*

Hepatocyte transplantation. (1994, November/December) *Science & Medicine.*

Intestinal oligopeptide transporter: From hypothesis to cloning. (1996, June) *News in Physiological Sciences* 11: 133-137.

Intestinal iron absorption: Cellular mechanism and regulation. (1997, August) *News in Physiological Sciences* 12: 184-188.

Neuronal inhibition determines the gastrointestinal motor state. (1996, April) *News in Physiological Sciences* 11: 67-72.

New concepts in hepatocellular transport and metabolism of bilirubin. (1995, February) *News in Physiological Sciences* 10: 35-41.

Novel kinase signaling cascades in pancreatic acinar cells. (1997, June) *News in Physiological Sciences* 12: 117-120.

Paige, D.M. and Bayless, T.M., eds. (1981) *Lactose Digestion: Clinical and Nutritional Implications.* Johns Hopkins University Press, Baltimore.

Peristalsis. (1994, November/December) *Science & Medicine.*

Rapid adaptation of intestinal sugar transport. (1994, April) *News in Physiological Sciences* 9: 84-87.

Regulation of ion transport in colonic crypts. (1997, April) *News in Physiological Sciences* 12: 62-66.

GI transporters

See *Annual Review of Physiology*, Vol. 55, 1993:

Molecular Basis of GI Transport	p. 571
Intestinal Na+/Glucose Cotransporter	p. 575
Facilitated Glucose Transporters	p. 591
Cystic Fibrosis Transmembrane Conductance Regulator	p. 609
Disorders of Intestinal Ion Transport	p. 631

Mammalian passive glucose transporters. (1993) *Biochimica et Biophysica Acta* 1154: 17-49.

Glucose transporter gene expression. (1995) *Pharmacology & Therapeutics* 66 (3): 465-505.

Facilitative glucose transporters. (1994) *European Journal of Biochemistry* 219 (3): 713-725.

Olestra

Olestra and the FDA. (1996, Apr 11) *New England Journal of Medicine* 334 (15): 984-986.

Are we ready for fat-free fat? (1996, Jan 8) *Time* 147 (2): 52.

Letting the chips fall where they may. (1996, March) *Tufts University Diet & Nutrition Letter* 14 (1): 2.

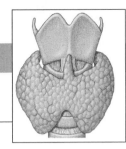

SUMMARY

This chapter is absolutely full of information! I hope you're not trying to do this one at the last minute. Here are some main points from this chapter:

- Anabolic and catabolic metabolic states are balanced to ensure that the body (especially the brain) has an adequate supply of fuel.
- Control of metabolic states is largely under endocrine control. Hour-to-hour control of metabolism is determined by the ratio of insulin to glucagon. Long-term metabolic control also involves cortisol, catecholamines, growth hormone, and the sex hormones.
- Abnormalities in insulin production or activity can lead to a condition called diabetes mellitus. Allow plenty of time to learn and understand the pathways in this condition!! Diabetes involves almost every pathway you've encountered throughout this textbook.
- Growth hormone, along with insulin, thyroid hormones, and sex hormones, controls growth of soft tissue and bone.
- Bone is dynamic tissue – always being formed and resorbed. This primarily to ensure adequate levels of free calcium.

Metabolism is the sum of the chemical reactions in your body that extract and use energy or store it for later use. There are two general types of metabolic reactions: anabolic (synthesis) and catabolic (breakdown). Normally, these reactions are balanced so that energy intake equals energy output. Energy is used to fuel biological work such as transport work, mechanical work, and chemical work. Body temperature regulation is also closely linked with metabolism.

Metabolism is divided into two states: fed (absorptive) and fasting (postabsorptive). The fed state is predominantly anabolic and the fasting state is predominantly catabolic. During the fed state, energy storage compounds are made. These include glycogen (liver, skeletal muscles) and fat (adipose). When the body enters the fasting state, these compounds are broken down to provide energy. Reactions involved in glycogen processing are: glycogenesis (glucose glycogen) and glycogenolysis (glycogen glucose). Glucose can also be synthesized from noncarbohydrate precursors. Fats and proteins are converted into glucose through a process called gluconeogenesis. Different enzymes control the forward and reverse reactions in these processes, adding an additional level of control to metabolism called push-pull control. The goal of all this is to maintain plasma glucose concentrations so that the brain receives enough glucose.

The hour-to-hour control of glucose concentration depends on the ratio of insulin to glucagon. These are hormones secreted by islets of Langerhans cells in the endocrine pancreas: beta cells produce insulin and alpha cells produce glucagon. Insulin and glucagon work antagonistically (remember hormone interactions from Chapter 6?) to maintain plasma glucose. Insulin dominates during the fed state, and glucagon dominates during the fasted state. Insulin lowers glucose levels by facilitating glucose uptake and utilization primarily in liver, adipose, and skeletal muscle cells. Glucagon has the opposite effect from that of insulin. It raises plasma glucose by initiating glycogenolysis and gluconeogenesis and increasing hepatic glucose output.

Diabetes mellitus is a collection of diseases marked by abnormal secretion or activity of insulin. BEWARE: This one topic integrates almost every control pathway you have learned in this course! Be sure to give yourself plenty of time to learn and understand every pathway involved. Once you have mastered the integration of diabetes, you can rest assured that you have reached a new level in understanding human physiology. Insulin-dependent diabetes (IDDM) is the more severe of the diabetes mellitus conditions. With IDDM, despite high plasma glucose concentrations, lack of insulin prevents most cells from taking up glucose. Therefore, the body is tricked into thinking it is in a constant state of fasting. The cells are breaking down proteins and fats, leading to muscle wasting and ketoacidosis. Suprathreshold glucose concentrations lead to glucosuria, which in turn causes osmotic diuresis and polyuria. Fluid loss leads to dehydration, which triggers homeostatic renal and cardiovascular reflexes. If all these problems aren't resolved, coma and death are inevitable. However, insulin injections and electrolyte therapy can correct the imbalances and prevent death. Study Figure 21-16 on p. 631 to see the integration of diabetes control. All the pathways are there, but there are some details left out (on purpose). See if you can supply additional detail to this map.

Four other hormones play roles in regulating metabolism. They are cortisol (from the adrenal cortex), catecholamines (adrenal medulla), thyroid hormones (thyroid gland), and growth hormone (anterior pituitary). The body uses these hormones for long-term control of metabolism.

Normal growth requires adequate amounts of growth hormone, thyroid hormones, insulin, and sex hormones. Growth hormone controls the release of somatomedins from the liver, which in turn stimulate bone and soft tissue growth. Bones increase in length by depositing new matrix at the epiphyseal plates.

The extracellular matrix of bone contains large amounts of calcium phosphate. Calcium is extremely important to our physiology, so the calcium stored in bones serves as a reservoir that can be tapped when free plasma calcium levels drop. Bone is constantly being formed and resorbed depending on calcium needs. Hormone control of calcium levels involves parathyroid hormone, calcitrol, and calcitonin.

TEACH YOURSELF THE BASICS

ENERGY BALANCE AND METABOLISM

• Define metabolism. _____

• What is the distinction between anabolic and catabolic pathways? _____

_____

Energy Input Equals Energy Output

• The first law of thermodynamics states that _____

• Energy input into the body comes in the form of _____.

• Energy output by the body takes one of two forms: _____.

• List the three kinds of biological work and give an example of each. [∫ p. 74]

1. _____

2. _____

3. _____

Body Temperature Is Regulated by Balancing Heat Production, Gain, and Loss

• How are body fat stores related to metabolic efficiency? _____

_____

Body temperature in humans
• What are homeothermic animals? _____

• What is the normal range for human body temperature range (°C)? _____

• List three factors that alter body temperature throughout the day.

_____

• When is body temperature normally highest? _____ lowest? _____

Heat gain and loss
• List two sources of internal heat production. (Fig. 21-2) _____

• Describe two types of thermogenesis. _____

_____

• What is the difference between radiant and conductive heat gain? _____

_____

• Heat loss takes place by four different routes. List and describe them.

1. _____

2. _____

3. _____

4. _____

• Which of the four forms of heat loss above is usually most significant? _____

Regulation of body temperature
• What is the thermoneutral zone? _____

_____

• What is the temperature range for the human thermoneutral zone? _____ °C (_____ °F)

• Above this range, heat production (> / < ?) heat loss; below this range, heat loss (> / < ?) production.

• The greatest physiological challenge to homeostasis comes from (hot / cold ?) temperatures.

• The integrating center for control of body temperature resides in the _____.

• The body's thermoreceptors monitor temperature in what locations? _____

• Heat loss is achieved by what two mechanisms? _____

• Heat generation is achieved by what two mechanisms? _____

Cutaneous Blood Flow

• Superficial cutaneous blood vessels lose heat when cutaneous blood flow (increases / decreases ?).

• The primary regulation of cutaneous blood flow comes from (local chemical controls / neural regulation ?).

• Sympathetic cholinergic neurons cause cutaneous (vasoconstriction / vasodilation ?) and heat (loss / retention ?).

• According to another hypothesis, sympathetic neurons cause sweat glands to release _____,

 which in turn causes (vasoconstriction / vasodilation ?).

• What are arteriovenous anastomoses and what role do they play in the regulation of cutaneous blood flow?

_____

_____

Sweat

• How does sweat promote surface heat loss? _____

• Describe the histological organization of sweat glands. _____

• What kind of neuron innervates sweat glands? _____

• Secretion and reabsorption of water and NaCl in the sweat glands are similar to what other processes

 you have studied? [∫ p. 554] _____

• Sweat is (hyper / hypo / iso ?) osmotic to body fluids.

• Why is it more difficult to lose body heat in humid environments? _____

_____

Heat Production
• Give two examples of unregulated heat production in the body. _____

_____

• Give two examples of regulated heat production in the body. _____

_____

• Heat production in brown fat cells in nonshivering thermogenesis results from what two metabolic events?

_____

• Nonshivering thermogenesis is promoted by the hormone _____ and by

efferent_____ neurons.

• What is thermogenin? _____

☛ *The body's responses to high and low temperatures are summarized in Fig. 21-4.*

Abnormal Temperature Regulation
• Physiological temperature variations, such as fever, are thought to be a result of what mechanism?

_____

• What are pyrogens? _____

• Explain how abnormally high Ca^{2+} concentrations in skeletal muscle cytoplasm are linked to excessive heat production in the condition malignant hyperthermia.

_____

• Why is hypothermia dangerous? _____

• Why do surgeons sometimes induce a state of hypothermia in patients? _____

_____

Energy Balance Is Reflected by an Individual's Oxygen Consumption

Measurement of metabolic rate
• Explain direct calorimetry. _____

_____

• The energy content of food is measured in what units? _____

• This unit is defined as _____.

• True or false? The metabolic energy content of food is equal to the content measured by direct calorimetry.

• Metabolic energy content of proteins and carbohydrates equals _____ kcal/g; that of fats

    is _____ kcal/g.

• What is the most accurate way to measure a person's metabolic rate? _____

_____

• What is the more practical way to measure a person's metabolic rate? _____

• This method assumes that for each 4.5-5 kcal of energy metabolized, _____ will be consumed.

• What is indirect calorimetry? _____

• What is the respiratory quotient (RQ) and why does it vary with diet? _____

_____

Factors that influence metabolic rate
• Define the basal metabolic rate (BMR). _____

• List seven factors that influence metabolic rate. _____

_____

• Males have an average BMR of _____ kcal/hr/kg body weight, while in women the average is _____.

• Why is the rate lower in women? _____

• Name two reasons metabolic rate increases above the BMR. _____

_____

• What is diet-induced thermogenesis? _____

Daily energy requirements
• The daily caloric intake of the average man is _____ kcal.

• Animals store glucose in the form of _____. [∫ p. 27]

• Most energy in animals is stored in the form of _____.

• Explain why glycogen is more efficient than glucose and fat more efficient than glycogen for energy storage.

_____

_____

• What is the advantage of keeping some energy stored in glycogen rather than fat?

_____

Energy for Work May Come from Absorbed Nutrients or from Storage

• Human metabolism is divided into what two states? _____

_____

• Which state is (net) anabolic and which is catabolic? _____

_____

• Ingested biomolecules meet one of three fates in the body. List them.

_____

_____

• Most fats enter energy production pathways as _____ acids and _____ (Fig. 21-5).

• The products of the citric acid cycle are _____ , _____, and _____

• Most carbohydrates are absorbed in the form of _____. Most proteins are absorbed

in the form of _____.

• The concentration of what plasma biomolecule is the most closely regulated? _____

• Define the following terms:

glycogenesis _____

glycogenolysis_____

gluconeogenesis _____

• What is the primary fate of ingested amino acids? _____

Metabolic Pathways Are Regulated by Changes in Enzyme Activity

☞ *Figure 21-6 is a summary of the important biochemical pathways for energy production. [∫ p. 91] It shows how different pathways for carbohydrates, proteins, and fats intersect.*

• What is push-pull control of metabolism? _____

Anabolic Metabolism Dominates in the Fed State

• In the fed state, what are the primary fates of absorbed nutrients? (Table 21-1)

_____

Carbohydrates
• In the intestine, absorbed glucose enters the (lymph system / blood ?) and is transported to the _____.

• How does glucose enter cells? [∫ p. 589] _____

• What happens to most absorbed glucose? _____

• What happens to unused glucose? _____

• Where are the body's glycogen stores? _____

Proteins
• In the intestine, absorbed amino acids enter the (lymph system / blood ?) for transport to the _____.

• How do amino acids enter cells? [∫ p. 589] _____

• What happens to most absorbed amino acids? _____

• What happens to unused amino acids? _____

Fats

• In the intestine, absorbed fats enter the (lymph system / blood ?) in what form? _____

• From the intestine, absorbed fats are transported to the [∫ p. 593] _____.

• What is the function of lipoprotein lipase and where is it found? (Fig. 21-8)

_____

• How do fatty acids enter cells? [∫ p. 589] _____

• What happens to most absorbed fats? _____

• Where are the body's fat stores? _____

• What are chylomicron remnants? _____

• How do chylomicron remnants enter cells? _____

• In what form is cholesterol transported in the blood? _____

• Contrast HDL- and LDL-cholesterol. _____

_____

• Why is LDL-cholesterol sometimes called "bad cholesterol?" _____

_____

Metabolism in the Fasted State

• What is the primary goal of the fasted state? _____

• What biomolecule and what organ provide most glucose during the fasted state? (Fig. 21-9)

_____

• How can skeletal muscle help raise plasma glucose? _____

• What happens to the amino groups removed from amino acids? _____

• In adipose tissue, triglycerides are broken down into _____.

• How does the body metabolize glycerol? _____

• How does the body metabolize fatty acids? [∫ p. 96] _____

• Why is excess ketone production dangerous? _____

ENDOCRINE CONTROL OF METABOLISM: PANCREATIC HORMONES

• The hour-to-hour regulation of metabolism depends on what two hormones? _____

• What glands secrete insulin and glucagon? _____

The Endocrine Pancreas Secretes Several Hormones

• Most pancreatic tissue produces and secretes _____.
   [∫ p. 584]

• What peptide is secreted by the following islet of Langerhans cells? (Fig. 21-10)

   β cells: _____   α cells: _____

   D cells: _____   PP (or F) cells: _____

The Ratio of Insulin to Glucagon Is a Key to Metabolic Regulation

• Insulin and glucagon are (synergistic / permissive / antagonistic ?) hormones. [∫ p. 149]

• True or false? Only insulin is present in the fed state, while only glucagon is present in the fasted state.

• Mark the following metabolic pathways as belonging to fed state or fasted state metabolism. (Fig. 21-11)

gluconeogenesis _____   glycogenolysis _____

glycolysis _____   ketone production _____

lipogenesis _____   lipolysis_____

protein breakdown _____   protein synthesis_____

• Compare plasma glucagon and insulin concentrations over the course of a day. (Fig. 21-12)

_____

Insulin Is the Hormone of the Fed State

• Insulin is a typical (amine / steroid / peptide ?) hormone. [∫ p. 179] (Table 21-2)

• The insulin receptor is located where in/on the target cell? _____

• What is the second messenger system for insulin? _____

• List four stimuli for insulin release. _____

_____

• What three tissues are the primary targets of insulin? _____

Metabolic effects of insulin
• Match the actions of insulin with the tissue(s) in which they take place.

1. Increased glucose uptake from blood _____   a) liver

2. Increased glycolysis _____   b) brain

3. Increased glycogenesis _____   c) skeletal muscle

4. Increased protein synthesis _____   d) adipose tissue

5. Increased triglyceride synthesis _____   e) kidney epithelium

• What is the significance of the fact that liver cells do not have GLUT 4 transporters?

_____

• Insulin is considered to be an (anabolic / catabolic?) hormone.

Diabetes mellitus
• Explain the etiology (cause) of IDDM. _____

_____

• What is the primary treatment for IDDM? _____

• Describe the typical patient with NIDDM. _____

• Most patients with diabetes have (IDDM / NIDDM ?).

• Describe the glucose tolerance test used to diagnose diabetes. _____

_____

Physiology of Insulin-Dependent Diabetes Mellitus [∫ p. 223, 443, 533]
• Explain the physiology behind each of the symptoms of untreated insulin-dependent diabetes mellitus (IDDM). (Fig. 21-16)

glucose in the urine _____

ketone production _____

muscle wasting _____

excessive urination _____

metabolic acidosis _____

excessive thirst _____

polyphagia _____

osmotic diuresis _____

hyperglycemia _____

increased ventilation _____

lactic acidosis _____

Physiology of Non-Insulin-Dependent Diabetes Mellitus
• Why don't untreated patients with NIDDM usually have ketosis?

_____

• What are the major complications of NIDDM? _____

_____

Glucagon Raises Blood Glucose Concentrations (Table 21-3)

• Glucagon is a typical (amine / steroid / peptide ?) hormone.

• When plasma glucose < 100 mg/dL, glucagon secretion _____.

• The primary target of glucagon is the _____. (Fig. 21-17)

• What effect does glucagon have on this tissue? _____

• Name another chemical stimulus for glucagon release. _____

NEURALLY MEDIATED ASPECTS OF METABOLISM

• List three ways the brain influences metabolism. _____

_____

• In the fed state, (sympathetic / parasympathetic ?) influence over metabolism is dominant.

• What effect do catecholamines released by stress have on metabolism?

_____

Regulation of Food Intake

• Name the two hypothalamic centers that control food intake and explain their interaction.

_____

_____

• Explain the glucostatic theory of control of eating.

_____

_____

• Explain the CCK theory of control of eating. [∫ p. 598] (Fig. 21-18)

_____

_____

• Why is the mouse model of obesity associated with leptin deficiency a poor model for human obesity?

_____

• What is enterostatin? [∫ p. 593] _____

• What is the difference between hunger and appetite? _____

_____

• Explain the lipostat theory for control of food intake. _____

_____

LONG-TERM ENDOCRINE CONTROL OF METABOLISM

• List four groups of hormones involved in long-term control of metabolism; what gland secreted each?

1. _____ from _____

2. _____ from _____

3. _____ from _____

4. _____ from _____

• Be able to draw the complete control and feedback pathways for these hormones. [p. 185]

The Adrenal Glands Secrete Both Amine and Steroid Hormones

• The adrenal glands are embryologically distinct glands that merged during development: an inner

portion called the _____ that secretes _____,

and the outer portion called the _____ that secretes three major groups of hormones.

• Why is the adrenal medulla considered a modified sympathetic ganglion? [∫ p. 311]

_____

• Name the three layers (zones) of the adrenal cortex and tell which hormones each layer secretes. (Fig. 21-19):

Outer layer = _____

Middle layer = _____

Inner layer = _____

• How do the glucocorticoids get their name? _____

• Cortisol is a typical steroid hormone. Fill in the blanks below. [∫ p. 180] (Table 21-5)

Made in what organelle of the endocrine cell? _____

Stored? (explain) _____

How is it transported in the blood? _____

Where on/in the target is its receptor? _____

What is the general cellular response to hormone-receptor activation? _____

• In the space below, draw the control pathway for cortisol. [∫ Fig. 7-17, p. 191]

• Why is it important to know what time of day a blood sample was drawn if you are assessing cortisol level?

_____

• List five significant effects of cortisol on metabolism. _____

_____

_____

• Give one reason cortisol is essential for life. _____

_____

Thyroid Hormones Are Important for Quality of Life

• Describe the anatomical location of the thyroid gland. (Fig. 21-21)

_____

• Name the two cell types of the thyroid gland and the hormones secreted by each.

_____

• Describe the structure of a thyroid follicle. Where does hormone synthesis take place?

_____

_____

• What is unique about thyroid hormone structure? (Fig. 21-22) _____

• In the space below, map thyroid hormone synthesis, beginning with amino acids and iodide. Be sure to show where each step takes place. Include the following terms: amino acids, colloid, diiodotyrosine, enzymes, follicle, I-ATPase, iodide, monoiodotyrosine, tetraiodotyrosine, T_3, T_4, thyroglobulin, thyroid-binding globulin, thyroxine, triiodotyrosine

• Explain the structural and functional relationship between T_3 and T_4. _____

_____

• What enzyme for thyroid hormone metabolism is found in the peripheral tissues? _____

• Although thyroid hormones are classified as (peptide / steroid / amine ?) hormones, at their target tissues they behave most like (peptide / steroid ?) hormones. [∫ p. 180]

• Thyroid hormones (are / are not ?) essential for life.

• Describe the metabolic effects of thyroid hormones and relate these effects to the symptoms exhibited by people with hyperthyroidism or hypothyroidism.

_____

_____

_____

_____

• People with hyperthyroidism often have a rapid heartbeat. Explain how thyroid hormones cause this.

_____

• In the space below, draw the complete control pathway for thyroid hormones. Tell what gland secretes each hormone. (Fig. 21-24)

• The trophic effect of excessive TSH can cause an enlarged thyroid gland or goiter. Explain why knowing only that the thyroid gland is enlarged does not tell you whether a person is hyperthyroid or hypothyroid. (Fig. 21-25a)

_____

_____

_____

ENDOCRINE CONTROL OF GROWTH

• Name four factors that are necessary for normal growth in children.

1. _____

2. _____

3. _____

4. _____

Growth Hormone Is Essential for Normal Growth in Children

• In the space below, draw the control pathway for growth hormone, including its trophic action. Tell what gland or tissue secretes each hormone. (Fig. 21-26)

• Explain the actions of growth hormone and somatomedins. _____

_____

• True or false? Growth hormone secretion ceases when growth stops.

• Growth hormone is a typical (steroid / peptide ?) hormone. Fill in the blanks below.

Made in what organelle of the endocrine cell? _____

Stored? (explain) _____

How is it transported in the blood? _____

Is this form of transport typical? Explain. _____

Where on/in the target is its receptor? _____

What is the general cellular response to hormone-receptor activation? _____

• Hypersecretion of GH in children causes the condition _____. Hyposecretion

   in children causes the condition _____.

Tissue Growth Is under the Control of Several Hormones

• Soft tissue growth requires what three hormones? _____

• What is the difference between hypertrophy and hyperplasia? _____

_____

• Explain what is meant by the statement that thyroid hormones and insulin are permissive for growth
   hormone. [∫ p. 189]

_____

Bone Growth Requires Adequate Amounts of Calcium in the Diet

• Describe the structure of the extracellular matrix of bone. _____

_____

• What does "resorbed" mean? _____

• Name the two forms of bone. (Fig. 21-27) _____ (dense); _____ (spongy)

• Explain the difference between osteoblasts, osteoclasts, and osteocytes. _____

_____

_____

• What are the epiphyseal plates of long bones? (Fig. 21-28) _____

• What is the relationship between epiphyseal plates and the development of acromegaly in adults
  who are over-secreting growth hormone?

_____

_____

Calcium Concentrations in the Blood Are Closely Regulated

• What compartment or tissue of the body contains the largest amount of Ca^{2+}? _____

• List eight important functions of Ca^{2+} in the body. (Table 21-8)

_____

_____

_____

• Extracellular Ca^{2+} is (more concentrated / less concentrated / the same as ?) intracellular Ca^{2+}, so
  if Ca^{2+} channels in the cell membrane open, Ca^{2+} will flow (into / out of ?) the cell.

• If Ca^{2+} concentrations move outside the normal range, which of the functions above is most

  likely to create significant problems? _____

• Describe how osteoclasts affect bone. (Fig. 21-19) _____

_____

Endocrine control of calcium balance
• List the three hormones that regulate Ca^{2+} homeostasis and name the cell/gland that secretes each.
  (Fig. 21-30)

_____

• What is the stimulus for parathyroid hormone release? _____

What is the effect of PTH action on plasma Ca^{2+}? _____

• Describe the three ways PTH alters plasma Ca^{2+}. _____

_____

Vitamin D hormones
• 1,25-dihydroxycholecalciferol is also known as _____.

• Where does the body get vitamin D_3? _____

• What organs are involved in the conversion of vitamin D_3 to calcitrol? (Fig. 21-32)

_____

• Describe the action of calcitrol on each of the following organs:

bone _____

intestine _____

kidney _____

Put a star next to the most important function of calcitrol above.

Calcitonin
• Calcitonin is a (steroid / peptide ?) hormone produced by what tissue/cell? _____

• Compare the role of calcitonin in humans to that of PTH. _____

_____

Other hormones and calcium
• Name two other hormones linked to Ca^{2+} balance and bone metabolism.

_____

• What is osteoporosis? _____

TALK THE TALK

1,25-dihydroxycholecaliferol (1,25(OH)$_2$D$_3$)

A cell

absorptive state (fed state)

acetoacetic acid

acini

acromegaly

active vasodilation

adipose tissue

adipsin

adrenal cortex

adrenal gland

adrenal medulla

adrenocorticotropic hormone (ACTH)

aldosterone

amino acids, metabolism of

amylin

anabolism

androgen

anorexia nervosa

apolipoprotein

apoprotein B

apoprotein E

appetite

arteriovenous anastomoses

basal metabolic rate (BMR)

beta cell

beta-adrenergic receptor

bomb calorimeter

bradykinin

brown adipose tissue

C cells (thyroid gland)

calcitonin

calcitrol

calcium phosphate

carbohydrates, as energy source

catabolism

catecholamine

CCK theory of feeding

chemical work

chondrocyte

chromium

chylomicron

chylomicron remnant

circadian rhythm, cortisol

colloid

compact bone

conductive heat gain

convective heat loss

corticotropin releasing hormone (CRH)

cortisol

cortisone

cretinism

cutaneous blood flow

D cell

deiodinase

diabetes mellitus

diabetic ketoacidosis

diaphysis

diet-induced thermogenesis

diiodotyrosine (DIT)

direct calorimetry

dwarfism

energy balance

enterostatin

epiphyseal plate

epiphysis

essential amino acid

estrogen

estrogen replacement therapy

evaporative heat loss

F cell

failure to thrive

fat, as energy source

fatty acid

feedforward

feeding center

fever

first law of thermodynamics

follicle, thyroid

giantism

glucagon

glucocorticoid

gluconeogenesis

glucose tolerance factor

glucose tolerance test

glucose-dependent insulinotropic peptide (GIP)

glucostatic theory of feeding control

glucosuria

GLUT transporter

glycogen, as energy source

glycogenesis

glycogenolysis

goiter

Graves' disease

growth

growth hormone

growth hormone binding protein

growth hormone-inhibiting hormone

growth hormone-releasing hormone (GHRH)

heat index

high density lipoprotein (HDL)

homeothermy

hot flashes

hydroxyapatite

hydroxybutyric acid

hypercalcemia

hypercholesterolemia

hyperglycemia

hyperplasia

hyperthyroidism

hypertrophy

hypocalcemia

hypoglycemia

hypothalamic thermostat

hypothermia

hypothyroidism

indirect calorimetry

insulin

insulin-dependent diabetes mellitus (IDDM)

insulin-like growth factor

iodine

islet amyloid polypeptide

islets of Langerhans

ketoacidosis

ketone body

ketosis

kilocalorie (kcal)

Langerhans, Paul
leptin
lipoprotein lipase
lipostat theory
low density lipoprotein (LDL)
malignant hyperthermia
mechanical work
metabolic acidosis
metabolism
microtubule assembly
mineralocorticoid
monoiodotyrosine (MIT)
non-insulin dependent diabetes
    mellitus, or NIDDM
non-shivering thermogenesis
obese (ob) gene
osmotic diuresis
osteoblast
osteoclast
osteocyte
osteoporosis
oxidation
oxygen consumption
Paget's disease
pancreatic polypeptide
parathyroid glands

parathyroid hormone (PTH)
polyphagia
polyuria
post-absorptive state (fasted
    state)
postprandial period
PP cell
push-pull control
pyrogen
radiant heat loss
receptor-mediated endocytosis
renal threshold for glucose
resorption of bone
respiratory quotient (RQ)
satiety center
shivering thermogenesis
somatomedin
somatostatin
somatotropin
sweat
sweat gland
sympathetic cholinergic
    vasodilator neuron
thermogenin
thermography
thermoneutral zone

thermoreceptor
thermoregulation
thyroglobulin
thyroid gland
thyroid-stimulating
    immunoglobulin (TSI)
thyrotropin, thyroid-stimulating
hormone (TSH)
thyrotropin-releasing hormone
    (TRH)
thyroxine-binding globulin
    (TBG)
trabecular bone
transport work
triiodothyronine (T_3)
tyrosine
tyrosine kinase
uncoupling protein
up-regulation
very low density lipoprotein
    (VLDL)
vitamin D_3
wind chill factor
zona fasciculata
zona glomerulosa
zona reticularis

ERRATA

p. 636, section header in left column should read: "Adrenal glands secrete both <u>amine</u> and steroid hormones."

p. 636, right column, item #3, last word should be "cheeks," not "checks."

ETHICS

Should athletes be allowed to take synthetic growth hormone? Should children who are genetically short but have normal growth hormone secretion be given growth hormone? What are the potential risks of taking hormones if a person's own hormone production is normal?

QUANTITATIVE THINKING

Body Mass Index: The **body mass index** (BMI) has been shown to correlate well with how much body fat a person has, and it can be calculated without special equipment or testing. The 1995 NIH guidelines define a BMI below 25 as healthy.

To calculate: $BMI = w/h^2$

w = weight in kilograms or weight in pounds divided by 2.2
h = height in meters or height in inches divided by 39.4

Calculate your BMI: weight = _____ lbs. ÷ 2.2 = _____ kg

height = _____ in. ÷ 39.4 = _____ m

$BMI = weight (kg) \div height^2$

PRACTICE MAKES PERFECT

1. Circle the letter of each pair representing an INCORRECT cause:effect relationship:

    a) epinephrine : increased glycogenolysis in the liver
    b) insulin : increased protein synthesis
    c) glucagon : decreased gluconeogenesis

2. The active hormone of the thyroid gland is:

    a. thyroglobulin
    b. thyroxine
    c. diiodotyrosine

3. TRUE/FALSE: The heat produced by an organism is one way of defining (or measuring) metabolism.

4. Classify each of the following hormones as anabolic (A) or catabolic (C).

cortisol _____   growth hormone _____   glucagon _____   insulin _____

5. If you go on a no-carbohydrate diet, why doesn't the brain starve to death for lack of glucose?

6. Compare insulin secretion when glucose is given orally to insulin secretion after the same amount of glucose is given intravenously.

7. Generally insulin and glucagon are released by opposing stimuli and have opposing effects on metabolism. However, *both* hormones are released by the stimulus of an increase in blood amino acids. Circle all the answers below that explain correctly why this occurs.

     a) Glucagon will prevent hypoglycemia following ingestion of a pure protein meal.
     b) Both insulin and glucagon promote amino acid absorption at the small intestine.
     c) Amino acids are present in the blood during both anabolism and catabolism.
     d) Glucagon release is part of a positive feedback loop.
     e) Amino acids stimulate the release of insulin that stimulates the release of glucagon.
     f) Glucagon stimulates transcription and translation of amino acid transporters at the cell membrane.

8. Analyze the food label below for fat content (% of total calories) and critique the labeling. Is this a "good" food according to current guidelines on recommended fat intake?

Calories per serving:	190
Total fat	7 g
Total carbohydrate	25 g
Total protein	9 g

9. Mice with leptin deficiency become obese, but obese humans have elevated levels of leptin in their blood. Can you think of an alternate explanation for leptin-related obesity in humans that fits these findings?

10. True or false? Explain your reasoning.

a) When we hear of athletes and body-builders taking "anabolic steroids," we know that they are taking glucocorticoids.

b) A child who has a vitamin D deficiency will develop rickets (poor bone formation) because vitamin D plays a major role in the precipitation of calcium phosphate into bone.

c) Glucagon promotes glycogenolysis, gluconeogenesis, and ketogenesis.

d) Insulin promotes transport of glucose into liver.

e) Glucose transport in all cells of the body is via a mediated transport system that exhibits saturation.

11. Why are body-builders wasting their money if they take amino acid supplements in addition to a balanced diet?

12. The diagram below represents calcium balance in the human body. The labeled arrows represent calcium movement of calcium between the compartments. Answer the questions below that refer to the letters on the diagram.

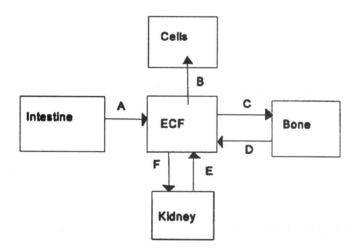

a) What hormone(s) directly or indirectly regulate movement at arrow A? _____

b) Describe *as specifically as possible* the calcium movement at arrow F. _____

c) What is the hormonal control over calcium movement at arrow F? _____

d) Name two physiological functions of the calcium movement shown by arrow B.

_____

e) Name the hormone that controls calcium movement at arrow C. _____

13. On axis (a) below, plot the effect of plasma parathyroid hormone concentration on plasma Ca²⁺ concentration. On axis (b) below, plot the effect of plasma Ca²⁺ concentration on plasma parathyroid hormone concentration. Be sure to label the axes of each graph!

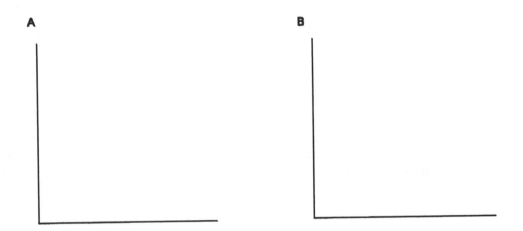

14. A patient comes in with a diagnosis of alpha-cell hyperplasia due to a pancreatic tumor. You draw a blood sample and send it out for analysis. Before it comes back, predict what changes you expect to see in the following parameters (up, down, no change) and explain your rationale.

Plasma concentration	Change: ↑, ↓, or N/C	Rationale for your answer
glucose		
insulin		
amino acids		
ketones		
K⁺		

MAPS

1. Create pathway maps for each hormone, adding details at the cellular level where appropriate.

2. Map in detail the physiological complications arising from insulin-dependent diabetes mellitus. Include all homeostatic controls, all hormones, neurotransmitters, receptors, integrating centers, effector tissues, etc. Be very SPECIFIC. This map will require a large sheet of paper or a poster board.

BEYOND THE PAGES

TRY IT
Can you tie a bone in a knot? Place the leg bone of a chicken in commercial white vinegar and monitor what happens. The vinegar will dissolve the calcium phosphate of the bone in vinegar, leaving behind the soft organic matrix. This is similar to the way osteoclasts secrete acid to dissolve bone in the body.

READING

Assessment of insulin resistance. (1994, September/October) *Science & Medicine.*

The atherogenic lipid profile. (1997, September/October) *Science & Medicine.*

Body weight regulation in humans: The importance of nutrient balance. (1993, December) *News in Physiological Sciences* 8: 273-275.

Diet and cancer. (1987, November) *Scientific American.*

Encapsulated cell therapy. (1995, July/August) *Science & Medicine.*

Endocrinology: People and Ideas, (1988) American Physiological Society.

Fetal nutrition and adult diabetes. (1994, July/August) *Science & Medicine.*

The fish oil puzzle. (1996, September/October) *Science & Medicine.*

Functional role of P_i transport in osteogenic cells. (1996, June) *News in Physiological Sciences* 11: 119-125.

Gaining on fat. (1996, August) *Scientific American.*

Glucose transport and glucose homeostasis: New insights from transgenic mice. (1995, February) *News in Physiological Sciences* 10: 22-29.

How does the fetus cope with thermal challenges? (1996, April) *News in Physiological Sciences* 11: 96-100.

The insulin factory. (1988, September) *Scientific American.*

Insulin-like growth factors. (1996, March/April) *Science & Medicine.*

Malignancy-associated hypercalcemia. (1995, September/October) *Science & Medicine.*

A metabolic sensor in action: News from the ATP-sensitive K^+-channel. (1997, December) *News in Physiological Sciences* 12: 255-262.

The molecular basis of insulin resistance. (1997, May/June) *Science & Medicine.*

Mouse models of human obesity. (1997, May/June) *Science & Medicine.*

Nature's K_{atp}-channel knockout. (1997, October) *News in Physiological Sciences* 12: 197-202. The glucose-sensing mechanism from beta cells is linked to these channels.

New concepts concerning the regulation of renal phosphate excretion. (1997, October) *News in Physiological Sciences* 12: 211-213.

The obese gene (ob) and human obesity. (1996, June) *News in Physiological Sciences* 11: 147-148.

Prolactins: Novel regulators of angiogenesis. (1997, October) *News in Physiological Sciences* 12: 231-237.

Reaching out: Nutrition quackery. (1997, May/June) *Science & Medicine.*

Species specificity of the primate growth hormone receptor. (1996, August) *News in Physiological Sciences* 11: 157-160.

Temperature and proteins: Little things can mean a lot. (1996, April) *News in Physiological Sciences* 11: 72-76.

Treating diabetes with transplanted cells (1995, July).

Vitamin E is nature's master. (1994, March/April) *Science & Medicine.*

What causes diabetes? (1990, July) *Scientific American.*

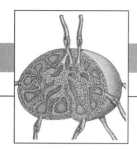

SUMMARY

This is another chapter packed with information. In other words, don't wait until the last minute to learn this. Our suggestion is to break the material up into types of responses, then types of cells involved in those responses, then cellular components involved in those responses, and so forth. But first, take the time to learn the vocabulary. Here are the main topics presented in this chapter:

• The immune system is spread throughout the body. It consists of primary and secondary lymphoid tissue and immune cells.
• There are five types of immune cells, and each plays a different role in the immune system.
• There are two types of immune responses: innate and acquired. Innate is nonspecific; acquired is highly specific.
• There are two types of acquired responses: humoral immunity and cell-mediated immunity. Humoral immunity involves antibodies from B cells; cell-mediated immunity is under the direction of T cells.
• Secondary responses are stronger and more rapid than primary responses.
• The immune system handles different pathogens in different ways.
• The immune system recognizes self from nonself. Major histocompatibility complexes play an integral role in determining self and nonself.
• The immune system is linked to the nervous and endocrine systems.

Immunity is the ability of the body to protect itself from pathogens. Anything that elicits an immune response is called an antigen, while a pathogen is anything that creates a pathophysiological condition. As you're probably beginning to see, the first step to understanding the immune system is conquering its vocabulary.

The immune system is spread throughout the body. The primary lymphoid tissues are the thymus and the bone marrow. These are places where immune cells are produced and mature. Secondary lymphoid tissues include encapsulated tissues (spleen and lymph nodes) and unencapsulated tissues (tonsils, GALT, skin and respiratory clusters). There are several types of immune cells: eosinophils, basophils, neutrophils, monocytes, lymphocytes, and dendritic cells. Learn how each plays a different role in the immune response.

There are two types of immune responses: innate immunity and acquired immunity. Innate immunity is a nonspecific immune response that reacts to all foreign particles. If an antigen should get past the body's physical defenses, then it will initiate an inflammatory response and ultimately meet one of the following fates: it could be phagocytized by tissue macrophages or neutrophils, or it could be destroyed via the complement system. The inflammatory response is the hallmark of the innate immune response. At the site of antigen entry, chemicals are released by local cells that result in the attraction of immune cells and the formation of a warm, red, swollen area. You should learn the steps and chemicals involved in the inflammatory response (Table 22-2).

Acquired immunity involves two responses: humoral immunity brought about by B lymphocytes and cell-mediated immunity brought about by T lymphocytes. B lymphocytes (B cells) secrete antibodies (immunoglobulins) that recognize specific sites (epitopes) on specific antigens. B cells that have never been exposed to an antigen are called naive B cells. Once naive cells have encountered their antigen, they divide and produce memory cells and plasma cells in a process called clonal expansion. Memory cells stick around in the body and are responsible for the phenomenon of secondary immune response. Plasma cells secrete antibodies (Ab) into the "humors" or body fluids (hence the name humoral immunity). The primary function of Ab's is to bind B cells and antigens and cause the production of more Ab's, but Ab's also have many secondary functions that target an antigen for destruction (Fig. 22-13). Ab's are Y-shaped protein structures: the arms recognize the antigen, and the stem can attach to cell membranes. There are five classes of Ab's: IgG, IgE, IgD, IgM, IgA.

T cells develop in the thymus. There, T cells that bind self proteins are destroyed. Mature T cells must have contact with their target cells to initiate cell-mediated immunity. T cell receptors (related, but not equivalent to Ab's) must bind to a major histocompatibility complex (MHC) bearing a specific antigen fragment on a target cell surface. MHCs display protein fragments (either self or nonself) on the cell surface. If a T cell recognizes a foreign protein displayed in an MHC, it will initiate a response that either kills that cell or activates other immune cells. There are two types of MHC proteins: MHC-I is found on all nucleated cells; MHC-II is found only on antigen-presenting cells (APCs). There are three subtypes of T cells: cytotoxic T cells, helper T cells, and natural killer cells (possibly also suppressor T cells). Cytotoxic T cells kill a target cell by apoptosis when they recognize an MHC-antigen complex. When helper T cells recognize an MHC-antigen complex, they secrete cytokines that activate other T and B cells. Natural killer cells are actually a distinct cell line. If they recognize an MHC-antigen complex on a cell, they will kill that cell. However, they can also kill an Ab-coated cell through a nonspecific process called antibody-dependent cell-mediated cytotoxicity.

That's a mouthful! We suggest making charts that show the differences between 1) innate and acquired immunity, and 2) humoral and cell-mediated immunity. Be sure to include information about the different types of cells and Ab's involved.

Beginning on p. 672, the different types of immune response pathways are discussed. Learn how the immune system responds differently to different types of antigens (bacterial, viral, allergenic).

The same processes we discussed earlier also apply to the body's recognition of self and nonself tissues. With RBC recognition, it involves surface antigens (because RBCs don't have MHC proteins). With other cases, like organ transplants, foreign MHCs and/or MHC-antigen complexes can cause an immune response that results in tissue rejection.

Finally, we are learning that the immune system is closely linked with the nervous and endocrine systems. It seems that some cytokines, neuropeptides, and hormones are secreted by some or all of these systems. Therefore, they all exert an integrated control over each other. Neuroimmunomodulation is the field that studies the brain-immune system interaction.

TEACH YOURSELF THE BASICS

• Define immunity. _____

• List three major functions of the immune system: _____

_____

_____

PATHOGENS OF THE HUMAN BODY

• List the types of infections common in the USA.

_____

• List the types of infections common in the world.

_____

Bacteria and Viruses Require Different Defense Mechanisms

• Fill in the following table on the differences between bacteria and viruses (Table 22-1):

	Bacteria	Viruses
Structure		
Living conditions		
Reproduction		
Susceptibility to antibiotics		

Viruses Must Reproduce Inside Host Cells

• Briefly describe ways that virus particles can enter a host cell. _____

_____

• Next, describe how a virus takes over host cell resources (Fig. 22-1b). _____

_____

• Describe two ways that viruses are released from host cells. _____

_____

• Give two examples of viral damage to host cells. _____

_____

THE IMMUNE RESPONSE

• List the four basic steps of all internal immune responses.

1._____   2. _____

3._____   4. _____

• What are the two categories of the human immune response? Describe them.

1._____

2._____

• A distinguishing feature of the immune system is that it uses what kind of signaling?

_____

• What are cytokines? [∫ p. 148] _____

_____

ANATOMY OF THE IMMUNE SYSTEM

Lymphoid Tissues are Distributed Throughout the Body

• Name the 2 primary lymphoid tissues (See Focus box, p. 670 and Fig. 22-2a).

_____

• List the two types of secondary lymphoid tissues. _____

_____

• Describe the following encapsulated lymphoid tissues:

Spleen (Fig. 22-2, 22-3) _____

Lymph nodes (Fig. 22-2b) [∫ lymphatic circulation, p. 439-440] _____

_____

• Identify and describe diffuse lymphoid tissues (Fig. 22-2): _____

_____

• What is GALT? _____

Leukocytes are the Primary Cells of the Immune System

• List four ways that leukocytes (WBCs) differ from RBCs. [∫ p. 454]

_____

_____

• List and briefly describe the six basic groups of immune cells:

1. _____

2. _____

3. _____

4. _____

5. _____

6. _____

• How are immune cells distinguished from one another in stained tissue samples?

_____

Terminology

• Know the classification systems for immune cells (Fig. 22-4).

• What are granulocytes? Describe the different types of granulocytes.

_____

_____

• What are immunocytes and what distinguishes them from other immune cells?

_____

• What are phagocytes? What cells belong to this group? _____

_____

• Explain the following terms:

cytotoxic cell _____

antigen-presenting cell (APC)_____

reticuloendothelial system _____

_____

• What is the modern term for the reticuloendothelial system? _____

Eosinophils

• Describe eosinophils. Include their relative abundance, location in the body, and their role(s) in the immune system. What is the physiological significance of their granules?

_____

_____

_____

Basophils

• Describe basophils. Include their relative abundance, location in the body, and their role(s) in the immune system. What is the physiological significance of their granules?

_____

_____

_____

Neutrophils

• Do the same for neutrophils that you did for the other groups.

_____

_____

_____

Monocytes and macrophages

• Describe monocytes and macrophages as you have been doing for the other groups.

_____

_____

_____

Lymphocytes and plasma cells

• Describe lymphocytes and plasma cells as you have done for the other lymphocytes.

_____

_____

_____

Dendritic cells
• Finally, do the same for dendritic cells

_____

_____

INNATE IMMUNITY

• Are innate immune defenses considered specific or nonspecific? _____

Physical and Chemical Barriers are the Body's First Line of Defense

• Describe three examples of physical barriers against foreign invaders. [∫ p. 66]

_____

_____

_____

• Describe the specialized physical barriers of the:

respiratory system [∫ p. 483] _____

stomach_____

tears_____

• Lysozyme can only attack which type of bacterial cell wall?_____

• What type of immune response takes place if the physical barriers of the body are breached?

_____

• What are chemotaxins? Give some examples. _____

_____

Phagocytes Recognize Foreign Material and Ingest It

• What are the primary phagocytic immune cells? _____

• What is diapedesis? _____

• Describe the process of phagocytosis. [∫ p. 126] _____

_____

• Some bacteria have evolved ways to "hide" from phagocytes. How have they accomplished this?

_____

• What are opsonins? _____

• What happens to particles ingested by phagocytes? _____

_____

• Why do sites of bacterial infection develop pus? _____

_____

Chemical Mediators Create the Inflammatory Response

• What signs and symptoms suggest an inflammation? _____

• What cells create the inflammatory response? _____

• What role do cytokines play in this response? (Fig. 22-2) _____

_____

Acute phase proteins

• When do we see an increased presence of acute phase proteins? _____

• What are acute phase proteins and where do they originate? Include examples.

_____

_____

• What happens to the level of acute phase proteins after an acute infection? _____

• What happens to the level of acute phase proteins in cases of chronic infection?_____

• What is the significance of serum amyloid A? _____

Histamine

• Describe the chemical structure of histamine. [∫ p. 147] _____

• Where do we find histamine in the body? _____

• Describe the histamine reaction. _____

• What is the purpose of the histamine reaction in the larger immune response?

_____

• How do antihistamines work? _____

_____

Interleukins
• What are interleukins? What is their role in the immune response?

_____

• As an example, interleukin-1 (IL-1) is secreted by _____.

Its overall role is _____

• Four specific actions of IL-1 include:

1. _____

2. _____

3. [∫ p. 615] _____

4. _____

Kinins
• What are kinins? _____

• What is the physiological action of bradykinin? _____

Complement
• What is included in the term "complement?" _____

• Briefly describe the complement pathway. [∫ coagulation cascade, p. 465] _____

_____

• What is membrane attack complex, and how does it cause cell lysis? (Fig. 22-8)

_____

ACQUIRED IMMUNITY

• How does acquired immunity differ from innate immunity? _____

_____

• Which immune cells primarily mediate the acquired response? _____

Lymphocytes are the Primary Cells Involved in the Acquired Immune Response

• What gives different lymphocytes specificity for specific pathogens? (Fig. 22-9) _____

• What are lymphocyte clones? _____

• Outline or map the lymphocyte life cycle, using the following terms: antigen, clonal expansion, effector cells, naive lymphocytes, memory cells, primary response, secondary response.

• How do memory cells differ from effector cells? _____

_____

• List examples and briefly describe the action of some lymphocyte cytokines. _____

_____

• There are three main types of lymphocytes: _____ lymphocytes secrete antibodies while _____

lymphocytes and _____ attack and destroy foreign cells.

B Lymphocytes Secrete Antibodies

• Where do B lymphocytes (B cells) develop? _____

• The primary function of B cells is the humoral immune response. Briefly, how does the humoral response work?

_____

• What do mature B cells insert into their membranes and what is the function of these structures? (Fig. 22-9)

_____

• Outline the events that take place after B cell activation.

_____

_____

_____

• What role do memory cells play? _____

• Draw and label a graph that shows plasma antibody concentration during a primary response and during a secondary response.

• Based on your graph and what you know so far, why do immunizations help fight against infection?

_____

Antibodies Are Glycoproteins Secreted by Plasma Cells

• Antibodies (Ab's) make up _____% of plasma proteins in a healthy individual?

• There are five general classes of Ab's: IgG, IgA, IgE, IgM, IgD. Match the antibody group to its functions.

      1. Ig _____: Allergic response
          a. Mast cell + Ig___ + antigen = degranulation and histamine release

      2. Ig _____: B lymphocyte surface; an unclear physiological role

      3. Ig _____: Found in external secretions (saliva, tears, mucus, breast milk)
          a. Disables pathogens before body entry

      4. Ig _____: Produced in secondary immune responses
          a. 75% of adult serum Ab's
          b. Can cross placenta and provide the initial immunity for the infant

      5. Ig _____: B lymphocyte surface; primary immune responses; reacts to blood group antigens

• Define gamma globulins. _____

Antibody structure
• Sketch a picture of the typical antibody molecule. Label all the parts and briefly describe the relevance of each part.

Antibody functions

• What is the primary function of an antibody? Where are they most effective?

_____

• How are antibodies attached to a B cell? Sketch and label a picture of an antibody attached to a B cell membrane.

• On the drawing you just made, show how the B cell could be activated. Be sure to label what you've done (Fig. 22-13, #1). What changes are brought about by B cell activation?

_____

• Explain the role of antibodies in the following secondary Ab functions. How does the Fc region function in each of these activities? (Fig. 22-13, #2-6)

Enhance phagocytosis _____

_____

Activate complement _____

_____

Trigger mast cell degranulation _____

_____

Activate cytotoxic cells _____

_____

T Lymphocytes Must Form Direct Contact with Their Target Cells

• Where do T lymphocytes (T cells) form? _____

• T cells are responsible for which type of immunity? _____

• What must T cells do before acting against an antigen? _____

• T cells bind to targets using what type of receptor? (Fig. 22-14) _____

• Structurally and functionally, how are antibodies related to T-cell receptors? _____

_____

• What happens when T-cell receptors bind to antigen? _____

_____

Major histocompatibility complex
• Define and describe major histocompatibility complex (MHC). _____

_____

• Where do we find MHCs? _____

• What is the function of MHC molecules? _____

• Why do MHC proteins vary from person to person? What about MHCs in identical twins?

_____

_____

• We see the importance of MHCs with transplants and in some autoimmune diseases. What roles
  do MHC proteins play in these two cases?

_____

_____

• Describe how MHCs work. (Figs. 22-5, 22-13, 22-14) _____

_____

_____

• Name the two types of MHC molecules and describe the differences between the types.

_____

_____

Subtypes of T lymphocytes
• What are the three major subtypes of cells that develop from T cell precursors? (Fig. 22-16)

_____

• Describe the action of cytotoxic T cells. Include the following terms: apoptosis, granzymes, perforin, channels.

_____

_____

• Describe the action of helper T cells. _____

• List four cytokines secreted by helper T cells. _____

Natural Killer Lymphocytes

• What is another name for natural killer cells? _____

• How do NK cells work? _____

• What is antibody-dependent cell-mediated cytotoxicity and how is it different from the mechanism used by cytotoxic T cells?

_____

_____

IMMUNE RESPONSE PATHWAYS

☛ *Once you've worked through these pathways the first time, go back and see if you can describe them completely without looking.*

• Describe how the body responds to 4 challenges:
  1. Extracellular bacterial infection
  2. Viral infection
  3. Allergic response to pollen
  4. Transfusion of incompatible blood
☛ *These pathways show how the innate and acquired immune responses are interconnected processes.*

Inflammation is the Typical Response to Bacterial Invasion

• Outline the integrated response to bacterial entry into the ECF. (Fig. 22-17), [∫ p. 465] Feel free to do this as a map, but don't directly copy the one in the book (this is for your own good). Be sure to use all the following terms:

acute phase proteins	chemotaxin	mast cell
antibodies	complement	membrane attack complex
antigen	diapedesis	memory B cells
B lymphocytes	histamine	opsonins
bacteria, encapsulated	leukocyte	phagocytes
bacteria, not encapsulated	lyse	plasma cells
capillary permeability	lysozyme	plasma protein

Intracellular Defense Mechanisms are Needed to Fight Viral Infections
☛ *Before viruses enter host cell, innate and humoral defenses both control infection.*

• After viral entry into a host cell, what cell type is the main line of defense? _____

• Map the steps of viral infection, assuming previous viral exposure/presence of Ab's (Fig. 22-18). Again, don't directly copy the figure in the book. Be sure to use all the following terms:

γ -interferon	cytotoxic T cell	MHC-II
acute phase proteins	granzymes	NK cell
antibodies	helper T cell	opsonins
antigen	host cell	perforin
apoptosis	lyse	phagocytosis
B lymphocytes	macrophage	T cell receptor
cytokines	MHC-I	virus

Antibodies against viruses
• Why might Ab's from one viral infection not be effective against subsequent viral infections? Give some examples of viruses for which this is true.

_____

_____

Allergic Responses are Inflammatory Responses Triggered by Specific Antigens

• Define allergy and allergen. _____

_____

• How do immediate hypersensitivity and delayed hypersensitivity reactions differ?

_____

_____

• What kinds of molecules can be allergens? Name some common allergens.

_____

• How are people exposed to allergens? _____

• Map the steps in allergy development. (Fig. 22-19) Include the following terms:

allergen	first exposure	memory cell
antibodies	helper T cell	MHC-I
antigen-processing cell	histamine	plasma cell
B lymphocyte	Ig __?__	re-exposure
complement	inflammation	T cell receptor
cytokines	mast cell	

• Name three effects of massive histamine release that cause the condition known as anaphylactic shock or anaphylaxis.

_____

Recognition of Foreign Tissue

• Which cell surface markers determine tissue compatibility? _____

• Use blood transfusion as example of compatibility. Make a chart or table with the four blood groups, tell what antigens and antibodies a person with each group will have, and tell which groups will be compatible and incompatible for transfusions.

Recognition of Self is an Important Function of the Immune System

• Define self-tolerance. _____

_____

• How does self-tolerance arise? _____

_____

• What happens if self-tolerance fails? _____

_____

Immune Surveillance Allows the Body to Remove Abnormal Cells

• Briefly describe the immune surveillance theory. _____

_____

• If this theory is accurate, how can cancer cells escape detection and destruction? _____

_____

INTEGRATION BETWEEN THE IMMUNE, NERVOUS, AND ENDOCRINE SYSTEMS

• Define neuroimmunomodulation and describe some examples of current research into the area.

_____

_____

_____

• Describe the three known links between the immune, nervous, endocrine systems. (Fig. 22-21)

_____

_____

_____

The Interaction between Stress and the Immune System

• Define stress and describe the stress response. [∫ p. 636]

_____

_____

• Where are most physical and emotional stressors integrated? _____

• The nervous system responds to acute stress. Hormonal responses attend to chronic stresses. Based on what you learned in Chapter 6, why does this make sense?

_____

• Why is study of the stress response difficult? _____

_____

TALK THE TALK

ABO blood group
acquired immunity
ACTH
acute phase protein
adrenal gland
agglutinin
AIDS (acquired immune
   deficiency syndrome)
allergen
allergy
amyloid protein
anaphylaxis
antibiotic
antibody
antibody-dependent cell-
   mediated cytotoxicity
antigen
antigen-presenting cell
antihistamine
apoptosis
autoimmune disease
B lymphocyte
bacteria
basophil
beta adrenergic receptor
blood transfusion
bone marrow
bradykinin
brown adipose tissue
cancer
capsule, bacterial
cell-mediated immunity
Chédiak-Higashi syndrome
chemokine
chemotaxin
clonal expansion
clone
complement
corticotropin releasing hormone
   (CRH)
cytokines
cytotoxic cells
cytotoxic T cell
Dalkon shield
degranulation
delayed hypersensitivity reaction

dendritic cell
diapedesis
diffuse lymphoid tissue
diphtheria
DNA
edema
effector cell
electrophoresis
encapsulated lymphoid tissue
envelope, viral
eosinophil
Fab region
Fc region
fight-or-flight reaction
gamma globulin
gamma-interferon (interferon-)
general adaptation syndrome
glucocorticoid
granulocyte
granzyme
growth hormone
gut-associated lymphoid tissue
   (GALT)
helper T cell
hemolysin
heparin
herpes virus
histamine
histiocyte
histoplasmosis
human leukocyte antigen (HLA)
humoral immunity
hypersensitivity
hypochlorous acid
IgA
IgD
IgE
IgG
IgM
immediate hypersensitivity
   reaction
immune surveillance
immunity
immunization
immunocyte
immunoglobulin

inflammation
innate immunity
insulin-dependent diabetes
   mellitus
interleukin
interleukin-1 (IL-1)
keratinocyte
killer T cell
kinin
Kupffer cell
large granular lymphocyte
leukocyte
life cycle of a virus
lymph node
lymphatic circulation
lymphocyte
lymphoid tissue
lymphoma
lysozyme
macrophage
major histocompatibility
   complex
malaria
mast cell
melatonin
membrane attack complex
memory cell
MHC class I molecule
MHC class II molecule
microbe
microglia
monocyte
mononuclear phagocyte system
mucus escalator
naive lymphocyte
natural killer cell (NK cell)
neuroimmunomodulation
neutrophil
oncogenic virus
opiate receptor
opsonin
osteoclast
parasite
pathogen

perforin	pus	stress
peroxide	pyrogen	substance P
phagocyte	recognition of self	T lymphocyte (T cell)
phagocytosis	reticuloendothelial cell	T-cell receptor
phagosome	reticuloendothelial system	thymopoietin
pineal gland	retrovirus	thymosin
placebo effect	reverse transcriptase	thymulin
plasma cell	Rh antigen	thymus gland
polymorphonuclear leukocyte	RNA	thyrotropin (TSH)
precipitin	secondary immune response	tonsil
primary immune response	self-tolerance	transplantation of organs
prolactin	Selye, Hans	valley fever
psychosomatic illness	spleen	virus

RUNNING PROBLEM -AIDS

How HIV defeats the immune system. (1995, August) *Scientific American*.

Hurdles in the path of an HIV-1 vaccine. (1995, May/June) *Science & Medicine*.

Intracellular antibodies for HIV-1 gene therapy. (1996, May/June) *Science & Medicine*.

The pathogenesis of AIDS. (1997, March/April) *Science & Medicine*.

ETHICS IN SCIENCE

A mother brings her ill child to the doctor, who diagnoses a viral illness and sends the child home with instructions for his care. The mother is convinced that only antibiotics will cure her son, so she keeps calling and pestering the doctor's staff until the doctor finally prescribes antibiotics, despite knowing that they will not be effective. How is this scenario related to the development of drug-resistant strains of bacteria? What should doctors and the public be doing to prevent the development of more drug-resistant strains?

PRACTICE MAKES PERFECT

1. From the following list, place one appropriate letter in each blank:

a) neutrophil   b) eosinophil   c) basophil   d) erythrocyte   e) monocyte   f) lymphocyte

_____ produces antibodies; has large round nucleus with very little surrounding cytoplasm

_____ phagocytic; has nucleus with 3-5 segments, pale pink granules in cytoplasm

_____ releases histamine and heparin; dark blue-staining granules in cytoplasm

2. Match the cell surface markers/receptors with the cells on which they are found.

a) MHC-I   b) MHC-II   c) T-cell receptors
d) glycoprotein markers   e) antibodies   f) no receptors

_____ macrophage   _____ red blood cell

_____ B lymphocyte   _____ natural killer cell

_____ liver cell   _____ cytotoxic T cell

_____ plasma cell   _____ helper T cell

3. List three functions of macrophages.

4. Can a mother with blood type A and a father with blood type B have a baby with blood type O? Explain. (Remember that each parent carries two alleles for the RBC surface antigens.)

5. A technician runs an ABO blood type test on Aparna's blood. Her blood agglutinates with anti-B serum but not with anti-A serum.

a. What is Aparna's blood type? _____

b. To what other ABO groups can she donate blood? _____

c. From what ABO groups can she receive blood? _____

6. You are walking barefoot through the cool spring grass, and you unconsciously step on a plant to which you are allergic. Your feet swell up in response. When you consider the physiology behind this reaction, you realize that there are seven kinds of leukocytes involved in this hypersensitivity reaction. Name these blood cells.

7. Where do B lymphocytes develop? _____ T lymphocytes? _____

8. Your friend Elizabeth knows that she is ABO blood type O. Why is the blood bank always calling her to ask her for a blood donation?

9. You are the microscopic ace reporter for the news station Plasma 1, and you have just been alerted to a skin breach at the index finger (let's say, a nasty paper cut . . .). As you arrive on the scene, describe the events you observe.

10. What is the hallmark of the innate response?

11. Peng-Chai is feeling bad and thinks he is coming down with the flu virus that has been going around school, so he takes some old antibiotics that he has left over from a previous sinus infection (not something anyone should ever do!). Will the antibiotics help his viral infection? Explain.

12. What role do acute phase proteins play in an innate response?

13. What is the result of the terminal step in the complement system?

14. What is the difference between the Fc region and the Fab region of an antibody?

15. What is a major histocompatibility complex (MHC)? Describe the differences between the functions of type I MHC and type II MHC.

16. How do the body's white blood cells know to attack and destroy old red blood cells but not new ones?

MAPS

1. Create a map showing the process of clonal expansion.

2. Map the different groups of white blood cells, their structure and their function.

BEYOND THE PAGES

FOCUS ON PHYSIOLOGY

Biotechnology Focus: Stealth® Liposomes

Early in their history, liposomes seemed to be the ideal vehicle for drug delivery: "magic bullets" that could deliver drugs wherever they were needed. But the first animal trials were disappointing, since liposomes injected intravenously disappeared rapidly from the circulation without reaching their target cells. The body's immune system recognized them as foreign even though they were composed of biological phospholipids, and macrophages gobbled them up and digested them. The solution that occurred to researchers was to make the liposomes invisible to the immune system somehow so that they could slip by the macrophages and reach their intended targets. The first type of "invisible" liposome relied on carbohydrate groups attached to the liposome exterior that made the liposomes resemble red blood cells. The next generation of invisible liposomes was trademarked Stealth® liposomes. They are coated with a polyoxyethylene polymer that allows them to slip into the body undetected and remain in the circulation for as long as a week, delivering their encapsulated drugs to the targeted tissues.

See: Stealth® Liposomes. D. Lasic and F. Martin, editors. (1995) CRC Press, Boca Raton. 289 pp.

READING

A is for...: Histocompatibility. (1996, November/December) *Science & Medicine.*

Adenoviruses as vectors for gene therapy. (1997, March/April) *Science & Medicine.*

Adhesion molecules in leukocyte emigration. (1995, November/December) *Science & Medicine.*

Alternative cancer treatments. (1996, September) *Scientific American.*

Anoikis: Assisted cell suicide. (1997, April) *News in Physiological Sciences* 12: 96-97.

Autoimmunity. (1994, May/June) *Science & Medicine.*

Bacterial virulence factors. (1995, May/June) *Science & Medicine.*

Biological basis of the stress response. (1993, April) *News in Physiological Sciences* 8: 69-73.

Cell suicide in health and disease. (1996, December) *Scientific American.*

Discovering the benign traits of the mast cell. (1997, September/October) *Science & Medicine.*

Emergence of tick-borne diseases. (1997, March/April) *Science & Medicine.*

Emerging viruses. (1995, October) *Scientific American.*

Evolution of infectious disease. (1996, April) *News in Physiological Sciences* 11: 83-89.

Fever: How may circulating pyrogens signal the brain? (1997, February) *News in Physiological Sciences* 12:1-8.

Helicobacter binds to blood group antigens. (1994, September/October) *Science & Medicine.*

How breast milk protects newborns. (1995, December) *Scientific American.*

How killer cells kill. (1988, January) *Scientific American.*

How cells present antigens. (1994, August) *Scientific American.*

How interferons fight disease. (1994, May) *Scientific American.*

How the immune system learns about self. (1991, October) *Scientific American.*

Immunology and the invertebrates. (1996, November) *Scientific American.*

Immunotherapy for cancer. (1996, September) *Scientific American.*

Immunotherapy for cocaine addiction. (1997, February) *Scientific American.*

Influenza virus control of protein synthesis. (1994, July/August) *Science & Medicine.*

Interleukin-2. (1990, March) *Scientific American.*

Killing mechanisms of cytotoxic T lymphocytes. (1998, February) *News in Physiological Sciences* 13.

Macrophages in the brain: Friends or enemies? (1994, April) *News in Physiological Sciences* 9: 80-84.

MHC polymorphism and human origins. (1993, December) *Scientific American.*

Perennial allergens and the asthma epidemic. (1996, May/June) *Science & Medicine.*

Perforin: Mediator of cellular defense. (1995, February) *News in Physiological Sciences* 10: 51-52.

Psychoneuroimmunology: Brain and immunity. (1995, November/December) *Science & Medicine.*

Radiolabeled antibodies. (1994, March/April) *Science & Medicine.*

Selection and survival of activated lymphocytes. (1995, March/April) *Science & Medicine.*

Sharks and the origins of vertebrate immunity. (1996, November) *Scientific American.*

Silicone immunopathology. (1996, September/October) *Science & Medicine.*

Solving the mysteries of viral hepatitis. (1994, March/April) *Science & Medicine.*

Special issue: The immune system. (1993, September) *Scientific American.*

Suicide genes for cancer therapy. (1997, July/August) *Science & Medicine.*

Teaching the immune system to fight cancer. (1993, March) *Scientific American.*

TGF-β. (1995, January/February) *Science & Medicine.*

T_H1 and T_H2 subsets of CD^{4+} lymphocytes. (1994, May/June) *Science & Medicine.*

The T cell and its receptor. (1986, February) *Scientific American.*

The immunology of inflammatory arthritis., (November/December 1996) *Science & Medicine.*

Tumor suppressor genes. (1995, January/February) *Science & Medicine.*

Understanding non-Hodgkin's lymphoma. (1997, May/June) *Science & Medicine.*

Xenotransplantation. (1996, July/August) *Science & Medicine.*

CHAPTER 23

INTEGRATIVE PHYSIOLOGY III: EXERCISE

SUMMARY

After Chapters 20 and 21, this chapter should feel relatively light! There are only a few main points:

- ATP for muscle contractions comes from aerobic metabolism, phosphocreatine, and glycolytic metabolism.
- During exercise, the catabolic hormones (glucagon, cortisol, catecholamines, and growth hormone) dominate. Insulin levels remain low.
- Exercise intensity is reflected by oxygen consumption.
- The respiratory and cardiovascular systems alter their activity to ensure that efficient oxygen and nutrients are delivered to exercising tissues.
- Moderate exercise can improve your immune system and decrease the risk of certain health problems.

Exercise is defined as any muscular activity that generates force and disrupts homeostasis. ATP for muscle contraction comes from several sources: aerobic metabolism, phosphocreatine, and anaerobic (glycolytic) metabolism. Glucose and fats are the primary substrates for energy production, but fats can only be metabolized in aerobic conditions.

Remember from Chapter 21 that glucagon is a catabolic hormone. The catabolic hormones participating in exercise metabolism include glucagon, cortisol, catecholamines, and growth hormone. The action of these hormones raises the plasma glucose concentration. However, insulin concentrations do not rise during exercise. Remember that active skeletal muscle doesn't require insulin for glucose uptake. Therefore, low plasma insulin concentrations prevent other cells from taking glucose that could be used by the muscles.

The intensity of exercise is indicated by oxygen consumption. Oxygen consumption increases rapidly at the onset of exercise and remains elevated after exercise as oxygen consumption returns to its resting level. The ability of the muscle cells to consume oxygen is possibly a limiting factor in exercise capacity. This is reflected by the fact that mitochondria can increase in size and number with endurance training. Cardiovascular activity is the major factor limiting maximal exertion.

Respiratory and cardiovascular systems make adjustments in response to exercise. Feedforward signals and sensory feedback initiate exercise hyperventilation. Exercise hyperventilation maintains nearly normal P_{O_2} and P_{CO_2} by steadily increasing alveolar ventilation in proportion with exercise level. Cardiovascular responses include increased cardiac output, vasodilation and increased blood flow in skeletal muscles, and a slight increase in mean arterial blood pressure. The cholinergic vasodilator system plays a feedforward role in causing vasodilation in skeletal muscle arterioles.

Exercise generates heat. In fact, most of the energy released during metabolism is not converted to ATP but is released as heat. There are two mechanisms by which the body regulates temperature during exercise: sweating and increased cutaneous blood flow.

Moderate exercise has been shown to affect health positively. It can alleviate or prevent the development of high blood pressure, strokes, and diabetes mellitus. However, a J-shaped curve (Fig. 23-9) relates immune function to exercise: only moderate exercise improves immune function. There has been no sound evidence to support the theory that exercise improves the immunity of immunocompromised individuals.

TEACH YOURSELF THE BASICS

METABOLISM AND EXERCISE

• Muscles require ATP for contraction. What are the sources of this ATP? (Fig. 23-1) [∫ p. 336]

_____

• Without new ATP production, a muscle has enough ATP to supply energy for how many seconds

of contraction? _____

• What macromolecules are the primary substrates for energy production?

_____

• The most efficient ATP production is a result of the glycolysis-citric acid cycle pathway. [∫ p. 91]
Briefly describe or map these pathways in the presence and absence of oxygen. (Fig. 23-1) [∫ p. 90]

• Discuss the advantages and disadvantages of anaerobic muscle metabolism over aerobic metabolism (Fig. 23-2).

_____

_____

• Where does muscle obtain glucose for ATP production (Fig. 23-1) [∫ p. 97]?

_____

• True or false? Aerobic exercise first uses glucose for ATP production, then turns to fatty acid metabolism.

• Beta oxidation [∫ p. 96] is (faster / slower?) than glycolysis.

Hormones Regulate Metabolism During Exercise

• List four hormones that affect glucose and fat metabolism during exercise and briefly describe their action.

1. _____

2. _____

3. _____

4. _____

• What happens to insulin secretion during exercise? Give the physiological mechanism and the adaptive significance for this pattern of insulin secretion.

_____

_____

Oxygen Consumption is Related to Exercise Intensity

• Exercise intensity is quantified by measuring oxygen consumption (Vo_2). Define oxygen consumption.

_____

• What is indicated by the maximal rate of oxygen consumption (Vo_{2max})?

_____

• Increased O_2 consumption persists even after activity ceases. (Fig. 23-3) Why?

_____

Several Factors Limit Exercise

• List three possible factors that might affect the muscle fibers' exercise capacity.

_____

_____

• Which of these factors is the most significant?

_____

VENTILATION RESPONSES TO EXERCISE

• How do total pulmonary ventilation, alveolar ventilation, rate and depth of breathing change in response to exercise [∫ p. 494]?

_____

• Exercise breathing is influenced by feedforward signals from the CNS and sensory feedback from peripheral receptors. Which peripheral receptors are most likely involved?

_____

• Fill the gaps in the following pathways:

　　1. Exercise begins muscle _____ and proprioceptors send signals to motor cortex.

　　2. Motor cortex signals respiratory control center in the _____ to increase ventilation.

　　3. As exercise continues, sensory feedback from what peripheral receptors to the respiratory control center ensures that O_2 use and ventilation are matched? _____

• How does exercise hyperventilation affect arterial oxygen and carbon dioxide levels? (Fig. 23-4)

_____

• Summarize the postulated stimuli for exercise hyperventilation.

_____

_____

CARDIOVASCULAR RESPONSES TO EXERCISE

• The cardiovascular control center (CVCC) responds to exercise by (increasing / decreasing ?)

 sympathetic output. What effect does this have on cardiac output? _____

 peripheral arterioles? _____

Cardiac Output Increases During Exercise

• Cardiac output increases dramatically with strenuous exercise. Cardiac output is influenced by what three factors [∫ p. 417]?

_____

• Explain the relationship between increased venous return and increased heart rate. _____

_____

• Describe the changes in the autonomic nervous system that alter exercise heart rate [∫ p. 404].

_____

• How is venous return affected by skeletal muscle contraction? [∫ p. 484] _____

Peripheral Blood Flow Redistributes to Muscle During Exercise

• During exercise, 88% of blood flow is diverted to exercising muscles (Fig. 23-5). How is this different from resting muscle blood flow? _____

• Describe the local and reflex processes that influence how the body redistributes blood flow during exercise.

_____

_____

Blood Pressure Rises Slightly During Exercise

• What factors determine peripheral blood pressure? [∫ p. 387] _____

• Explain what happens to blood pressure as these factors increase and decrease.

_____

• Total peripheral resistance decreases as exercise intensity increases (Fig. 23-6a). This would be

 expected to (raise / lower ?) arterial blood pressure.

• What is the net result of the changes in cardiac output and peripheral resistance during exercise? (Fig. 23-6b)

_____

The Baroreceptor Reflex Adjusts to Exercise

• Normally, increased blood pressure triggers a homeostatic decrease in blood pressure. During exercise, though, there is no homeostatic decrease in BP. Why is this? Outline the possible mechanisms behind the absent baroreceptor reflex during exercise.

_____

_____

_____

FEEDFORWARD RESPONSES TO EXERCISE

• Feedforward responses play a significant role in exercise physiology. For example, ventilation

(increases / decreases ?) upon beginning exercise, despite normal P_{CO_2} and P_{O_2}. (Fig. 23-7)

• What is the sympathetic cholinergic vasodilator system and what role does it play in exercise?

_____

_____

TEMPERATURE REGULATION

• What happens to most of the energy released during metabolism? _____

• Endurance exercise events can create core body temperatures of _____ °C.

• How does the body respond to this rise in temperature? [∫ p. 615]

_____

• Both homeostatic responses to increased temperature can disrupt other homeostatic conditions. Describe the potential threats posed by:

sweating [∫ p. 560] _____

increased cutaneous blood flow _____

• Faced with maintaining either blood pressure or body temperature, which will the body select?

_____

• Describe the changes that take place with acclimatization to exercise in hot environments.

_____

_____

EXERCISE AND HEALTH

• Exercise can improve several pathological conditions. List two of these conditions.

_____

Exercise Lowers the Risk of Cardiovascular Disease

• There is a relationship between exercise and cardiovascular disease. How does exercise affect

BP? _____ plasma triglycerides? _____ HDL levels?_____

☛ *Even mild exercise can have significant health benefits.*

Non-Insulin-Dependent Diabetes Mellitus May Improve with Exercise

☛ *p. 693, Fig. 23-8a: The line colors were switched. The top line should be red, the middle line green, and the bottom line blue.*

• Chronic exercise can improve type II diabetes mellitus. Briefly describe how this is so.

_____

Exercise, Stress, and Immunity

☛ *Exercise is associated with a reduced incidence of disease and improved longevity.*

• No solid evidence confirms that exercise boosts immunity, prevents cancer, or helps HIV-positive people fight AIDS. In fact, strenuous exercise can be detrimental. What physiological mechanism would explain this?

_____

• How does the literature show that exercise and depression are related? Is this relationship proved by the evidence or overstated?

_____

_____

TALK THE TALK

acclimatization
adipose tissue
aerobic metabolism
alveolar ventilation
anaerobic pathways
ATP
baroreceptor reflex
beta cells of the pancreas
beta-oxidation
carbohydrate
cardiac output
cardiovascular control center
carotid body chemoreceptors
catecholamine
central chemoreceptor
cholinergic vasodilator system
citric acid cycle
contractility
convective heat loss
cortisol
dehydration
diabetes mellitus
endurance
epinephrine
evaporative cooling
exercise
exercise and depression
exercise and immunity
exercise hyperventilation
fats
fatty acid
feedforward responses to exercise
force of contraction
glucagon
glucose
glucose transporters
glycogen
glycolysis
glycolytic metabolism
growth hormone

heart rate
high blood pressure
homeostasis
hyperpnea
insulin
K^+
lactic acid
limbic system
liver
maximal rate of oxygen consumption
      (V_{O_2}max)
metabolic acidosis
mitochondria
motor cortex
muscle contraction
norepinephrine
oxidative phosphorylation
oxygen consumption
paracrines
parasympathetic output
P_{CO_2}
peripheral blood flow
peripheral resistance to blood flow
phosphocreatine
physiological integration
P_{O_2}
proprioceptors
pulmonary stretch receptors
pyruvate
respiratory control center of the medulla
Starling's Law of the heart
sweating
sympathetic output
thermoregulation
vasoconstriction
vasodilation in skeletal muscle arterioles
venous return
ventilation
V_{O_2}

PRACTICE MAKES PERFECT

1. Which would you expect change more during exercise: diastolic or systolic blood pressure?

2. TRUE/FALSE: During exercise, cardiac output, stroke volume, and oxygen debt all should be greater in a trained (physically fit) person than in an average person.

3. How is cardiorespiratory endurance (aerobic fitness) usually measured?

4. During exercise, inspiratory and expiratory reserve volumes decrease. Why do you think that is?

5. List two beneficial effects of exercise.

6. What physiological factors can limit exercise capacity?

7. Concisely describe the relationship between exercise and immunity.

MAPS

Design a map that integrates cardiovascular and respiratory responses to exercise. Include receptors, chemicals, calculations, etc.

BEYOND THE PAGES

READING

Carbohydrate ingestion during prolonged exercise: Effects on metabolism and performance. (1991) *Exercise and Sports Sciences Reviews* 19:1ff.

Exercise, Infection, and Immunity. (1994) *International Journal of Sports Medicine* 15: S131-S141.

Prevention of type II diabetes by physical training. (1992) *Diabetes Care* 15 (suppl. 4):1794-1799.
This paper is one of a collection written for a World Health Symposium, Diabetes & Exercise '90, all published in Diabetes Care 15 (Suppl. 4), beginning on pg. 1676.

A review of the control of breathing during exercise. (1995) *European Journal of Applied Physiology* 71:1-27.

Synovial fluid hydraulics. (1996, September/October) *Science & Medicine*.

Venular endothelium: Metabolic sensors? (1995, February) *News in Physiological Sciences* 10: 50-51.
Paracrines released from venular endothelium may alter blood flow to exercising muscle.

Wilmore, J.H. and Costill, D.L. (1994) *Physiology of Sport and Exercise*. Human Kinetics, Champaign IL, p 549 .
An excellent text on exercise physiology that includes normal physiology, environmental effects on exercise performance, exercise in special populations, and exercise in health and disease.

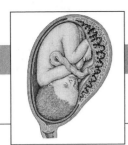

SUMMARY

OK, you got your break in Chapter 23, now it's back to work for one last big chapter. Most of this chapter is spent discussing the differences between male and female reproductive development. Therefore, when you're studying, make a big chart that compares these differences (be sure to include hormone action). The discussion of pregnancy and parturition provides a perfect opportunity to make a detailed flowchart. Start with the human sex act and go all the way through milk production.

Here are some key points:

• Autosomes are diploid; sex cells are haploid. Meiotic division [Appendix B] creates haploid cells.
• Sexual differentiation occurs during embryonic development. The presence of a Y chromosome initiates male reproductive development. The absence (or inactivity) of a Y chromosome initiates female reproductive development.
• Gametogenesis begins during embryonic development, stops before or just after birth, then resumes at puberty in males. New gamete production in females ceases before birth.
• The female menstrual cycle is complex and involves extensive hormonal control (Fig. 24-13).
• Fusion of egg and sperm leads to procreation. Pregnancy ends with parturition (the birthing process).

On a very general level, the human body contains two types of cells: diploid autosomal (somatic or body) cells and haploid sex (or germ) cells. The diploid number of chromosomes for humans is 46, and the haploid number is therefore 23 (exactly $\frac{1}{2}$ the diploid number of chromosomes). In haploid cells, there are 22 autosomal chromosome pairs that control autosomal cell development and one pair of sex chromosomes that control germ cell development. The male sex cells, sperm, are produced by the testes; female sex cells, ova, are produced by the ovaries. Gametogenesis is the process by which sex cells are produced. The diploid primary sex cells (spermatogonia, oogonia) become secondary sex cells, and after the second meiotic division, these become haploid germ cells. While the process is similar for both male and female, the timing of gametogenesis is very different. Spermatogenesis begins in the embryo and stops just after birth. It resumes at puberty and continues throughout the male's lifetime. Oogenesis also begins in the embryo but stops before birth. At puberty, ovulation begins and occurs in a cyclical fashion until menopause. Both spermatogenesis and oogenesis are under hormonal control. Hormones involved include gonadotropins (FSH and LH from the anterior pituitary), sex hormones (androgens, estrogens, progesterone), inhibins, and activins.

Sexual differentiation takes place during the seventh week of embryonic development. Before differentiation, the gonadal tissue is considered bipotential. If the embryo has a Y chromosome, the SRY gene produces testis-determining factor that signals the testicular Sertoli cells to secrete Müllerian inhibiting substance. Müllerian inhibiting substance causes the degeneration of the Müllerian ducts. Leydig cells then begin to secrete testosterone and DHT, which cause the Wolffian ducts to develop into male accessory structures. Testosterone and DHT also create the external male genitalia. On the other hand, if the embryo does not have a Y chromosome, then none of the above happens, and the Müllerian ducts develop into female reproductive structures.

Male reproductive anatomy consists of the testes, accessory glands and ducts, and external genitalia. The urethra, which runs through the penis, is the common duct for sperm and urine movement (though not concurrently). Female reproductive anatomy consists of the ovaries, Fallopian tubes, uterus, vagina, labia (majora, minora), and clitoris. The female urethra is completely separated from the reproductive structures.

The female reproductive cycle is perhaps one of the most complex patterns in physiology. It is under extensive hormonal control (Fig. 24-13) and can be influenced by many emotional and physical factors. The menstrual cycle is composed of two concurrent cycles: the ovarian cycle and the uterine cycle. The ovarian cycle is further divided into phases: the follicular phase, ovulation, and the luteal phase. The uterine cycle is also divided into phases: menses, the proliferative phase, and the secretory phase. Follow Fig. 24-13 and note the hormones involved in these cycles.

During unprotected sex, the sperm and egg have the possibility of joining to form a zygote. Sperm must first be capacitated before they can make their way to the egg. If a sperm reaches an egg and successfully fertilizes it, only then will the egg complete its second meiotic division. Only one sperm is allowed to fertilize an egg, as the cortical reaction prevents polyspermy. Additionally, fertilization is species specific, meaning that only sperm and egg of compatible species can initiate procreation. If fertilization is successful, the two haploid germ cells create a diploid zygote that begins mitotic division to produce an embryo. The developing embryo attaches itself to the mother's uterine wall and parasitizes the mother's nutrient supply until birth. The placenta of the developing embryo secretes hormones that alter the mother's metabolism to ensure adequate nourishment for development.

After 38-40 weeks of pregnancy, the baby is born in a process called parturition. This process is under hormonal control, though the specific signaling is still unclear. The mother's mammary glands are activated under hormonal control and begin to secrete milk to feed the baby.

TEACH YOURSELF THE BASICS

• What is a zygote? _____

SEX DETERMINATION

• Define the structures found in both male and female sex organs, then give the sex-specific names for each.

genitalia _____

gamete _____

germ cell _____

• Review of genetics [∫ Appendix B].

How many sets of chromosomes do nucleated body (somatic) cells have? (Fig. 24-1) _____

How many autosomal pairs? _____ How many sex chromosome pairs? _____

How many autosomes do gametes have? _____ How many sex chromosomes? _____

How is the X sex chromosome different from the Y? _____

The Genetic Sex of an Individual is Determined by the Sex Chromosomes

• What determines the genetic sex of an individual? (Fig. 24-2) _____

• XX individuals are usually phenotypic _____ and XY individuals are usually _____.

• There are many documented cases of abnormal sex chromosome distribution in humans. What are the roles of X and Y in determining the physical sex of an individual? What happens in the absence of the Y chromosome?

_____

_____

• What are Barr bodies? How and why are they formed?

_____

Sexual Differentiation in the Embryo Takes Place in the Second Month of Development

• Before the seventh week of development, it is morphologically difficult to determine an embryo's sex. What are the bipotential components of the external genitalia present at this stage? Into what components of the male and female genitalia do each later develop? (Fig. 24-3) (Table 24-1)

1. _____

2. _____

3. _____

4. _____

• Name the two ducts associated with the bipotential gonadal tissue and tell what structure(s) each duct develops into in the male and female.

1. _____

2. _____

• Explain the role of the following in sexual development of the male embryo:

Y chromosome _____

SRY gene _____

testis determining factor _____

Müllerian inhibiting substance (MIS)_____

Sertoli cells _____

Leydig cells _____

testosterone and DHT _____

• In the absence of TDF, MIS, and testosterone, what happens to the bipotential structures in the female embryo?

_____

_____

• What effect, if any, do sex hormones have on sexual behavior and gender identity?

_____

BASIC PATTERNS OF REPRODUCTION

• Define gametogenesis. _____

• The timing of gametogenesis is different for males and females. Briefly compare these timing differences. [∫ Fig. 3-2, p. 42]

_____

_____

Gametogenesis Begins in Utero and Resumes During Puberty
☛ *For a summary of male and female patterns of gametogenesis, see Fig. 24-5.*

• Gametogenesis has some similar steps in both sexes. List the steps of gametogenesis, describing them by whether DNA has replicated, whether the cell divides, and the number of chromosomes in the cell.

1. _____

2. _____

3. _____

4. _____

• Compare the timing of these events in gametogenesis for males and females.

_____

_____

• The first meiotic division creates two secondary gametes. Compare the fates of those gametes in males and females.

_____

• What are polar bodies? _____

• Define ovulation. _____

Hormonal Control of Reproduction is Directed by the Brain

• There is a basic pattern for hormonal control of reproduction (Fig. 24-6). Sketch a flowchart or describe this pattern, including the hormones, their targets, and their action.

• Structurally, how are the male and female sex hormones similar? (Fig. 24-7)

_____

• Name the three basic groups of sex steroids and identify which are predominately found in males and which in females.

_____

• What is the function of the following?

aromatase _____

inhibin _____

activins _____

Feedback pathways

• Sex hormones follow general short-loop and long-loop feedback patterns [∫ p. 187]. Describe the feedback effect of each of the following:

androgens _____

low estrogen levels _____

sustained elevated estrogen _____

The hypothalamus and gonadotropin release

• Describe the tonic, pulsed GnRH release in both sexes. What is the physiological mechanism for this pulsing?

_____

_____

MALE REPRODUCTION

• Male anatomy (Fig. 24-8):

    The external genitalia are the _____ and _____.

    What is the common passageway for urine and sperm? _____

    What two tissues form the erectile tissue? _____

    The tip of the penis is called the _____ penis. It is covered by tissue called the _____

    unless this tissue has been surgically removed in a procedure known as _____.

• What is the function of the scrotum? _____

• What is cryptorchidism and why is it usually corrected? _____

_____

• List the three male accessory glands. _____

_____

Sperm Production Takes Place in the Testes

• Human testes (Fig. 24-9):

    The tough outer fibrous capsule encloses the _____ tubules.

    What structures/cells are found between these tubules? _____

    What is the epididymis (Fig. 24-9b)? _____

    What is the vas deferens? _____

The seminiferous tubules
• Describe the two cell types that compose the seminiferous tubules (Fig. 24-9d,e).

_____

_____

• What is the function of the basement membrane that surrounds the outside of the tubule (Fig. 24-9c)?

_____

• List the three functional compartments in the testes.

    1. _____

    2. _____

    3. _____

• Describe the composition of the fluid in the lumen of the seminiferous tubules.

_____

Sperm production
• Describe the anatomical arrangement of the Sertoli cells and the spermatogonia.

_____

• Outline or map the process of sperm maturation. (Fig. 24-10)

• Describe the parts and their functions in a mature sperm. _____

_____

The Sertoli cells
• What role do the Sertoli cells play in sperm maturation? _____

_____

• List the substances made or secreted by the Sertoli cells.

_____

• What is the function of androgen-binding protein? Why is it necessary?

_____

Leydig cells
• What is the primary function of the Leydig cells?

_____

• Leydig cells are active in the fetus but inactive after birth until puberty. What role do they play in
the fetus?

_____

• The bulk of the body's testosterone is produced in the Leydig cells. What other hormones can testosterone be converted into?

_____

Spermatogenesis Requires Gonadotropins and Testosterone

• What is the target tissue of FSH and what effect does the hormone have on this tissue? (Fig. 24-11)

_____

_____

• What is the target tissue of LH and what effect does the hormone have on this tissue?

_____

Male Accessory Glands Contribute Secretions to Semen

• What is semen?_____

• What chemicals are found in semen? Give their source and their function. (Table 24-3)

_____

_____

• What percentage of semen is from the accessory glands? _____ List the accessory glands.

_____

Androgens Influence Primary and Secondary Sex Characteristics

• List the primary sex characteristics of males. _____

• List the secondary sex characteristics of males. _____

_____

• Androgens are (anabolic / catabolic ?) steroids.

FEMALE REPRODUCTION

The Female Reproductive Tract Consists of the Ovaries, Accessory Structures Such as the Uterus, and the External Genitalia

• List the parts of external female genitalia that make up the vulva (or pudendum). (Fig. 24-12a)

_____

_____

• Describe the internal anatomy by following the path of sperm.

_____

_____

• How do the Fallopian tubes move an egg to the uterus? _____

_____

The Ovary, Like the Testis, Produces Both Gametes and Hormones

• Describe (or draw) the anatomy of an ovary and its follicles. (Fig. 24-12d)

• What is an oocyte? _____

Female Reproductive Cycles Last about One Month

• How often does a menstrual cycle occur? _____

• What is the purpose of the menstrual cycle? _____

• The menstrual cycle is described according to changes in what two structures?

_____

• Name the three phases of the ovarian cycle. (Fig. 24-13)

_____

• Name the phases of the uterine cycle (Fig. 24-13).

_____

• What hormones control the ovarian and uterine cycles? (Fig. 24-13)

_____

• Which hormone dominates the follicular phase? _____

• Which hormones dominate the luteal phase? _____

Early follicular phase
• What event marks day 1 of a cycle? _____

• Hormones:

What happens to FSH and LH secretion just before the cycle begins? _____

What changes does FSH cause? _____

_____

Which hormones do the granulosa cells and theca secrete? (Fig. 24-14a, p. 719)

_____

Why do FSH levels decline as the follicular phase progresses? _____

• Follicle development. (Table 24-4) Explain the role of the following:

granulosa cells _____

thecal cells_____

• What is atresia? _____

• What is the antrum? _____

• When menstruation ends, what happens to the endometrium?

_____

Late follicular phase
• What happens to ovarian estrogen levels as the follicular phase progresses? _____

• What cells are the granulosa secreting at this point? (Fig. 24-14b) _____

• How does this affect the feedback loop to the pituitary?

_____

• Ovulation requires a surge in which hormone? _____

• Meiosis resumes just before ovulation. Review: When did meiosis pause for this oocyte?

_____

• What is the result of this meiotic division? _____

• Antral volume now reaches its maximum or minimum? _____

• How do the high estrogen levels prepare uterus for pregnancy?

_____

Ovulation
• At what point of the cycle does ovulation occur? _____

• What enzyme does a mature follicle secrete that promotes ovulation? _____

• The resulting breakdown of collagen causes what kind of reaction? _____

• How is the egg released from the follicle?

_____

• Describe luteinization. _____

_____

• What happens to estrogen secretion at ovulation? _____

Early to mid-luteal phase
• Describe what happens to progesterone and estrogen secretion following ovulation.

_____

• How do these hormones affect FSH and LH production? (Fig. 24-14c)

_____

• What are the effects of progesterone on the uterus?

_____

_____

Late luteal phase and menstruation
• What happens to the corpus luteum in the absence of pregnancy? (Fig. 24-14d)

_____

• When progesterone secretion decreases, what change does this initiate in the endometrium?

_____

• What is a corpus albicans? _____

• Describe menstruation. _____

_____

Estrogens and Androgens Control Secondary Sex Characteristics in Women

• What are some female secondary sex characteristics? _____

_____

• Which part of the adrenal gland secretes androgens? _____

• What effect do these androgens have in women? _____

_____

PROCREATION

• What types of structures and behaviors have evolved to ensure reproductive success in humans
  and many other terrestrial vertebrates?

_____

• List the two stages of the male sex act. _____

The Human Sexual Response Has Four Phases

• Describe the four stages of the human sex act (coitus). _____

_____

_____

The Male Sex Act is Composed of Erection and Ejaculation

• What happens physiologically in the penis to allow erection?

_____

• Draw the erection reflex. (Fig. 24-15)

• What is the climax of the male sexual act? What does this accomplish?

_____

• The average semen volume is 3 mL. What percentage of that volume is sperm? _____

• Explain the process of emission. _____

• What structure prevents the mixing of sperm and urine? _____

• What are some factors involved in impotence? _____

_____

Contraceptive Measures are Designed to Prevent Pregnancy

• List some common methods of contraception (birth control). Briefly discuss their effectiveness.

_____

_____

_____

_____

_____

• Why have male hormonal contraceptive efforts failed thus far? _____

_____

• Give an example of a potential new form of contraception. _____

Infertility is the Inability to Conceive

• Define infertility. _____

• List some potential causes of infertility. _____

_____

• What number of all pregnancies terminate spontaneously? _____

• Describe the process of _in vitro_ fertilization. _____

_____

_____

PREGNANCY AND PARTURITION

Fertilization

• Define capacitation. What has to happen in order for capacitation to occur? _____

_____

• For how long following ovulation can the egg be fertilized? _____

• How long is sperm viable in the female reproductive tract? _____

• Where does fertilization take place? (Fig. 24-16) _____

• Describe fertilization. (Fig. 24-17) Include the following terms: acrosome, acrosomal reaction, capacitation, cortical granules, cortical reaction, enzymes, granulosa cells, meiosis, polyspermy, second polar body, sperm-binding receptor, sperm nucleus, zona pellucida, zygote nucleus.

_____

_____

_____

_____

_____

_____

_____

• Fertilization creates a zygote with a (diploid / haploid ?) set of chromosomes.

The Developing Zygote Implants in the Secretory Endometrium

• When does the dividing zygote move into the uterine cavity? (Fig. 24-18) _____

• The embryo is in what developmental stage at this point? _____

• The outer blastocyst becomes what structure? (Fig. 24-19a) _____

• The inner blastocyst becomes what four structures? _____

_____

• How long after fertilization does implantation usually take place? _____

• This is equivalent to about what day of the woman's menstrual cycle if ovulation was on day 14?

_____

• Describe the process of implantation. _____

• What are chorionic villi and what is their function? _____

_____

• Explain the relationship between the mother's blood supply and that of the fetus.

_____

• Why is the abnormal separation of the placenta from the endometrium a medical emergency?

_____

The Placenta Secretes Hormones During Pregnancy

• When an embryo has implanted, what prevents menstruation from occurring about day 28 of the cycle?

_____

Human chorionic gonadotropin (hCG)
• What other hormone is hCG structurally related to? _____

• How long into gestation does the corpus luteum secrete estrogen and progesterone? _____

• Give two functions of hCG. _____

_____

Human chorionic somatomammotropin (hCS)
• What is the older name for hCS? _____

• hCS is structurally related to what two other hormones? _____

• What are the postulated roles for hCS during pregnancy? _____

_____

Estrogen and progesterone
• What effect does continuous secretion of estrogen and progesterone during pregnancy have on

FSH and LH secretion? _____

• What are the functions of estrogen and progesterone during pregnancy? _____

_____

• What is relaxin? _____

Pregnancy Ends with Labor and Delivery

• How long is gestation in humans? _____

• What is parturition? _____

• What is labor? _____

• What hormonal changes take place around the time of parturition? _____

_____

_____

• Describe the changes in the uterus that take place during labor and delivery. _____

_____

_____

• The fetus is normally in what orientation at the time of labor? (Fig. 24-20a) _____

• What two hormones promote uterine contractions? _____

• Describe the positive feedback loop of parturition. What outside event stops the cycle?(Fig. 24-21)

_____

_____

• What happens to the placenta when the baby is born? _____

The Mammary Glands Secrete Milk during Lactation

• Describe mammary gland structure. (Fig. 24-22)

_____

• Breasts develop during puberty under the control of which hormone? _____

• How do the breasts change during pregnancy? What hormones induce these changes?

_____

_____

• Estrogen and progesterone (inhibit / stimulate ?) milk secretion by the mammary gland epithelium.

• Prolactin [∫ p. 186] controls milk production. Draw the prolactin control pathway.

• Compare the composition of colostrum and breast milk. _____

_____

• What causes milk production to increase after pregnancy? _____

_____

• How does suckling act as a stimulus for milk production and release? Describe or draw the reflex. (Fig. 24-23)

• How is milk ejection accomplished? _____

• What is a postulated role for prolactin in men and non-nursing women?

_____

GROWTH AND AGING

Puberty Marks the Beginning of the Reproductive Years

• When does puberty begin for girls? _____

      For boys? _____

• Puberty requires the maturation of which control axis? What is one explanation for how this happens?

_____

Menopause and Aging

• What changes happen in the female reproductive system after about 40 years of menstrual cycles? Why?

_____

_____

• This process causes an absence of estrogen. What changes can accompany this lack of estrogen?

_____

• What are some pros and cons of hormone replacement therapy in women?

_____

• Do men experience a "male menopause" in the same way that women do? Explain.

_____

TALK THE TALK

abstinence
acrosomal reaction
acrosome
activin
adrenal cortex
allantois
amnion
anabolic steroid
androgen
androgen-binding protein
anencephaly
anterior pituitary
anti-Müllerian hormone
antrum
aromatase
atresia
autosome
Barr body
barrier method of contraception
bipotential gonad
birth control
birth control pill
blastocyst
blood-testis barrier
breast development
breast-feeding
bulbourethral (Cowper's) gland
capacitation
centromere
cervical mucus
cervix
chorion
chorionic villi
ciliated epithelium
circumcision
coitus
condom
contraception
corpora cavernosa
corpus luteum
corpus albicans
corpus spongiosum
cortical granule of ovum
cortical reaction
cryptorchidism

decapacitation factor
diaphragm
dihydrotestosterone (DHT)
diploid number
down-regulation, GnRH
ductus deferens
ejaculation
emission
endometrium
epididymis
erection reflex
erogenous zone
erotic stimuli
estradiol
estrogen
excitement phase
extraembryonic membrane
Fallopian tube
fimbriae
follicle stimulating hormone
    (FSH)
follicular phase
foreskin
gamete
gametogenesis
genital tubercle
genitalia
germ cell
gestation
glans penis
GnRH agonist
gonad
gonadotropin releasing hormone
    (GnRH)
gonadotropin
granulosa cell
hot flash
human placental lactogen (hPL)
human chorionic gonadotropin
    (hCG)
human chorionic
somatomammotropin (hCS)
hymen
hyperglycemia of pregnancy
implantation

impotence
in vitro fertilization
infertility
inhibin
internal fertilization
intrauterine device (IUD)
labioscrotal swelling
labor
lactation
let-down reflex
Leydig cell
LHRH: see GnRH
libido
long-loop negative feedback
luteal phase
luteinization
luteinizing hormone (LH)
lysozyme
mammary gland
meiosis
melatonin
menarche
menopause
menses
menstrual cycle
menstruation
mifepristone
mitosis
Müllerian duct
Müllerian inhibiting substance
myometrium
oocyte
oogonia
orgasm
osteoporosis
ovarian cycle
ovary
oviduct
ovulation
ovum
oxytocin
parasympathetic vasodilation
parturition
penis
placenta

plateau phase	resolution phase	sustentacular cell
polar body, first and second	RU 486	testes (testis)
polyspermy	scrotum	testis determining factor
positive feedback loop	secondary spermatocyte	testosterone
prepuce	secondary oocyte	theca
primary oocyte	secondary sex characteristic	transforming growth factor
primary spermatocyte	secretory phase	tubal ligation
primary follicle	secretory epithelium	tubal pregnancy
primary sex characteristic	semen	urethral groove
progesterone, thermogenic ability	seminal vesicle	urethral fold
progesterone	seminiferous tubule	uterine cycle
prolactin	Sertoli cells	uterus
prolactin inhibiting hormone	sex chromosome	vagina
proliferative phase	short loop negative feedback	vasectomy
prostaglandin	sister chromatid	vulva
prostate gland	sperm (spermatozoa)	Wolffian duct
pseudohermaphrodite	spermatid	X-linked disorder
puberty	spermatogonia	zinc
pudendum	SRY gene	zona pellucida
pulse generator	sterilization	zygote
relaxin	stroma	

ETHICS IN SCIENCE

In recent years, the use of donor eggs and hormonal therapy has made it possible for post-menopausal women to bear children through *in vitro* fertilization. For example, in California a 53 year-old woman gave birth to quadruplets. When these children are twenty, their mother will be 73 years-old, if she is still alive. Older men, have been having children with younger women for years. Should an age limit be placed on women who desire *in vitro* fertilization?

PRACTICE MAKES PERFECT

1. Anatomical structures are considered **homologous** if they have the same origin and **analogous** if they are similar in function but do not have the same origin. Using the information in Fig. 23-4, pair the following parts of the male and female reproductive tract and mark them as homologous or analogous. Not every part will have a corresponding part in the opposite sex. In that case, mark them "unique."

Male	Corresponding female part?	Analogous or homologous?	Female
bulbourethral gland	_____	_____	a) clitoris
ductus deferens	_____	_____	b) Fallopian tube
penis	_____	_____	c) ovary
prostate gland	_____	_____	d) labia majora
scrotum	_____	_____	e) labia minora
seminal vesicle	_____	_____	f) uterus
testis	_____	_____	g) vagina

2. Which set of terms below corresponds to the blanks in the sentence?

The _____ later develops into the _____ which secretes _____

until the _____ takes over the role of maintenance of pregnancy.

    a. ovary, corpus luteum, estrogen and progesterone, endometrium
    b. follicle, corpus luteum, luteinizing hormone, placenta
    c. ovary, placenta, follicle stimulating hormone, corpus luteum
    d. ovary, corpus luteum, estrogen and progesterone, placenta
    e. follicle, corpus luteum, estrogen and progesterone, placenta

3. You are a researcher who has just discovered the hypothalamic-pituitary control axis for the ovary. Now you have conducted some experiments to see if menopause is due to a failure of the ovarian cells themselves, or a failure of one of the trophic hormones. Your working hypothesis is that menopause is a result of pituitary failure. If this is true, what results do you expect to obtain in your tests?

a) Levels of pituitary gonadotropins will be   *elevated*   *normal*   *decreased*

b) Estrogen levels will be   *elevated*   *normal*   *decreased*

c) Administering estrogen   *will restore normal menstrual cycles*   *will have no effect on cycles*

d) Administering GnRH   *will restore normal menstrual cycles*   *will have no effect on cycles*

e) Administering FSH and LH   *will restore normal menstrual cycles*   *will have no effect on cycles*

d) Administering progesterone   *will restore normal menstrual cycles*   *will have no effect on cycles*

4. Explain the function of the corpus luteum during early pregnancy.

5. What is the adaptive value of androgen-binding protein?

6. Fill in the following chart on reproduction.

	Male	Female
What is the gonad?		
What is the gamete?		
Cell(s) that produce gametes		
What structure has the sensory tissue involved in the sexual response?		
What hormone(s) controls development of the secondary sex characteristics?		
Gonadal cells that produce hormones and the hormones they produce		
Target cell/tissue for LH		
Target cell/tissue for FSH		
Hormone(s) with negative feedback on anterior pituitary		
Timing of gamete production in adults		

7. BRIEFLY but specifically explain the mechanism by which the oral birth control pill works in woman.

8. Name the hormone(s) in women that is(are) the primary control for the following events:

Proliferation of the endometrium _____

Initiates development of follicle(s) _____

Ovulation _____

Development of endometrium into a secretory structure _____

Keeps corpus luteum alive in early pregnancy _____

Keeps endometrium from sloughing (i.e., menstruating) in early pregnancy _____

9. If a child who was genetically female had an aromatase deficiench, how would this affect the child's sexual development?

MAPS
1. Create a map showing the events determining the sex of an embryo, starting with the sex chromosomes.

2. Outline the uterine and ovarian cycles and include the major hormones active during each phase.

3. Draw a reflex map that shows how the female birth control pill that contains estrogen and progesterone works to stop the production of ova.

BEYOND THE PAGES

READING

Animal sexuality. (1994, January) *Scientific American.*

Barriers to sexually transmitted diseases. (1996, March/April) *Science & Medicine.*

Biology of aggression. (1995, January/February) *Science & Medicine.*

Can environmental estrogens cause breast cancer? (1995, October) *Scientific American.*

Debate: Is homosexuality biologically induced? (1994, May) *Scientific American.*

The dilemmas of prostate cancer. (1994, April) *Scientific American.*

Does screening for prostate cancer make sense? (1996, September) *Scientific American.*

The epididymis as a chloride-secreting organ. (1994, February) *News in Physiological Sciences* 9: 31-34.

Estrogen actions in arteries, bone, and brain. (1994, July/August) *Science & Medicine.*

Estrogen receptors in breast cancer. (1996, January/February) *Science & Medicine.*

Fatness and fertility. (1988, March) *Scientific American.*

Fertilization in mammals. (1988, December) *Scientific American.*

Future contraceptives. (1995, September) *Scientific American.*

The gravid kidney: Clues to preeclampsia? (1996, August) *News in Physiological Sciences* 11: 192-193.

The history of synthetic testosterone. (1995, February) *Scientific American.*

Hormonal manipulation to prevent breast cancer. (1995, July/August) *Science & Medicine.*

How breast milk protects newborns. (1995, December) *Scientific American.*

Human Reproductive Ecology: Interactions of environment, fertility, and behavior. (1994) *Annals of the New York Academy of Sciences*, Volume 709.

Is hormone replacement therapy a risk? (1996, September) *Scientific American.*

Menopause: Aging of pacemakers. (1997, June) *News in Physiological Sciences* 12: 143-144.

The neurobiology of mammalian puberty: Has the contribution of glial cells been underestimated? (1994, February) *Journal of NIH Research* 6: 51-56.

Pre-term birth. (1996, March/April) *Science & Medicine.*

RU 486. (1990, June) *Scientific American.*

Sexually transmitted diseases in the AIDS era. (1991, February) *Scientific American.*

The "stress" of being born. (1986, April) *Scientific American.*

Testicular cancer. (1995, January/February) *Science & Medicine.*

Why do we age? (1992, December) *Scientific American.*

ANSWERS TO WORKBOOK QUESTIONS

CHAPTER 1- Practice Makes Perfect
1. Physiology is the study of body function; anatomy is the study of body structure.
2. a) The hypothesis might be "Pravistatin lowers cholesterol in rats."
   b) An appropriate control would be to administer an inert substance in the same fashion that the Pravistatin is administered.
3. As the concentration of nerve growth factor (NGF) increases, the number of migrating cells increases. NGF has its minimum effect at a concentration of 10-5 M and its maximum effect at a concentration of 10-3 M.
4. The independent variable, which is controlled by the experimenter, is the extracellular concentration of glucose; it goes on the x-axis. The dependent variable, which is measured by the experimenter, is the intracellular concentration of glucose; it goes on the y-axis. The data do not begin at low values, so the origin is not given the value of 0,0. On the x-axis, the first heavy line is given the lowest value (80), with each heavy line after that representing an increase of 10 mM. The x-axis data are evenly spaced. The y-axis values are not evenly spaced, so the person constructing the graph must select a range of values that includes all points. The lowest data point is 52, so the first heavy line on the y-axis is given a value of 50, with each heavy line after that representing an increase of 10 mM (60, 70, etc.). The y-axis data points can then be plotted according to their actual values. The data points do not fall exactly on a line, so a best-fit line should be drawn (see Fig. 1-7d, p. 9). The x-axis is not time, so individual points should not be connected. The graph flattens out at x = 130 and the line changes slope from diagonal to horizontal. Summary of results: The intracellular concentration of glucose increases linearly as extracellular glucose concentration increases, up to an extracellular concentration of 100 mM. At that concentration, the intracellular concentration of glucose reaches a maximum value of about 100 mM.

CHAPTER 2
p. 2-5: To make a 1 molar (1 mole/L) solution of NaCl, weight out 58.5 g NaCl and add water until the total volume is 1 liter. One mole of magnesium ions (Mg^{2+}) contains 2 equivalents.
p. 2-6: A 10% (wt/vol) solution contains 10 g of solute per 100 mL solution. To make 250 mL of 10% NaCl, weigh out 25 g of NaCl and add water until the final volume is 250 mL.
   A solution with 200 mg NaCl/dL = 200 mg/100 mL. This is equal to 2000 mg/1000 mL, or 2 g/L.

Quantitative Thinking
Task 2: 6 moles NaCl + 6 moles glucose = 12 moles total solute in 3 L volume = 4 mol/L = 4 M.
Task 3: 300 mM glucose = 0.3 moles glucose/liter. Mol. weight glucose = 180.
   180 g/mole = ? g/0.3 mole = 54 g
   54 g glucose/1 liter = ? g/0.6 L = 32.4 g glucose into 600 mL solution.

Practice Makes Perfect
1. The carbon atom has 6 protons and 6 neutrons in the nucleus. There are 2 electrons in the first electron shell and 4 electrons in the outer shell. See Fig. 2-6b, p. 20. The oxygen atom has 8 protons and 8 neutrons in the nucleus. There are 2 electrons in the first electron shell and 6 electrons in the outer shell. See Fig. 2-6c, p. 20.
2.

$$\overset{..}{\underset{..}{O}} :: C :: \overset{..}{\underset{.}{O}}$$

3. B, D, A, C, F, G, H

4.

ELEMENT	SYMBOL	AT. #	PROTONS	ELECTRONS	NEUTRONS	AT. WT.
Calcium	Ca		20	20		
Carbon	C	6			6	
Chlorine	Cl	17		17		
Cobalt		27			32	
Hydrogen	H	1	1			1
Iodine			53	53		127
Magnesium	Mg	12				24
Nitrogen			7		7	
Oxygen		8		8	8	
Sodium	Na	11		11	12	
Zinc	Zn	30		30		65
Copper	Cu	29	29			64
Iron			26	26	30	
Potassium	K			19		39

5.　Water = H2O. (2 H × at. wt. 1 = 2) + (1 O × at. wt. 16 = 16) = 18 daltons

6.　F, E, B, G, A, C

7.　Both A and C are correct. They have the correct number of carbon, hydrogen, oxygen and nitrogen atoms. The COO⁻ and 4 electrons in the outer shell. See Fig. 2-6b, p. 20. The oxygen atom has 8 protons and 8 neutrons in the nucleus. There are 2 electrons bon instead of 4.

8.　a) Covalent　b) Hydrogen　c) Water is the only solvent in biological systems. d) The charged ions will be attracted to the partial charges in the polar regions of the water molecule. This allows the ions to dissolve in the water.

9.　If an oxygen atom gains a proton, it becomes a fluorine atom.

10.　Chlorine has 7 electrons in its 8-place outer shell; potassium, like sodium, has only one electron in its outer shell. Thus, chlorine takes an electron from potassium, creating Cl⁻ (chloride) and K⁺. Both ions are stable as their outer shells are now filled. See Fig. 2-9, p. 23.

11.　Polar molecules are hydrophilic because their regions of partial charge will interact with the polar regions of water, allowing the polar solute to dissolve. Nonpolar molecules are hydrophobic and will not dissolve because they have no regions of charge to disrupt the hydrogen bonds between adjacent water molecules.

12.　The molecular weight of sodium chloride (NaCl) is 58.5. One mole weighs 58.5 grams, so 0.5 moles will weight 29.25 g.

13.　A 0.5 M NaCl solution has 0.5 moles/L, or 29.25 g/L. To make 0.5 L, put 14.625 g into 500 mL solution.

14.　A 0.1 M solution has 100 mmoles/L.

15.　A 50 mM solution contains 50 mmoles glucose/liter, or 5 mmoles/100 mL. One mole contains 180 g, so 1 mmole contains 0.180 g. Therefore, 5 mmoles = 0.9 g.

16.　To make 100 mL of a 3% glucose solution, take 3 g glucose and add water to give 100 mL final volume. For molarity: 1 mole/180 g = ? mole/3 g = 0.017 moles. 0.017 mol/100 mL = 0.17 mol/L = 170 mM.

17.　Mixed solution contains 2 L with 400 mmoles glucose and 800 mmoles NaCl. Total concentration is 1200 mmol/2 L or 600 mmol/L = 0.6 M. The NaCl concentration is 400 mmol/2L or 200 mmol/L = 0.2 M. The glucose concentration is 800 mmol/2L or 400 mmol/L = 0.4 M.

18. Na$^+$ contains 1 mEq/mmol because the ion has a single charge. Therefore, 142 mEq/L = 142 mmol/L.

19. Ca^{2+} contains 2 mEq/mmol because the ion has a charge of 2+. Therefore, 5 mEq/L = 2.5 mmol/L.

20. C; D and E; A; F, B and D

21. On the left side of the reaction, water (H_2O) is the acid because it donates an H$^+$, and the amine is the base. On the right side of the reaction, the amine donates the H$^+$ and acts as the acid, while the hydroxide ion (OH the reaction, th22. H_2CO_3 is carbonic acid, which dissociates into H+ and bicarbonate ion, HCO_3^-, which acts as a base.

23. A 0.5 M solution equals a 500 mM solution, which is more acidic than a 50 mM solution.

24. C, B, E, A, F

25. D; A and E; B and C; C; F

26. See Fig. 2-19 on p. 34 for one possible map. Your map does not have to match it exactly, but it should have the same links. If you have put in links that are not shown in Fig. 2-19, ask your instructor if they are correct.

CHAPTER 3 - Practice Makes Perfect

1. Mitochondria would have the highest probability of existing independently and evolving because they have their own DNA with which they can reproduce and make new proteins. Mitochondria also contain the enzymes and proteins needed to make ATP.

2. Compartmentation of the nucleus allows the cell's control center to operate without being greatly affected by conditions in the cytoplasm. For example, cytoplasmic enzymes cannot enter the nucleus.

3. In the absence of a cytoskeleton, intestinal cells could not link to each other at cell junctions. They would also not have microvilli that increase the surface area for absorption of nutrients.

4. Many epithelial cells are exposed to chemical and mechanical stress, so they are constantly undergoing mitosis to make new cells. Many environmental chemicals can damage chromosomes, leading to abnormal (cancerous) daughter cells. In addition, the constant reproduction of these cells increases the probability of genetic mutations during cell division.

5. You would expect the solutions to be different because tight junctions are used to create a barrier between the compartments on each side of the cell layer.

6. The surface layer of the epidermis is composed of mats of keratin fibers and extracellular matrix that are left behind when keratinocytes die. This layer acts as a waterproof layer to prevent loss of water and heat.

7. C, D, A, E, G, F

8. Under skin- a, b, c, d. Sheaths- a. cartilage - a. adipose - e; tendons and ligaments - a; blood - d. lungs and blood vessels - a, b, c, d. bones - a.

9. packages proteins - c. modifies proteins - c. series of tubes - c. protein synthesis - a.

10. a) See pp. 53-55. b) See pp. 47-48. c) See pp. 43, 47. d) See pp. 57-58. e) See pp. 42-43. f) See p. 48. g) See pp. 51-52. h) See pp. 49-50. i) See p. 57. j) See p. 58. k) See Fig. 3-17, p. 59. l) See p. 60.
m) See p. 60. n) See pp. 62-63 and Fig. 3-24, p. 65. o) See p. 62. p) See Fig. 3-22, p. 64.

CHAPTER 4

p. 4-14: (f) The scale is balanced again with 6 blocks on the left and 12 blocks on the right. (i) The scale is balanced again with two blocks on the left and 4 blocks on the right. If more A is added to the reaction, A, C and D all increase. If C and D are converted into E, the amount of A will decrease.

Practice Makes Perfect

1. entropy - D. potential energy - A, E. kinetic energy - B. exergonic - F. endergonic -C, G.

2. Vitamins and ions act as cofactors or coenzymes. Cofactors must bind to the enzyme before substrates will bind to the active sites. Coenzymes act as receptors or carriers for atoms or functional groups.

3.

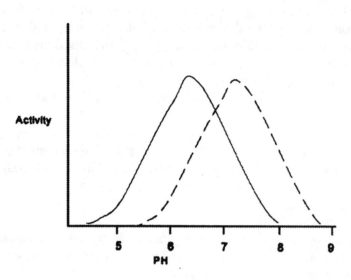

4. a) Graph A is exergonic and graph B is endergonic.   b) Graph A is more likely to go in the forward direction because it has a lower activation energy.   c) In graph A, the products have a lower free energy (stored energy) than the substrates; therefore, more energy was released. d) Graph A belongs with reaction 2. Graph B belongs with reactions 1, 3, and 4.

5. C, B, A, D, G, H, F

6. a) and c)

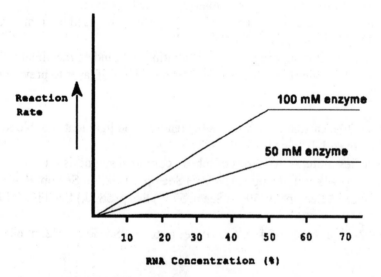

b)   When the reaction rate is maximal, all active sites on the enzymes are occupied with substrate; i.e., the enzyme is saturated. See Fig. 4-15 on p. 85.

7. NADH is oxidized (loses an electron) to become NAD^+. H^+ is reduced because it gains an electron to become an H that combines with carbon.

8. This is a dehydration reaction because water is removed.

9. D, A, G, F, K, C, B, H, E, I

10. a) NADH donates high-energy electrons. b) FADH2 donates high-energy electrons. c) Oxygen combines with H^+ and electrons to form water. d) ATP synthase transfers the kinetic energy of electrons moving down their concentration gradient to the high-energy bond of ATP. e) H^+ is concentrated in the intermembrane space, storing energy in its concentration gradient. f) The inner membrane proteins convert energy from high-energy electrons into either the work of moving H^+ ions against their concentration gradient or into heat.

CHAPTER 5 - Practice Makes Perfect

1. Time values are distance 2, or 1, 4, 9, and 16. [insert figure here]

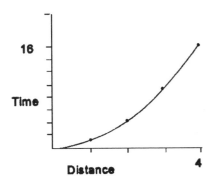

The graph has distance on the x-axis and evenly spaced values from 0 at the origin to 4 mm. The y-axis is time and the values range from 0-16, with evenly spaced tic-marks.

2. According to Fick's Law, the rate of diffusion $\dfrac{\text{surface area} \times \text{concentration gradient}}{\text{membrane resistance} \times \text{membrane thickness}}$

   So as membrane thickness increases by a factor of 2, the rate of diffusion will decrease by half.
3. See table on next page.
4. Na^+/K^+-ATPase - B, C, D. Na^+-glucose - A, D, E. Ca^{2+}-ATPase - C. $Na^+/K^+/2\,Cl^-$ - A, D, E. Na^+/H^+ - B, D, E.
5. The membrane permeability to Ca^{2+} appears to change with light. Light is a form of energy, so it could be doing one of several things: (1) opening a Ca^{2+} channel, (2) providing energy for the conformation change of a Ca^{2+} carrier protein, or (3) changing the structure of the membrane in some other way to make it permeable to Ca^{2+}. The first two explanations are more specific and relevant to the material in this chapter.
6. See Fig. 5-28, p. 131.
7. To compare your answer to another map, see Fig. 5-13, p. 117.
8. 1 OsM/1.86 = ? OsM/0.55 = 0.296 OsM or 296 mOsM plasma osmolarity. Intracellular osmolarity is always the same as plasma at equilibrium.
9. 1 mL sample/1 mg glucose = ? mL total/5000 mg glucose. The beaker has 5000 mL or 5 liters.
10. Absorption rate decreased as apical Na^+ concentration decreased, so transport appears to be Na^+-dependent. At the Curesall concentrations tested, transport rate was always proportional to concentration. The Na^+-dependence of the transport suggests that the transporter may be a secondary active transporter similar to the Na^+-glucose transporter even though transport did not show saturation.
11. Molecular weight of NaCl = 58.5. A 0.9% solution contains 0.9 g/100 mL solution or 9 g/liter. 1 mole NaCl/58.5 g NaCl = ? mole/9 g NaCl = 154 mmoles NaCl $\times$ 2 osmoles/mole = 308 mosmoles/L.
12. Molecular weight of glucose = 180. 5% solution = 5 g/100 mL or 50 g/liter. 1 mole glucose/180 g = ? moles/50 g = 0.278 moles $\times$ 1 osmole/mole = 0.278 osmoles or 278 mosmoles/liter.

Table for question 3:

METHOD	Movement relative to concentration gradient	Energy source	Rate membrane	Movement relative to structure	Exhibits Specificity?	Exhibits Competition?	Exhibits Saturation?	Examples
simple diffusion	down	[ ] gradient	[ ] dependent	diffuse across lipid bilayer	no	no	no	gases, small nonpolar molecules, urea
restricted diffusion	down	[ ] gradient	[ ] dependent	through open channel	yes	yes	yes	urea
facilitated diffusion	down	[ ] gradient	[ ] dependent has .maximum rate	on protein carrier	yes	yes	yes	glucose into cells from ECF
direct active transport	against	ATP	[ ] dependent has maximum rate	via protein carrier	yes	yes	yes	Na-K ATPase
indirect active transport	one or more down, one or more against	[ ] gradient, usually of Na^+	[ ] dependent	via protein carrier	yes	yes	yes	Na-glucose symport

[ ] = concentration. ECF = extracellular fluid

13. a) Rate is measured as mg or mmole per minute or second.   b) In graph #1, the concentration inside the cell reaches a maximum value over time, presumably showing that the system has come to equilibrium. If concentration inside equals concentration outside, the movement of glucose must be by diffusion. You cannot tell from this graph if the diffusion is simple diffusion across the phospholipid bilayer or if it is facilitated diffusion, a passive process. In graph #2, the rate of movement does not reach a maximum, suggesting that there are no transporters to become saturated. This graph would support a hypothesis of simple diffusion. However, it is possible that the extracellular glucose concentration did not reach a high enough value to cause saturation of the carriers in the artificial membrane. c) The line in graph #1 leveled off because there was no more glucose entering the artificial cell.

14.

solution A	Membrane	solution B	osmolarity of A relative to B
100 mM glucose	no net movement	100 mM urea	isosmotic
200 mM glucose	no net movement	100 mM NaCl*	isosmotic
300 mOsM NaCl	no net movement	300 mOsM glucose	isosmotic
300 mM glucose	$\Rightarrow$	200 mM $CaCl_2$**	hyposmotic

* 100 mM NaCl = 200 mOsM NaCl.   **200 mM CaCl2 = 600 mOsM CaCl2

15. The IV solution is isosmotic so there will be no change in the osmolarity of either the ECF or ICF. The infusion goes into the plasma (ECF), so ECF volume will increase initially. Because the solution is all NaCl, a nonpenetrating solute, the solution will all remain in the ECF. There is no concentration gradient to cause water to move into or out of the ICF, so ICF volume will remain unchanged.

16. The solution is isosmotic and hypotonic to the cell. The graph should be labeled with "time" on the x-axis and "cell volume" on the y-axis. Cell volume increases at the arrow, then levels off at a new, larger volume.

17. a) 340 mosmoles/L $\times$ 13 L = 4420 osmoles in ECF b) 160 mosmoles/L $\times$ 1 mmole NaCl/2osmoles = 80 mmoles c) new ECF volume = 13 L + 1 L = 14 L. Solute = 4420 + 160 mosmoles = 4580 osmoles. Osmolarity = 4580 osmoles/14 L = 0.327 OsM or 327 mOsM.

18. Cl will move from side 2 to side 1 because of a concentration gradient for Cl^-. As the ions move, side 1 will develop a net negative charge while side 2 will develop a net positive charge, creating a membrane potential. Net Cl^- movement will stop when the positive charge on side 2 that holds Cl^- in that compartment is equal in magnitude to the concentration gradient driving Cl^- into side 1.

19. When ECF K^+ increases, an equal amount of some anion has also been added to the ECF, so there is no change in the charge on the ECF. When ECF K^+ increases, the membrane potential of a liver cell depolarizes because less K^+ leaks out of the cell. Liver cells are not excitable and do not fire action potentials but can experience changes in membrane potential.

20. The membrane potential will slowly depolarize to zero as K^+ leaks out and Na^+ leaks into the cell. Normally the Na^+/K^+-ATPase removes ions that leak across the membrane.

CHAPTER 6 - Practice Makes Perfect

1. Local communication with neighboring cells is carried out by chemical communication (autocrines, paracrines, cytokines). Long distance communication is carried out by the nervous system, hormones, and cytokines. Local communication takes place by diffusion, which limits its speed. Nervous signals are the fastest means of communication. Long-distance chemical communication relies on the circulatory system.

2. Different types of receptors for one chemical signal allow different responses to a single signal. In addition, there are many different chemical signals, each with its own receptor or receptors. Receptor number is not constant. Cells can alter their receptor number by adding or withdrawing receptors (up- and down-regulation).

3. Cascades allow amplification of signals. A single event could not elicit as large a response without a cascade.

4. You would not expect the effects of growth hormone to exert negative feedback because then growth would not take place. On the other hand, without some form of control, growth would continue unchecked. By having growth hormone concentration act as the negative feedback signal, the body can keep the secretion of growth hormone in a desirable range.

CHAPTER 7 - Practice Makes Perfect

1.

	Peptide	Steroid
Transport in plasma	dissolved	bound to carrier proteins
Synthesis site in endocrine cell	rough endoplasmic reticulum	smooth endoplasmic reticulum
Method of release from endocrine cell	exocytosis from secretory vesicles	simple diffusion across phospholipid bilayer
General response of endocrine cell	modification of existing proteins	transcription, translation, and synthesis of new proteins

2. Connected by nerve fibers - P; connected by blood vessels - A; hormones made in hypothalamus - P; under influence of hypothalamic hormones - A; secretes peptide hormones - A, P.

3. e

4. 1- a, b; 2 - c, f; 3 - d; 4 - c, h; 5 - e.

5. False: The onset of steroid action takes at least 30 minutes, which is too slow for most fight-or-flight situations.

6. a) False: Steroid hormones cannot be stored in vesicles because they are lipid-soluble and would diffuse out across the vesicle membrane.   b) It is true that steroid-secreting cells have lots of smooth endoplasmic reticulum but false that they have lots of Golgi. The Golgi apparatus is used to package material into vesicles (see 6a).

7. a) 2;   b) 1. mRNA carries the code for peptide synthesis.

8. The tissue has lots of membrane-bound secretory vesicles, indicating that the cells synthesize and store peptides. Insulin is a peptide hormone, so the tissue is endocrine pancreas.

9. For the pathway, see Fig. 7-17, p. 191. Secondary hypocortisolism originating at the pituitary means that both ACTH and cortisol secretion are down. If you administer ACTH, her cortisol secretion should increase because there is nothing wrong with the adrenal cortex; it has simply lacked stimulation by ACTH.

10. The antibodies stimulate the thyroid gland, so you expect thyroid hormone (thyroxine) levels to be elevated. This eliminates patients A and C. The elevated thyroxine will have a negative feedback effect on the pituitary and shut off endogenous TSH production, therefore TSH should be below normal. Only patient B has elevated thyroxine and decreased TSH.

Graphs

The data for the patients in problem #10 are best shown with a bar graph. There are two parameters being measured for each patient, so you would need two sets of values on the y-axis. This can be done by placing the scale for thyroxine on the left side of the graph and the scale for TSH on the right side of the graph. On the x-axis, each patient is represented by a pair of bars, side by side, one bar for each hormone. Distinguish the hormones by using different colors or fill patterns in the bars.

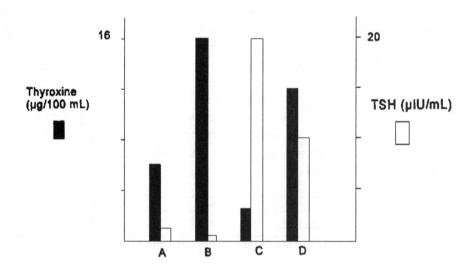

CHAPTER 8 - Practice Makes Perfect

1. speed

2. ECF (left to right) Na, Cl, K; ICF - K, Na, Cl

3. Dendrite ⇒ cell body ⇒ trigger zone (axon hillock) ⇒ axon ⇒ axon terminal ⇒ releases neurotransmitter by exocytosis ⇒ neurotransmitter diffuses across synaptic cleft to combine with receptor on postsynaptic cell

4. a) K^+ will leave the cell due to the electrical gradient (positive inside) repelling it. Na^+ will leave because of both the electrical gradient and the concentration gradient favoring its movement out.

   b) To convert 110 F to degrees Kelvin, see p. 736.

$ENa = \dfrac{RT}{Fz} \ln \dfrac{[Na]out}{[Na]in}$ R = 8.314, T = 316.48 K, F = 96,000, z = +1, [Na]out = 15, [Na]in = 175

   ln = 2.3 log10

   = -67 mV

5. a

6. d

7. b

8.

	Location of channel	Ion(s) that moves	Chem or volt gating?	Physiological process in which ion participates
NEURON				
	dendrite	Na^+	chemical	graded potentials
	axon	Na^+	electrical	action potentials
	axon	K^+	electrical	action potentials
	axon terminal	Ca^{2+}	electrical	exocytosis of neurotransmitter

9. The ground electrode is set to zero millivolts and the potential difference is measured between this electrode and the inside of the cell.

10. Stretching could destroy the integrity of the cell membrane or could pop open channels that would allow ions to move between the cell and the extracellular fluid.

11. See Fig. 8-14, p. 215.

CHAPTER 9 - Practice Makes Perfect

1. a) The cerebrum contains the motor cortices for voluntary movement (walking to class) and the cerebral cortex for higher brain functions (pondering physiology). It also contains the visual cortex for integrating visual information from the eyes, and association areas that integrate sensory information into the perception of tripping and falling. The basal ganglia are associated with control of movement. The amygdala of the limbic system is linked to the emotions of anger and embarrassment. b) The cerebellum coordinates the movement of walking.

2. An electroencephalogram measures the summed electrical activity of brain neurons. The wave patterns of the EEG vary with different states of consciousness and can be used to assess if brain activity is following normal patterns.

3. a) higher thought processes, b) centers for homeostasis, neurohormone production, c) centers for involuntary functions (breathing, heart rate) and eye movement, d) coordination and control of body movement

4. b

5. c

6. c

7. Stimulus: sight of *T. rex* $\Rightarrow$ (receptor) eyes $\Rightarrow$ sensory neurons to visual cortex $\Rightarrow$ association areas of cerebral cortex recognize *T. rex* as dangerous $\Rightarrow$

   Efferent pathway 1: amygdala responds with fear $\Rightarrow$ limbic system communicates with hypothalamus to initiate responses of fight-or-flight $\Rightarrow$ medullary centers for breathing and heart rate increase both

   Efferent pathway 2: Motor cortices communicate with cerebellum to coordinate skeletal muscles to flee

CHAPTER 10 - Practice Makes Perfect

1. Rapidly adapting phasic receptors adapt to constant pressure from clothing and stop sending information to the brain.
2. A salty solution cannot be tasted at the back of the tongue because there are no salt receptors there.
3. True. The layers of the retina are arranged so that light must pass through layers containing nerves and blood vessels before it strikes the photoreceptors.
4. e
5. In nearsightedness (myopia), the light focuses in front of the retina. In astigmatism, the cornea is not a perfect dome, so the light does not focus evenly on the retina.
6. b
7. Loss of air conduction but not bone conduction suggests a problem with the middle ear or a plugged ear canal.
8. The spot of bright light appears black on the paper. Light causes retinal to release from the opsin portion of rhodopsin and be transported out of the photoreceptor. The slow recovery results from time needed to transport the retinal back into the rod and also the time needed for the rod membrane potential to return to its former state.
9. The horizontal canal is involved in movement that changes the left-right position of the head.
10. Rods are responsible for low-light monochromatic vision, so a nocturnal animal would have more rods and fewer cones.
11. d

CHAPTER 11 - Practice Makes Perfect

1. a
2. a) antagonist,   b) agonist
3. Autonomic neurotransmitters (NTs) are released from varicosities along the axon as well as at the axon terminal, so their release is less specifically associated with receptors on the target cell. Autonomic NT release can be modulated by many chemical factors such as hormones and paracrines. In addition to being broken down by synaptic enzymes, autonomic NTs may diffuse away from the receptors or be transported intact back into the neuron. A major difference between autonomic and somatic motor NTs is that the amount of NT will vary the response of the target cell. Each action potential in a somatic motor neuron creates one muscle twitch.
4. Monoamine oxidase (MAO) is the enzyme that breaks down catecholamines at autonomic synapses, therefore inhibition of MAO will prolong or enhance target responses to autonomic signals.

CHAPTER 12 - Practice Makes Perfect

1. Muscle contraction requires Ca^{2+} and uses ATP.
2. Somatic motor neurons secrete acetylcholine.
3. True.
4. Na^+ entry is greater because Na^+ has both concentration and electrical gradients favoring its movement. K^+ has concentration favoring its efflux but an electrical gradient opposing efflux.
5. d
6. Contraction is more rapid because Ca^{2+} floods the cytoplasm as it moves down its concentration gradient. Removal of Ca^{2+} for relaxation requires that the cell use ATP to pump Ca^{2+} against its gradient.
7. Temporal summation in muscle results when multiple action potentials repeatedly stimulate a muscle fiber, causing an increase in its force of contraction, up to some maximum value. Temporal summation in a neuron occurs when multiple sub-threshold stimuli arrive at the trigger zone and create a suprathreshold signal. In the neuron, the result is an action potential of constant amplitude, in contrast to the increasing force observed in the muscle.
8. Fatigued muscles have less ATP with which to pump Ca^{2+} back into the sarcoplasmic reticulum.

9. If skeletal muscles had gap junctions, all fibers would contract simultaneously. This would prevent us from regulating which muscles are contracting and would prevent us from regulating the force of contraction.

10. True. Muscles shorten during isotonic contractions.

11. a

12. a

13. Motor proteins in the nervous system include the microtubules and foot proteins used for axonal transport and the cytoskeleton that helps growing neurons find their targets. Motor proteins in Chapter 3 include microtubules that move cilia and flagella, microtubules associated with centrioles for chromosome movement during cell division, fibers that help mobile white blood cells (phagocytes) move, and fibers that move vesicles and other organelles around the cytoplasm.

14. Fewer ACh receptors would mean that a muscle would not respond as strongly to neurotransmitter release. This would probably manifest itself as muscle weakness or paralysis because the muscle fiber would not fire action potentials and excitation-contraction coupling would not occur.

15. You would expect the latency period to be longer when the nerve is stimulated because the additional steps of neurotransmitter release, diffusion across the synapse, opening of ACh-gated channels, and ion movement must take place.

CHAPTER 13 - Practice Makes Perfect

1. The steps are stimulus, receptor, afferent pathway, integrating center, efferent pathway, tissue response, systemic response. Examples of reflexes in Chapter 13 include the muscle spindle and Golgi tendon reflexes, the flexion reflex, and the crossed extensor reflex.

2. a

3. In Fig. 13-10, the sensory neuron diverges to excite multiple interneurons. In Fig. 13-11, sensory neurons diverge, neurons converge on the thalamus and cerebral cortex, and multiple neurons converge on the somatic motor neuron.

CHAPTER 14

p. 14-3: Tube B has the larger pressure gradient, 65 mm Hg (75 - 10), so it will have the greatest flow. The pressure gradient in Tube A is only 50 mm Hg.

p. 14-8: The small change in the membrane potential between points 1 and 2 in Fig. 14-14 (p. 401) is due to K^+ leaving the cell through K^+ leak channels.

Quantitative Thinking

1. Resistance should decrease and radius increase.

2. Flow = $\Delta P/R$, where R ∝ L/r^4. So flow ∝ $\Delta P \times r4/L$
   Flow A = (50 × 16)/16 = 50. Flow B = (72 × 1)/2 = 36

3. CO = HR × SV so SV = CO/HR. Before Romeo, SV = 69.4 mL/beat. After Romeo, SV = 125 mL/beat.

Practice Makes Perfect

1. acetylcholine, atrioventricular, cardiac output, electrocardiogram, end-diastolic volume, end-systolic volume, sinoatrial, stroke volume.

2.

	Electrical Event	Mechanical Event
P wave	atrial depolarization	atrial contraction follows
QRS complex	ventricular depolarization	ventricular contraction and atrial relaxation
T wave	ventricular repolarization	ventricular relaxation
PQ segment	AV node delay	atrial contraction
ST segment	some parts of ventricle depolarizing, some repolarizing	end of ventricular contraction
TP segment	none	atrial and ventricular diastole

3. If heart rate speeds up, there is less time for passive (gravity-assisted) ventricular filling, so atrial contraction takes on increasing importance.

4. See Table 14-4 on pg. 403.

5. Kinetic energy is unchanged but the hydrostatic pressure decreases. See Fig. 14-3 on p. 387.

6. Several mechanisms to alter would include Ca^{2+} entry from the extracellular fluid, Ca^{2+} release from the sarcoplasmic reticulum, the regulatory protein phospholamban, the activity of the Ca^{2+}-ATPase, and anything that would alter binding of Ca^{2+} to troponin or the interaction of actin and myosin.

7. As venous return increases, the muscle fibers stretch more and therefore contract more forcefully (the Frank-Starling Law). This increases stroke volume and therefore increases cardiac output. Sympathetic input increases ventricular contractility and therefore stroke volume. Sympathetic input onto veins will cause venous constriction and increase venous return, also increasing stroke volume.

8. See Fig. 14-19 on p. 404.

9. Cardiac output by the right heart must equal cardiac output by the left heart (4.5 L) or blood will begin to collect on one side of the circulation.

10. At point A, the atrial pressure exceeds ventricular pressure. Ventricular pressure equals aortic pressure at point C. The highest pressure in the aorta is about the same as the highest ventricular pressure, 120 mm Hg.

11.

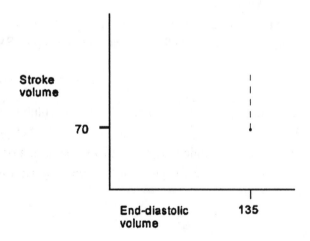

CHAPTER 15

Quantitative Thinking

1. a) mm Hg   b) At age 20, her pulse pressure is 40 mm Hg and her MAP is 83 mm Hg. At age 60, her pulse pressure is 43 mm Hg and her MAP is 96.   c) Postmenopausal women are more likely to develop atherosclerosis, which increases blood pressure when arteries stiffen and lose elastance.

2. If the radius goes from 2 to 3, radius^4 goes from 16 to 81, about a 5-fold increase. Resistance therefore falls by the same factor. As resistance decreases, blood flow increases.

3. Mean arterial pressure will increase.

4. CO = HR × SV. SV = EDV – ESV, or 110 mL. CO = 140 bpm × 110 mL/beat = 15,400 mL or 15.4 L.

Practice Makes Perfect

1. arteries - A, B; arterioles - A; capillaries - E; veins - A, D.

2. For physical characteristics, see Fig. 15-2, p. 426. Functions: arteries and arterioles - carry blood to tissues.

Arteries - store and release energy created by heart. Arterioles - site of variable resistance to regulate blood flow to individual tissues and to help maintain blood pressure. Capillaries - site of exchange between blood and cells. Veins - carry blood from tissues to heart and serve as a volume reservoir.

3. The cuff must be inflated to at least 130 mm Hg in order to stop flow through the artery. By the time blood reaches the cuff, the pressure will be less than 130 due to friction loss.

4. Healthy arteries expand during ventricular systole and store energy that they slowly release as they recoil during diastole. Hardened atherosclerotic arteries are unable to stretch, forcing the heart to work harder to eject the same amount of blood into them. Narrowing of the arteries also slows outflow so that diastolic pressures will be higher than normal. This will decrease pulse pressure.

5. Resistance to blood flow depends on the total cross-sectional area at any level of the circulatory system. Although individual capillaries are very narrow, their total cross-sectional area is very large, so their resistance is low.

6. Flow to different regions is altered by constricting or dilating the arterioles leading into the different tissues. Resistance of skeletal muscle arterioles decreases to increase flow, while flow to the digestive system is decreased by increasing arteriolar resistance.

7. SA node - D, E. Ventricle - E. Skeletal muscle capillary - F. Cardiac vasculature - A, C. Renal arterioles - A. Brain arterioles - F.

8. Blood flow through kidneys decreases. Mean arterial pressure increases. Blood flow through skeletal muscle increases. Cardiac output does not change. Total resistance increases. Blood flow through the venae cavae and lungs does not change.

9. False: Epinephrine comes from the adrenal medulla and combines with β_2 receptors to cause vasodilation. Epinephrine on 1 receptors will cause increased heart rate and force of contraction.

10. Calcium entering smooth and cardiac muscle is responsible for muscle contraction. By blocking the Ca^{2+} channels through which Ca^{2+} enters, you can relax arterioles and decrease stroke volume, both of which would decrease blood pressure.

11. Figure should include carotid and aortic baroreceptors, sensory neurons to the CVCC, parasympathetic neurons to the SA node (ACh on muscarinic receptors), sympathetic neurons to SA node and ventricles (norepi on β_2), sympathetic to α receptors on arterioles, epinephrine from the adrenal medulla to all adrenergic receptors.

12. See Fig. 15-20.

13. a) Blood flow in his legs is decreased because of increased resistance in the leg arteries. The pain comes from hypoxia when muscles are unable to get sufficient blood flow and oxygen. b) A peripheral vasodilator would have no effect in the legs because the arterioles that dilate would be beyond the obstructed arteries. If peripheral vasodilation decreased blood pressure, flow into the legs would decrease even more. c) Sympathetic nerves primarily affect arterioles, so there would be no effect for the same reason as in (b) above.

CHAPTER 16 - Practice Makes Perfect

1. Two main functions of blood are to transport substances throughout the body and to help protect the body from foreign invaders.
2. Leukocytes can be identified by the shape of the nucleus, the presence and staining color of granules in the cytoplasm, and by whether they can carry out phagocytosis.
3. Reticulocytes are immature red blood cells. Their presence in the circulation suggests that the bone marrow is responding to loss of circulating red blood cells by stepping up marrow production of new cells. This would mean that the marrow is responding normally but that something in the circulation is destroying the circulating red blood cells.
4. If hematocrit is 40%, then 60% of the sample is plasma. Sixty percent of a total blood volume of 4.8 L = 2.88 L plasma.
5. b
6. EPO is more accurately described as a cytokine because it is synthesized on demand rather than being stored in secretory vesicles. Because EPO is not made in advance and stored, it was only recently that the EPO-secreting cells in the kidney were identified.

CHAPTER 17
Quantitative Thinking

1. (approximate values) TLC = 4 L; VC = 2.75 L; ERV = 1 L; RV = 1.25 L.
   Total pulmonary ventilation = 3 breaths/15 sec $\times$ 60 sec/min $\times$ 500 mL/breath = 6000 mL/min
2. 720 mm Hg $\times$ 0.78 = 562 mm Hg P_{N2}
3. Pulmonary ventilation = tidal volume $\times$ rate = 300 mL/breath $\times$ 20 breaths/min = 6000 mL/min
   Alveolar ventilation = $(V_T$ – dead space) $\times$ rate = (300 mL – 150 mL/breath) $\times$ 20 breaths/min = 3000 mL/min
4. If you assume that both patients have dead space = 150 mL, patient A has alveolar ventilation of 350 mL/breath $\times$ 12 br/min = 4.2 L/min, while patient B has alveolar ventilation of 150 mL/br $\times$ 20 br/min = 3 L/min.
5. Total oxygen = amount dissolved + amount bound to hemoglobin. (See Fig. 17-24, p. 505).

   a) Dissolved in plasma = plasma volume $\times$ plasma concentration. If hematocrit is 38%, 62% of total blood volume is plasma, or 4.2 L $\times$ 0.62 = 2.6 L plasma $\times$ 0.3 mL O_2/100 mL plasma = 7.8 mL O_2 dissolved.

   b) Bound to hemoglobin (Hb) depends on amount of hemoglobin and the percent saturation of that amount. Maximum O_2-carrying capacity = O_2 carried by Hb at 100% saturation.

   13 g Hb/dL whole blood $\times$ 4.2 L blood = 546 g Hb in blood of this person.

   At 100% saturation, 1.34 mL O_2/g Hb $\times$ 546 g Hb = 731.64 mL O_2 carried on Hb. But this person's Hb is only 97% saturated: 731.64 mL O_2 $\times$ 0.97 = 709.7 mL O2 carried bound to Hb.

   709.7 mL O_2 carried bound to Hb + 7.8 mL O_2 dissolved = 717.5 mL O_2 in this person's blood.

   *This problem does not take into account the fact that venous blood will have a lower oxygen content than arterial blood.

6. The oxygen consumed by the tissues is extracted from the blood. Therefore, the difference in oxygen content between arterial and venous blood represents oxygen that went into the tissues. If you know how much oxygen is extracted per liter of blood, then you simply need to find how much blood must flow past the tissues to supply 1.8 L of oxygen per minute. Blood flow past the tissues is the cardiac output.

1.8 L O2/min = (190 – 143 mL O2/L blood) $\times$ blood flow past tissues (CO). CO = 38.3 L blood/min

Practice Makes Perfect

1. See Fig. 17-26, p. 507.

	Atmospheric air (dry)	Alveolar air	Arterial Blood	Venous Blood	Cells, resting	Expired Air
Oxygen	160	100	100	40	40	*120
Carbon dioxide	0.3	40	40	46	46	*27

*These values are not given in the text. However, you should have been able to say that the values would be between the values for alveolar and atmospheric air, as expired alveolar air will mix with "fresh" atmospheric air that has been in the dead space. See Fig. 17-15, p. 494.

2. 2,3-diphosphoglycerate; tidal volume; partial pressure of oxygen; residual volume; inspiratory reserve volume; hemoglobin
3. a) T   b) F   c) F
4. a) F   b) T   c) T
5. a) T   b) T   c) T   d) F   e) F
6. a, e
7. < - see Fig. 17-10, p. 486;   < - see Fig. 17-23a on p. 503;   < - The percent saturation of hemoglobin is essentially identical in these two people, so the amount of hemoglobin becomes the important factor for how much oxygen is being transported;   > - peripheral arterioles dilate when P_{O_2} decreases but pulmonary arterioles constrict; = - arterial P_{O_2} is determined by alveolar P_{O_2} and is not affected by the amount of hemoglobin in the blood; < - bronchioles have a much greater total cross-sectional area; > - surfactant decreases the surface tension of fluid lining the alveoli and therefore increases compliance.
8. Diaphragm and external intercostals are skeletal muscles, so answer is B. Bronchioles - C and D.
9. a) The P_{O_2} increases and P_{CO_2} decreases as fresh air comes in to the alveoli, but no gases exchange with the blood. b) Tissue P_{O_2} decreases and P_{CO_2} increases as there is no gas exchange.   c) The bronchioles constrict in response to decreased P_{CO_2} in an attempt to send ventilation to alveoli with better perfusion. Likewise, the pulmonary arterioles constrict in response to decreased P_{O2} in an attempt to send blood to better ventilated alveoli. The effectiveness of these responses will depend on the size of the affected area.
10. O_2 in air $\Rightarrow$ mouth, nasal cavity $\Rightarrow$ pharynx $\Rightarrow$ larynx $\Rightarrow$ trachea $\Rightarrow$ main bronchus $\Rightarrow$ secondary bronchi $\Rightarrow$ bronchiole $\Rightarrow$ alveolus $\Rightarrow$ crosses alveolar-capillary membranes $\Rightarrow$ pulmonary capillary $\Rightarrow$ pulmonary venule, small veins $\Rightarrow$ pulmonary vein $\Rightarrow$ left atrium $\Rightarrow$ mitral valve $\Rightarrow$ left ventricle $\Rightarrow$ aortic valve $\Rightarrow$ aorta $\Rightarrow$ hepatic artery, arteriole, capillary $\Rightarrow$ O_2 into liver cell $\Rightarrow$ CO_2 out of liver cell $\Rightarrow$ hepatic capillary, venule, vein $\Rightarrow$ inferior vena cava $\Rightarrow$ right atrium $\Rightarrow$ tricuspid valve $\Rightarrow$ right ventricle $\Rightarrow$ pulmonary valve $\Rightarrow$ pulmonary artery, arteriole, capillary $\Rightarrow$ crosses capillary- alveolar membranes $\Rightarrow$ [run backwards to air]
11. a) and b) P_{O_2} decreased because barometric pressure decreased due to increased altitude. c) Arterial P_{O_2} decreased because P_{O_2} of inspired air is down. d) Arterial P_{CO_2} is down because you

begin to hyperventilate. e) Arterial pH is up because of decreased P_{CO_2}. By law of mass action, the equilibrium between CO_2 and H^+/HCO_3^- is disturbed, converting more H^+/HCO_3^- to CO_2. f) They will have more hemoglobin because their state of chronic hypoxia triggered erythropoietin synthesis and new RBC synthesis. g) The best course of action is to take the person back to lower altitudes to remove the source of the stress (hypoxia). h) Initially the hypoxia triggers a hyperventilation response. But hyperventilation decreases arterial P_{CO_2} and increases arterial pH, both of which will tend to decrease ventilation and offset the hypoxic response. However, the central chemoreceptors are able to adapt to chronic changes in P_{CO_2}, so over the course of several days, the ventilation rate increases as the hypoxia effect is less opposed by the low P_{CO_2}. For a discussion of high altitude physiology, see the Focus feature on p. 504.

12. Hyperventilation will increase plasma P_{O_2} but have minimal effect on the total oxygen content because so little oxygen is carried dissolved in plasma. Hemoglobin is already carrying nearly maximal amounts of oxygen, and the saturation curve is nearly flat at these values of P_{O_2}, so no more oxygen will be transported by hemoglobin.

13. Expired air from the alveoli is mixing with atmospheric air in the dead space, increasing the P_{O_2} of the expired air and decreasing its Pco2.

CHAPTER 18
Quantitative Thinking

1. Clearance of = excretion rate of   plasma concentration of
   creatinine clearance = (276 mg/dL urine $\times$ 1100 mL urine/day)/1.8 mg/dL plasma = 168.7 L plasma/day
   GFR = creatinine clearance, so GFR = 168.7 L plasma/day

2. Filtration rate of = GFR $\times$ plasma concentration of
   1 mg X/mL plasma $\times$ 125 mL plasma/min = 125 mg X filtered /min. Same values for inulin. Excretion rate of inulin = filtration rate, or 125 mg inulin excreted/min. Cannot say what the excretion rate of X is because there is insufficient information. We must know whether X is reabsorbed or secreted by the tubule in order to estimate its excretion rate.

3. a) When reabsorption of glucose reaches the transport maximum, 100% of what is filtered is being reabsorbed. Therefore, the transport rate of glucose at the Tm is identical to the glucose filtration rate. The renal threshold is the plasma concentration at which the Tm is reached. By substitution:
   Filtration rate of = GFR $\times$ plasma concentration of   (or) Tm = GFR $\times$ renal threshold
   90 mg glucose/min = GFR $\times$ 500 mg glucose/100 mL plasma, or GFR = 18 mL plasma/min
   b) Creatinine clearance = GFR = 25 mL/min (*You cannot calculate this because you are not given plasma creatinine.)
   phenol red clearance = (5 mg/mL urine $\times$ 2 mL urine/min)/2 mg/mL plasma = 5 mL plasma/min

4. Graphing question. To calculate the points for the filtration graph of Z, multiply various plasma-concentrations in the range of 0-140 mg Z/mL plasma times the GFR. The line will be a straight line beginning at the origin and extending upward to the right. For secretion, you know that at a plasma concentration of 80 mg Z/mL plasma, secretion reaches its maximum rate of 40 mg/min. Plot that point. Draw the secretion line from the origin to that point. At plasma concentrations above the renal threshold, secretion rate does not change, so the line becomes horizontal. To draw the excretion line, add the filtration rate and secretion rate at a number of plasma concentrations of Z. The excretion line will extend upward with a steeper slope than that of the filtration line from the origin to the renal threshold. At that point, the slope of the line changes and the line runs parallel to the line of filtration rate.

Practice Makes Perfect

1. Proteinuria suggests that the filtration barrier in the glomerulus has been disrupted or that proximal tubule cells are no longer able to reabsorb the small filtered proteins.

2. Glucose is usually absent because 100% of what is filtered is reabsorbed.

3. See Fig. 18-2 on p. 522.

4.

	Hydraulic pressure	Fluid pressure	Osmotic pressure	Net direction of fluid flow
Glomerular capillaries	55	15	30	into Bowman's capsule
Peritubular capillaries	10	negligible	30	into capillaries
Systemic capillaries	32-15 (see p. 440)	negligible	25	out of capillaries into interstitial fluid
Pulmonary capillaries	< 14 (see p. 477)	negligible	25	into capillaries

5. a) Renal threshold: plasma concentration at which transport maximum for a substance is reached-mg/mL plasma. b) Clearance: volume of plasma cleared of a substance per unit time - mL plasma/min.

   c) GFR = glomerular filtration rate. Volume of plasma filtered into Bowman's capsule per unit time -
   mL plasma/min.

6. a) Both terms refer to a volume of plasma per unit time. Clearance is specific for a single substance that is removed from the plasma, while GFR is the bulk filtration of plasma and almost all its solutes.

   b) These terms are related (see definition in 5a above) but one deals with a transport rate while the other is the plasma concentration associated with that transport rate.

7. K^+ movement across the apical membrane is against the gradient, so it must be by some form of active transport. The $Na^+/K^+/2\,Cl$ symporter is one way to bring K^+ into the cell. On the basolateral side, K^+ can leave by moving passively down its concentration gradient. One way to do this would be to have K^+ leak channels on the basolateral membrane.

8. Capillary in hand $\Rightarrow$ venule, vein $\Rightarrow$ inferior vena cava $\Rightarrow$ right atrium $\Rightarrow$ tricuspid valve $\Rightarrow$ right ventricle $\Rightarrow$ pulmonary valve $\Rightarrow$ pulmonary artery, arteriole, capillary, venule, vein $\Rightarrow$ pulmonary vein $\Rightarrow$ left atrium $\Rightarrow$ mitral valve $\Rightarrow$ left ventricle $\Rightarrow$ aortic valve $\Rightarrow$ aorta renal artery $\Rightarrow$ afferent arteriole $\Rightarrow$ glomerulus $\Rightarrow$ Bowman's capsule $\Rightarrow$ proximal tubule $\Rightarrow$ loop of Henle $\Rightarrow$ distal tubule $\Rightarrow$ collecting duct $\Rightarrow$ renal pelvis $\Rightarrow$ ureter $\Rightarrow$ urinary bladder $\Rightarrow$ urethra $\Rightarrow$ leaves body in the urine

9. Inulin clearance (= GFR) at MAP = 100 will less than GFR at MAP = 200 because the kidney autoregulation of GFR is only effective to a MAP of 180.

10. a) The substance filters, then additional is secreted. The secretion line can be calculated by subtracting the filtration rate from the excretion rate at a series of plasma concentrations. The line goes from the origin to the point (0.05, 0.2), then changes slope and runs horizontally because the rate has reached its maximum. b) The slope changes because secretion has a transport maximum. c) If phenol red secretion is inhibited, less would be excreted and more would stay in the plasma. Thus, plasma clearance would decrease.

CHAPTER 19
Quantitative Thinking
<u>Osmotic diuresis:</u> A = 1.5 L, B = 0.5 L, and C = 0.125 L. With an additional 150 mosmoles of solute in the lumen, the volume will increase: A = 3 L, B = 1 L, and C = 0.25 L. The volume of urine doubled when the amount of unreabsorbed solute doubled.

Practice Makes Perfect

1. = - Urine osmolarity cannot be greater than the medullary interstitial osmolarity. With excreted glucose, the <u>volume</u> will increase; < - Aldosterone secretion is directly inhibited by high osmolarity, even if AGII is present; > - Respiratory acidosis is characterized by elevated P_{CO2}; > - Renal reabsorption of buffer will be greater in acidosis; < - Ventilation will be reduced in alkalosis and elevated in acidosis; < - Renin is secreted in response to low blood pressure; = - Filtration is not regulated and depends only on the plasma concentration of a substance and the GFR.

2. a) no change b) decrease c) decrease or no change d) no direct effect e) no change f) will decrease as a result of homeostatic compensation for elevated MAP

3. An increase in plasma osmolarity will cause an increase in vasopressin secretion, so the x-axis is osmolarity and the y-axis is vasopressin.

4. $CO_2 + H_2O$ $H^+ + HCO_3^-$. If the green fruit you ate represents bicarbonate or H^+, you have sent the equation out of balance relative to its equilibrium, making the CO_2 side appear to be too large. Convert some of the red fruit (CO_2) into green fruit to balance out the equation again.

5. Elevated P_{CO_2} and bicarbonate suggests a respiratory acidosis. Therefore, answers a, b, and c are all correct.

6. Normal pH is 7.4 (Mr. Osgoode is in acidosis) and normal arterial P_{CO_2} is 40. It appears that Mr. Osgoode is hyperventilating in an attempt to compensate for a metabolic acidosis; the hyperventilation is elevating his pH (decreasing his H^+) but also decreasing his P_{CO_2} and plasma HCO_3.

7. Ari is breathing through an extended dead space (the tube), so he is experiencing alveolar hypoventilation, resulting in respiratory acidosis. You would expect his pH to be lower than normal, and both HCO_3^- and P_{CO_2} to be greater than normal, due to the retention of CO_2.

Try It Box:
The baking soda and vinegar combination foams up, producing CO_2. This simple reaction has been used for centuries by cooks to make batters rise. If you taste the solution after adding baking soda, it will not be nearly as sour because the H^+ have been buffered and converted to water.

CHAPTER 20 - Practice Makes Perfect

1. Stomach - 2; ileum - 5; esophagus - 1; ascending colon - 6; pyloric sphincter - 3; duodenum - 4.
2. Salivary amylase is denatured by the acidic conditions in the stomach.
3. Pepsin is produced by the stomach and is active at low pH; trypsin is produced by the pancreas and is active at high pH.
4. The stomach prevents autodigestion by (1) secreting inactive enzyme, (2) secreting mucus, and (3) secreting a layer of bicarbonate under the mucus to neutralize acid.
5. a) False: Short-chain fatty acids are not incorporated into chylomicrons. All other parts are true.
   b) False: Most bile salts are reabsorbed and used again.
   c) False: Much more water is reabsorbed than ingested because large volumes of water are present in secreted fluids.
   d) True: Without oxygen, transporting epithelia cannot carry about aerobic metabolism and make enough ATP to support the secondary active transport of glucose into the cell.

6. a) See Fig. 20-9a on p. 588. b) Excess vomiting would cause metabolic alkalosis because of the bicarbonate that is reabsorbed which acts as a buffer.

7. The enteric nervous system can act as its own integrating center, without communicating with the CNS.

8. Gluten is a substance that can cause allergies if absorbed intact, and the intestines of infants have the ability to absorb intact molecules via transcytosis.

9. Receptors: Eyes (sight of test), ears (sound of people talking about the test)
Afferent path: sensory neurons
Integrating center: cerebral cortex, with descending pathways through limbic system
Efferent path: parasympathetic neurons to enteric nervous system
Effector: parietal cells
Cellular response: second-messenger initiated modification of proteins
Tissue response: acid secretion

10. Both systems secrete material from ECF into the lumen. Most of the GI secretion is reabsorbed while most of the renal secretion is excreted. Reabsorption in the kidney is equivalent to absorption in the GI tract. New material in the GI tract is absorbed, but many components of secretion are reabsorbed. Both systems excrete material to the external environment; GI excretion is primarily solid while urine is liquid. Movement through the systems is quite different. Filtration, created by hydraulic pressure, creates bulk flow of fluid through the tubule, unaided by muscle contraction until the urine has left the renal pelvis. Motility in the GI tract is completely dependent on coordinated smooth muscle contraction.

11. Test the pH optimum for activity of the two enzymes. Gastric lipase will be most active in low pH, while lingual lipase is more active at higher pH.

CHAPTER 21 - Practice Makes Perfect

1. (c) is incorrect.
2. b
3. True. See p. 618.
4. Cortisol and glucagon are catabolic, growth hormone and insulin are anabolic.
5. The body is able to make glucose from non-glucose precursors through the pathways of gluconeogenesis.
6. Insulin secretion will be greater when glucose is given orally because GIP will be secreted due to insulin in the intestine and because all absorbed glucose goes into the hepatic portal system. Glucose given by IV will be taken up by cells before it gets to the pancreas, so the pancreas will not sense as large an increase in blood glucose.
7. a
8. $(9 \times 7) = 63$ calories from fat. $63/190 = 33\%$, about the maximum recommended fat content for food.
9. Humans with obesity may have defective leptin receptors or a defective pathway for leptin action in target cells, so that they are actually leptin-deficient even though plasma levels of the protein are high.
10. a) False: Cortisol is net catabolic and causes muscle breakdown.
b) False: Vitamin D deficiency causes rickets because the children are unable to absorb dietary Ca^{2+}.
c) True: These are catabolic effects and can elevate blood glucose concentrations.
d) False: The glucose transporters of liver are not insulin-dependent. However, insulin does increase liver metabolism of glucose so that glucose utilization is increased.
e) True. All cells use a GLUT-family transporter to take up glucose, and all mediated transport systems show saturation.

11. If people ingest more amino acids than they need for protein synthesis, the excess is turned into glucose or fat.

12. a) Calcitrol b) Ca^{2+} is filtered into the tubule at the glomerulus if it is not bound to plasma proteins.

c) Renal reabsorption is enhanced by PTH and calcitrol.   d) Ca^{2+} entry into cells initiates exocytosis of vesicles and muscle contraction in smooth and cardiac muscle.   eP.H.TH; calcitrol to a lesser extent.

13. a) x-axis is PTH, y-axis is plasma Ca^{2+}. Line goes upward to the right.   b) x-axis is Ca^{2+}, y-axis is PTH. The line goes downward from left to right: as plasma Ca^{2+} increases, PTH secretion decreases.

14. Alpha cells secrete glucagon.

Plasma concentration	Change: , , or N/C	Rationale for your answer
Glucose		Glucagon increases plasma glucose.
Insulin		The increase in plasma glucose will decrease insulin secretion.
Amino acids		Glucagon stimulates conversion of amino acids into glucose.
Ketones	n/c	Although fats are broken down, the cells can use them and there is no excess ketone production.
K^+	n/c	These patients do not have an acid-base disorder and there is no other association of K^+ with glucagon.

CHAPTER 22 - Practice Makes Perfect

1. f, a, c
2. macrophage - a, b; red blood cell - d; B lymphocyte - a, b, e; natural killer cell - a, c;   liver cell - a; cytotoxic T cell - a, c; plasma cell - a; helper T cell - a, c
3. Macrophages ingest and destroy material; present antigens on MHC-II; secrete cytokines to initiate the inflammatory response and activate T cells.
4. If the mother has alleles AO and father has alleles BO, they could have a child with blood group O, alleles OO.
5. a) If Aparna's blood reacts to anti-B serum but not the anti-A, she has B antibodies on her cells and type B blood. b) She can donate to type B or type AB.   c) She can receive blood from type B or type O.
6. Antigen-presenting cells (macrophages, B cells), helper T cells, B cells, plasma cells, memory B cells, memory T cells, basophils (mast cells)
7. B cells develop in bone marrow; T cells develop in the thymus gland.
8. ABO blood type O has no antigens on the RBC membrane and will therefore not react with any antibodies in the recipient's plasma, so it is the universal donor. However, anti-A and anti-B antibodies in blood type O mean that type O patients can only receive type O blood, hence the high demand.
9. Your description should include the events shown in Fig. 22-17, p. 672.
10. Inflammation is the hallmark of the innate response.
11. Antibiotics will not help his viral infection because those drugs only act on bacteria. In addition, antibiotics stay in the extracellular fluid; viruses hide inside host cells.
12. Acute phase proteins act as opsonins and promote inflammation.
13. The complement cascade ends with production of membrane attack complex that causes pathogens to lyse.
14. The Fc region (stem) binds to immune cells; the Fab region recognizes and binds to antigen.
15. MHC is a family of membrane proteins that the body uses for presenting foreign antigens and for recognition of self. All nucleated cells have MHC-I but only antigen-presenting immune cells have MHC-II.
16. Old red blood cells lose the membrane markers that identify them as "self," which then leads the immune system to attack them.

Maps

1. See Fig. 22-10, p. 666.   2. See Fig. 22-4, p. 660.

CHAPTER 23 - Practice Makes Perfect

1. Systolic blood pressure, because the force of cardiac contraction is increased but peripheral resistance is decreased.
2. True
3. By the maximum rate of oxygen consumption.
4. Inspiratory and expiratory reserve volumes decrease because tidal volume increases.
5. Improved glucose tolerance, higher HDL and lower triglycerides, improved cardiovascular function, weight loss and/or improved muscle tone.
6. The factors that can limit exercise capacity include the ability of the muscle to provide ATP, the ability of the respiratory system to provide oxygen, and the ability of the cardiovascular system to supply the muscles with oxygen and nutrients.
7. Too much or too little exercise decreases immunity.

CHAPTER 24 - Practice Makes Perfect

1. <u>Male</u>   <u>Corresponding female part?</u>   <u>Analogous or homologous?</u>

Male	Corresponding female part?	Analogous or homologous?
bulbourethral gland	unique	
ductus deferens	(b)	analogous
penis	(a)	homologous
prostate gland	unique	
scrotum	(d)	homologous
seminal vesicle	unique	
testis	(c)	homologous

<u>Female</u>

(e) labia minora are homologous to parts of the penile shaft; f) uterus and g) vagina are unique.

2. e

3. a) decreased   b) decreased   c) will have no effect on cycles   d) will have no effect on cycles e) will restore normal menstrual cycles   f) will have no effect on cycles

4. The corpus luteum during early pregnancy secretes estrogen and progesterone to prevent the endometrium from sloughing.

5. Androgen-binding proteins keep androgens inside the tubule so that their concentration there is much higher than outside the tubules.

6. Fill in the following chart on reproduction.

	Male	Female
What is the gonad?	testis	ovary
What is the gamete?	sperm	ovum
Cell(s) that produce gametes	spermatogonia	follicle
What structure has the sensory tissue involved in the sexual response?	glans penis	clitoris
What hormone(s) controls development of the secondary sex characteristics?	androgens	estrogens and androgens
Gonadal cells that produce hormones and the hormones they produce	Sertoli: inhibin, activin Leydig: testosterone, DHT estradiol	Theca: androgens Granulosa: estrogens, progesterone, inhibin luteal: estrogen, progesterone, inhibin
Target cell/tissue for LH	Leydig cells	thecal cells; oocyte just before ovulation
Target cell/tissue for FSH	Sertoli cells	granulosa cells
Hormone(s) with negative feedback on anterior pituitary	inhibin, testosterone	inhibin, progesterone, sustained high estrogen
Timing of gamete production in adults	constant	cyclic until menopause

7. Proliferative endometrium - estrogen; initiates development of follicle(s) - FSH; ovulation - LH, estrogen; secretory endometrium - progesterone; keeps corpus luteum alive - human chorionic gonadotropin; keeps endometrium from sloughing (i.e., menstruating) in early pregnancy - estrogen and progesterone.